Monitoring Bathing Waters

Also available from E & FN Spon

Coastal Planning and Management
R. Kay and J. Alder

Coastal Recreation Management
The sustainable developemnt of maritime leisure
T. Goodhead and D. Johnson

Coastal and Estuarine Management
P. French

Determination of Organic Compounds in Natural and Treated Waters
T.R. Crompton

Ecological Effects of Waste Water, Second edition
Applied limnology and pollutant effects
E.B. Welch

Integrated Approaches to Water Pollution Problems
Edited by J. Bau, J.D. Henriques, J. de Oliviera Raposo and J.P. Lobo Ferreira

International River Water Quality
Pollution and restoration
Edited by G. Best, E. Niemirycz and T. Bogacka

Microbiology and Chemistry for Environmental Scientists and Engineers
J.N. Lester and J.W. Birkett

The Coliform Index and Waterborne Disease
Problems of microbial drinking water assessment
C. Gleeson and N. Gray

Water and the Environment
Innovative issues in irrigation and drainage
Edited by L.S. Pereira and J. Gowing

Water and Wastewater Treatment, Fourth edition
R. Bardolet

Water: Economics, management and demand
Edited by B. Kay, L.E.D. Smith and T. Franks

Water Policy
Allocation and management in practice
Edited by P. Howsam and R.C. Carter

Water Pollution Control
A guide to the use of water quality management principles
R. Helmer and I. Hespanhol

A Water Quality Assessment of the Former Soviet Union
Edited by V. Kimstach, M. Meybeck and E. Baroudy

Water Quality Assessments, Second edition
A guide to the use of biota, sediments and water in environmental monitoring
Edited by D. Chapman

Water Quality Monitoring
A practical guide to the design and implementation of freshwater quality studies and monitoring programmes
Edited by J. Bartram and R. Ballance

Water Resources
Health, environment and development
Edited by B. Kay

Water Wells: Monitoring, maintenance, rehabilitation
Proceedings of the International Groundwater Engineering Conference, Cranfield Institute of Technology, UK
Edited by P. Howsan

Monitoring Bathing Waters

A practical guide to the design and implementation of assessments and monitoring programmes

Edited by Jamie Bartram and Gareth Rees

London and New York

First published 2000 by E & FN Spon
11 New Fetter Lane, London EC4P 4EE

Simultaneously published in the USA and Canada
by E & FN Spon
29 West 35th Street, New York, NY 10001

E & FN Spon is an imprint of the Taylor & Francis Group

Printed and bound in Great Britain by TJ International Ltd, Padstow, Cornwall

Publisher's Note
This book has been prepared from camera-ready copy provided by the editors.

British Library Cataloguing in Publication Data
A catalogue record for this book is available from the British Library

Library of Congress Cataloging in Publication Data
A catalogue record for this book has been requested

ISBN 0-419-24380-1 (hbk)
0-419-24370-4

TABLE OF CONTENTS

FOREWORD

Coastal waters, rivers and lakes are used for a variety of recreational activities, including swimming, diving, fishing and sailing. If these activities are to be enjoyed safely, attention must be given to health hazards, as well as to the prevention of accidents.

Between 1993 and 1998, *Guidelines for Safe Recreational Water Environments* were developed by the World Health Organization (WHO) Headquarters in collaboration with the WHO European Centre for Environment and Health, Rome, Italy. These guidelines were released in the form of a draft for consultation in two volumes, *Coastal and Freshwaters* and *Swimming Pools, Spas and Similar Recreational Water Environments*. They comprise an assessment of the health risks associated with recreational use of water and outline linkages to monitoring and assessment and management practices. They are intended to provide guidance in identifying, characterising and minimising the risks to human health associated with recreational use of water and to promote the adoption of a risk-benefit approach to the management of such risks. The development of such an approach involves issues such as environmental pollution, conservation, and local and national economic development and may lead to the adoption of standards that can be implemented and enforced. To implement such an approach successfully requires considerable intersectoral co-operation and co-ordination at national and local levels as well as a coherent policy and legislative framework.

This book is a practical guide to the monitoring and assessment of freshwater and marine water used for recreation and builds upon the health risk assessment described in *Guidelines for Safe Recreational Water Environments*. It provides comprehensive guidance for the design, planning and implementation of assessments and monitoring programmes for water used for recreation. It addresses the wide range of hazards that may be encountered and emphasises the importance of linking monitoring programmes to effective and feasible management actions to protect human health. It also defines elements of good practice that together constitute the Code of Good Practice for the Monitoring and Assessment of Recreational Waters.

This book will be an invaluable source of information for anyone concerned with monitoring and assessing water used for recreation, including field staff. It will also be useful for national and regional government departments concerned with tourism and recreation, undergraduate and postgraduate students and special interest groups.

ACKNOWLEDGEMENTS

The World Health Organization wishes to express its appreciation to all those whose efforts made the production of this book possible. An international group of experts provided material and, in most cases, several authors and their collaborators contributed to each chapter. Because numerous contributions were spread over several chapters it is difficult to identify precisely the contribution made by each individual author and therefore the principal contributors are listed together below:

Nicholas Ashbolt, University of New South Wales, Sydney, Australia
Jamie Bartram, WHO, Geneva, Switzerland (formerly of the WHO European Centre for Environment and Health, Rome, Italy)
Juan Borrego, University of Malaga, Malaga, Spain
Raymond Briggs, Robens Centre for Public and Environmental Health, Guildford, Surrey, England
Maurizio Cavalieri, Local Agency for Electricity and Water Supply, Rome, Italy
Ingrid Chorus, Institute for Water, Soil and Air Hygiene, Federal Environmental Agency, Berlin, Germany
Karin Dubsky, Trinity College, Dublin, Ireland
Alfred Dufour, United States Environmental Protection Agency (US EPA), Cincinnati, OH, USA
Maria Jose Figueras, Rovira i Virgili University, Tarragona-Reus, Spain
Sylvie Goyet, World Wide Fund for Nature, Gland, Switzerland
Huw Jones, Department of the Environment, Transport and the Regions, London, England
David Kay, University of Aberystwyth, Aberystwyth, Wales
Bettina Menne, WHO European Centre for Environment and Health, Rome, Italy
Art Mittlestaedt, Recreation Safety Institute, New York, NY, USA
Eric Mood, Yale University, New Haven, CT, USA
Robin Phillipp, Bristol Royal Infirmary, Bristol, Avon, England
Edmund Pike, Consultant Microbiologist, Reading, Berks, England
Kathy Pond, WHO European Centre for Environment and Health, Rome, Italy and Robens Centre for Public and Environmental Health, Guildford, Surrey, England
Gareth Rees, Robens Centre for Public and Environmental Health, Guildford, Surrey, England
Ronnie Russell, Trinity College, Dublin, Ireland
William Robertson, Health Canada, Ottawa, Ontario, Canada

Henry Salas, PAHO/WHO Pan American Center for Sanitary Engineering and Environmental Sciences (CEPIS), Lima, Peru
Ann Storey, Robens Centre for Public and Environmental Health, Guildford, Surrey, England
Bert van Maele, European Commission, Brussels, Belgium
Allan Williams, Bath Spa University College, Bath, Avon, England
Adam Wooller, Surf Life-Saving Association of Great Britain, Plymouth, Devon, England

This book is based on the Code of Good Practice for the Monitoring and Assessment of Recreational Waters, which was prepared in co-operation with the European Commission. The Code was developed through a review process in which comments were received from 55 persons in 28 countries and reviewed at a meeting of an international group of experts.

Chapter 9 represents the conclusions of a meeting of experts organised by WHO in co-operation with United States Environmental Protection Agency (US EPA) and held in Annapolis, MD, USA, in November 1998. The meeting was attended by: Nicholas Ashbolt; Jamie Bartram; Rebecca Calderon, US EPA, Cincinnati, OH, USA; Joseph Cotruvo, NSF International, Washington, DC, USA; Alfred Dufour; Jay Fleisher, State University of New York Health Science Center at Brooklyn, Brooklyn, NY, USA; Walter Frick, USEPA, Cincinnati, OH, USA; Robert Gearheart, Humboldt State University, Arcata, CA, USA; Nancy Hearne, NSF International, Washington, DC, USA; Alan Jenkins, Institute of Hydrology, Wallingford, Oxon, England; Huw Jones; David Kay; Charles McGee, Orange County Sanitation District, CA, USA; Latisha Parker, US EPA, Cincinnati, OH, USA; Kathy Pond; Gareth Rees; David Rosenblatt, New Jersey Department of Environmental Protection, NJ, USA; Henry Salas; Stephen Schaub, US EPA, Cincinnati, OH, USA; Bert van Maele.

Chapter 10 was prepared by Ingrid Chorus and Maurizio Cavalieri, based on *Toxic Cyanobacteria in Water* (Chorus and Bartram, 1999). *Toxic Cyanobacteria in Water* was prepared by an international group of experts and was published by E & FN Spon on behalf of WHO.

Acknowledgements are also due to the following people, who reviewed the text: David Berry, Public Services Department, Jersey, Channel Islands; Robert Bos, WHO, Geneva, Switzerland; Anthony Bruce, Environmental Health Department, Jersey, Channel Islands; Enzo Funari, National Institute of Health, Rome, Italy; Richard Grainger, Environmental Health Department, Jersey, Channel Islands; Gerry Jackson, Public Services Department, Jersey, Channel Islands; Mihaly Kadar, National Institute of Hygiene, Budapest, Hungary; George Kamizoulis, WHO, Athens, Greece; Annette Prüss, WHO, Geneva, Switzerland; Mike Romeril, Chief Adviser's Office, Jersey,

Channel Islands; Regina Szewsyk, Federal Environmental Agency, Berlin, Germany; Rowena White, Consultant, Jersey, Channel Islands; and Terry Williams, Environmental Lawyer, Jersey, Channel Islands.

Thanks are also due to Deborah Chapman, University College Cork, Ireland, for editorial assistance, layout and production management and to Kathy Pond for editorial assistance. Deborah Chapman is the editor of the WHO-sponsored series of guidebooks on water resources management and was responsible for ensuring compatibility with *Water Quality Assessments*, *Water Quality Monitoring*, *Water Pollution Control* and *Toxic Cyanobacteria in Water*, four of the other books in the series. The contributions of the following people are also gratefully acknowledged: Leonard Chapman, Ann Willcocks and Lis Willcocks, who provided typesetting assistance; Stephanie Dagg, who prepared the index; and Alan Steel, who prepared the illustrations.

Special thanks are also due to the following institutions, which provided financial support for the preparation of the book and for the meetings at which the various drafts of the manuscript were reviewed: the European Commission; the Institute for Water, Soil and Air Hygiene of the Federal Environmental Agency, Germany; the Ministries of Environment and Health of Germany; the Ministry of Health of Italy; the States of Jersey; and the United States Environmental Protection Agency.

Chapter 1*

INTRODUCTION

From 1993 to 1998 the World Health Organization (WHO) worked on the progressive development of its *Guidelines for Safe Recreational-Water Environments* (WHO, 1998). The Guidelines comprise a health risk assessment of recreational water use to be published in two volumes (*Volume 1: Coastal and Fresh-Waters* and *Volume 2: Swimming Pools, Spas and Similar Recreational-Water Environments*). This present book is designed to complement the *Guidelines for Safe Recreational-Water Environments,* providing a practitioners' guide to the monitoring and assessment of coastal and freshwater recreational environments. It presents, in a methodological format, the information necessary to design and implement a monitoring and assessment programme for recreational water environments.

Surface and coastal waters are used for a variety of leisure and recreational activities, and for other purposes including transport, food production, hydroelectricity generation, as a transport medium and as a repository for sewage and industrial waste. Such activities are not always compatible with one another. Water and its recreational use have long been recognised as major influences on health and well being. The health benefits of bathing in saltwater were, and still are, promoted with enthusiasm. Sea water was once considered as an alternative medicinal treatment to spa water. Water-based recreation is an important component of leisure activities and tourism throughout the world. Tourists are responsible for the significant movement of economic resources both within and between countries. This may be typified by the annual influx of tourists from northern European countries to the countries surrounding the Mediterranean. A similar effect may occur within some countries where certain regions are favoured holiday destinations by those from other regions within the country.

Recreational use of the water environment may offer a significant financial benefit to the associated communities but it also has implications for health and for the environment. Visitors exert a variety of pressures on the very environment that attracts them. Water-based recreation and tourism can also expose individuals to a variety of health hazards, ranging from exposure

* *This chapter was prepared by G. Rees, J. Bartram, K. Pond and S. Goyet*

to potentially contaminated foodstuffs and potable water supplies, through to exposure to sunshine and ultra violet (UV) light and to bathing in polluted waters. Water, however clean, is an alien environment to humans and thus it can pose hazards to human health even when it is of pristine quality.

The varied nature of the hazards to human health and well-being posed by recreational waters demands a full audit of the relative importance of the resultant health effects and the resources required to mitigate those effects. Undoubtedly, the public health outcomes of accidents (including drowning and trauma associated predominantly with diving incidents) and potential infections acquired from contaminated waters are those that demand most attention world-wide.

All the trends indicate that leisure activities, including water-based recreation, will continue to increase. Thus the effects of the health hazards that face recreational water users are likely to gain more prominence in the future. Those responsible for monitoring the likely health impacts of recreational water use are going to face increasingly complex challenges as recreational uses diversify and the number of users increases.

1.1 Health hazards in recreational water environments

Amongst the unequivocal adverse health outcomes resulting from recreational water exposure are drowning and near-drowning. Such injuries account for a significant annual death toll, often associated with reckless behaviour and/or alcohol consumption. Unsafe diving into water bodies can lead to a range of traumatic injuries, including spinal injury, which ultimately may result in quadriplegia. More common, but less severe, incidents include those arising from discarded materials, such as glass, cans and needles on beaches and on the bottom of the bathing zone. Of particular concern is the presence of medical waste, particularly hypodermic needles. Chapter 7 addresses the dangers due to accidents and injuries including those associated with drowning and spinal injury.

The pathogenic micro-organisms that can be found in water bodies have a wide range of sources. These include sewage pollution, organisms naturally found in the water environment, agriculture and animal husbandry and the recreational users themselves. Sewage of domestic origin comprises a particularly unhealthy mixture of micro-organisms. The microbiological hazards encountered in water-based recreation include viral, bacterial and protozoan pathogens. Primary concern has usually been directed towards gastro-intestinal illnesses acquired from recreational waters, although acute febrile respiratory illness and infections of the eye, ear, nose and throat have all been identified as acquired through bathing. The link between recreational water use and more serious infections such as meningitis, hepatitis A, typhoid fever and poliomyelitis is difficult to determine unequivocally.

A key environmental effect of sewage discharges is nutrient enrichment largely, but not exclusively, attributable to phosphate and nitrate in the sewage. This nutrient enrichment can lead to localised eutrophication, which in turn is associated with more frequent or severe algal blooms. Prolonged and excessive eutrophication has also been responsible for algal blooms on a regional basis, such as those in the Adriatic and Baltic Seas in recent years.

A review of human health effects arising from exposure to toxic cyanobacteria, as well as discussion of the detailed analysis of toxic cyanobacteria in water and of their monitoring and management, is available in a companion volume in this series, *Toxic Cyanobacteria in Water* (Chorus and Bartram, 1999). Chapter 10 in this book deals with the monitoring and assessment of toxic cyanobacteria and algae in recreational waters and also provides a framework for assessing under what circumstances such organisms may pose a priority hazard.

A number of other health hazards may be encountered during recreational water use but which are typically local or regional in distribution. These include: chemical contaminants, arising principally from direct waste or wastewater discharge; non-venomous disease-transmitting organisms (e.g. mosquitoes as malaria and arboviral disease vectors and freshwater snails as intermediate hosts of the schistosomes that cause bilharzia or schistosomiasis); hazardous animals encountered near water (such as crocodiles and seals) and venomous invertebrates (such as sponges, corals, jellyfish, bristleworms, sea urchins and sea stars); and venomous vertebrates (catfish, stingrays, scorpionfish, weaverfish, etc.). The health risks associated with these hazards are outlined in the *WHO Guidelines for Safe Recreational-Water Environments* (WHO, 1998). Approaches to monitoring and management of these health hazards are often strongly influenced by local factors and for this reason the issue is dealt with here in generic terms, allowing the reader to make informed responses to circumstances where such hazards may arise (Chapter 11).

Excessive exposure to UV, although not exclusive to water-based recreation, may pose a significant health risk if recreational water users do not take appropriate care. Acute effects, such as the discomfort and injury associated with sunburn, or delayed effects (which may include malignant melanoma) are direct adverse health outcomes. Cold water is an important contributory factor in many cases of drowning, and excessive exposure to heat and/or cold can also be associated with adverse health outcomes. These issues are fully addressed in the *WHO Guidelines for Safe Recreational-Water Environments* (WHO, 1998). Hazards attributable to physical components, such as exposure to extremes of temperature or to excess UV radiation, are not included in this book due to the limited contribution that monitoring and assessment can make to risk management in this context.

1.2 Factors affecting recreational water quality

The health risks posed by poor quality recreational waters generally relate to infections acquired whilst bathing. A range of pollutants enters recreational waters from a number of sources — coastal waters can be regarded as the ultimate sink for the by-products of human activities. In terms of the quality of coastal recreational waters and the resultant impact on human health, the key sources of pollutants are riverine inputs of domestic, agricultural and industrial effluents and direct sewage discharges from the local population. Apart from regular discharges through short and long sea outfalls, irregular discharges may occur through storm water and overflow outfalls, and through unregulated private discharges. Freshwater bathing sites are subject to the same polluting sources as coastal sites, although the scale and extent of these sources may be easier to predict in more clearly delimited freshwater sites.

In both coastal and freshwaters the point sources of pollution that cause most health concern are those due to domestic sewage discharges. Diffuse outputs and catchment aggregates of such pollution sources are more difficult to predict. Discharge of sewage to coastal and riverine waters exerts a variable polluting effect that is dependent on the quantity and composition of the effluent and on the capacity of the receiving waters to accept that effluent. Thus enclosed, low volume, slowly-flushed water systems will be affected by sewage discharges more readily than will water bodies that are subject to rapid change and recharge.

Water-based recreation and leisure activities contribute relatively little to pollution inputs and associated adverse health outcomes when compared with other sources of aquatic pollution, such as sewage outfalls. Pollution originating from water-based recreation and leisure craft includes sanitation discharges, fuel spillages, the environmentally toxic effects of antifouling compounds and general debris. Because water-based recreational activities often occur in estuaries or embayments, any polluting effects may be exaggerated due to the enclosed nature of the system and the subsequent accumulation of pollutants. This is particularly evident in the large number of boating marinas that have been developed over recent years, where there is often a high density of craft and the associated crew, adjacent to bathing waters. Appropriate controls on sanitation discharges from pleasure craft are dependent on suitable holding tanks and port reception facilities. The contribution that such vessels make to the total sewage inputs may be small, but may become more significant in the situations described above where vessels aggregate.

Environmental hazards attributable to pleasure craft may also arise from fuel spillages or discharges. Oil, petrol and diesel may be spilt at filling barges and bilge waters may be discharged, adding to pollution of the coastal environment. Two-stroke outboard engines are thought to exert an annual polluting effect several times greater than that attributable to high profile oil-tanker disasters such as that of the Exxon Valdez (Feder and Blanchard, 1998). Apart from the fuels themselves, the emissions from the engines may be harmful and the oil and petrol mix in two-stroke fuels has been implicated in the tainting of fish and shellfish products. This type of pollution, therefore, can pose an indirect threat to health.

The toxic nature of antifouling paints applied to prevent the attachment and subsequent growth of organisms on pleasure craft hulls and on coastal installations defines them as environmental pollutants. They may thus have an effect on water quality, particularly where vessels are concentrated or the area is enclosed. Such effects are usually considered to affect the marine biota rather than human health.

Pleasure craft and their users undoubtedly contribute significantly to the load of marine debris. Plastics, fishing gear, packaging, food and other wastes are discarded overboard, even when there are controls to prevent this. About 70–80 per cent of marine debris comes from land-based sources and the rest comes from vessels and installations. Such materials rarely affect water quality and human health but do have environmental effects and contribute enormously to aesthetic pollution.

1.3 Effective monitoring for management

Particular types of recreational activities may be associated with certain hazards and therefore discrete patterns of action may be taken to reduce the risk of these hazards. For example, untreated sewage discharges will pose one type of risk — that of infection to bathers; glass discarded on a beach will pose a different type of hazard — injury to walkers with bare feet. Effective sewage discharge procedures can address the former and regular cleaning of the beach coupled with provision of litter bins and educational awareness campaigns can reduce the latter hazard. Therefore, each type of recreational activity should be subject to assessment to determine the most effective control measures. This assessment should include factors that may have a moderating effect on the particular type of risk, such as local features, seasonal effects and the competence of the participants in the activity where the risk is encountered.

The importance of effective use of information from monitoring must be stressed. There is little point in generating monitoring data unless they are to be used. The eventual use of the information products resulting from monitoring should guide and determine all the stages of the monitoring process from the setting of objectives through to design and implementation, reporting and to co-ordination of follow-up. The principal components of management of recreational water use areas for the protection of public health are described in Chapter 5.

For any monitoring and assessment programme to be effective there must be clear management outcomes from the use of the data produced. Effective monitoring requires collection of adequate quantities of data of the appropriate quality, and an understanding of the link between monitoring and management (and therefore to whom, in what format and when information would be best provided). Subsequent actions may be remedial, may provide public information or may inform planning. For example, microbiological monitoring data can be used to justify improved treatment of coastal sewage discharges or, alternatively, to indicate that a small investment in injury and prevention measures will yield more substantial public health benefit. The information links between regulators, government and industry, and those that can provide the financial support for remedial initiatives, must be based on good quality data.

In order for individuals to be able to make knowledgeable decisions about their ultimate recreational destinations, based on the existing facilities and the environmental quality of the available options, they must be aware and informed. Ideally, such judgements should be based on good quality, readily understandable and easily accessible information. Individual choice of site of recreational activity may indirectly result in improved levels of recreational water and bathing beach management by local and national governments. Furthermore, individuals can take responsibility for some important actions to protect their own health and well-being whilst involved in water-based recreation.

The *WHO Guidelines for Safe Recreational-Water Environments* (WHO, 1998) describe management actions that support improved safety in recreational water use in four broad areas under the umbrella of integrated coastal or basin management (Figure 1.1). All four broad areas rely on the output of sound monitoring and assessment processes for effective implementation. Several activity levels can be defined at international, national, regional and local levels. Actions that may be taken at international and national level consist primarily of the setting of standards, and such actions are the province of government. There are many local actions that can be undertaken and which may have a significant impact on the well-being of recreational water users. These include basic beach management schemes, comprising lifeguard provision, appropriate sanitation facilities, potable water, parking, medical facilities, beach cleaning, emergency communication and zoning of activities to avoid conflict. These initiatives are largely the domain of local municipalities or of the owners of private beaches. Although all these management measures have attributable direct costs, they may have major indirect benefits in the form of increased recreation and leisure at the location. Beach award schemes (see Chapter 6) harness the willingness of municipalities to provide such facilities.

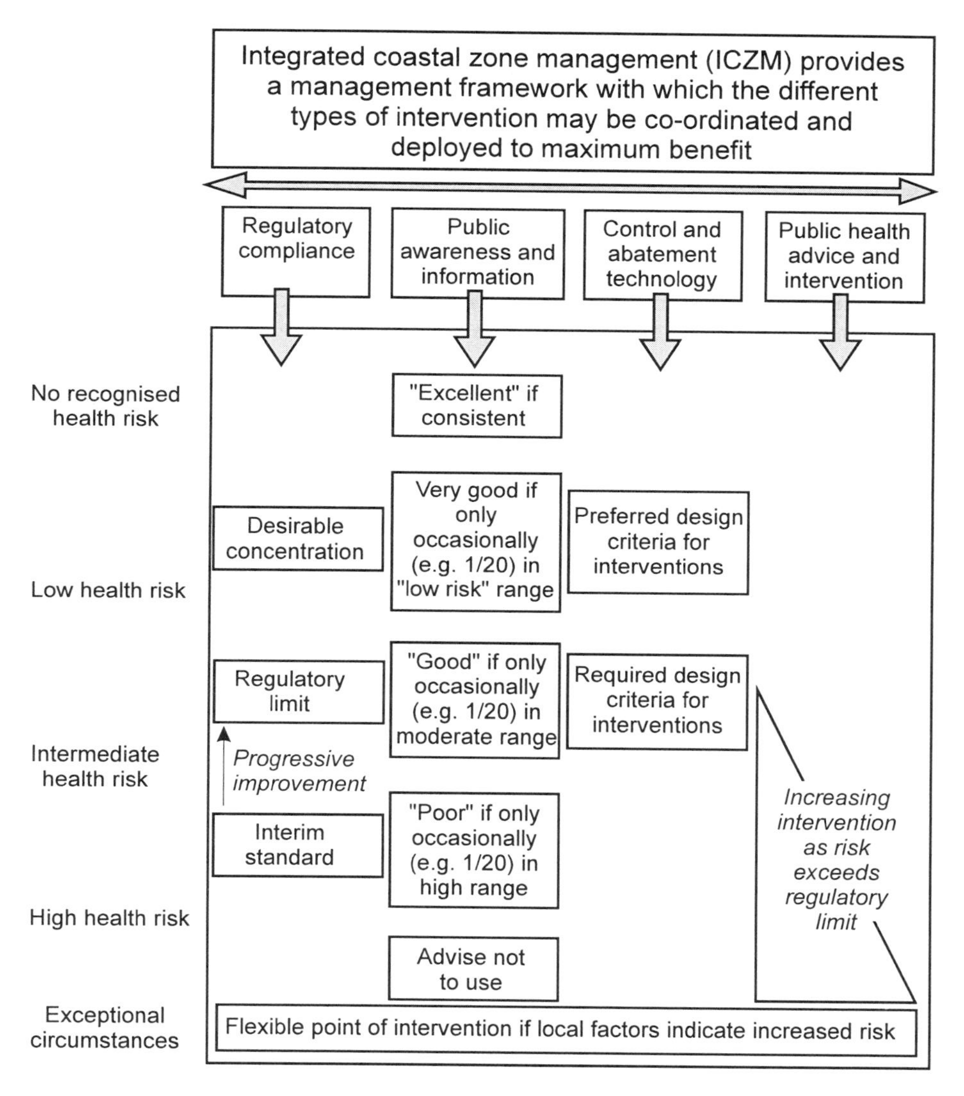

Figure 1.1 Management framework and types of intervention in relation to different types and degrees of hazard associated with recreational water use

Effective monitoring also requires the participation of all authorities, organisations, industries etc. with a vested interest. This implies that an agency involved in monitoring should ensure and maintain effective channels of communication with non-governmental organisations, industry (especially tourism), local and central government, trade associations, resort and tourism operators and elements of the media. Often monitoring and regulatory agencies are concerned with the quality and veracity of data, but are less aware of the need to package the information and display the results in ways that are easily understood by participating partners, i.e. the same partners with which they are trying to maintain links.

Many uses of the water environment have the capacity to conflict with each other. Such competing pressures must be monitored in a coherent fashion to minimise conflict and, where appropriate, risk to health. Conflict may arise between different groups of water users or between local users and those visiting an area. It is one of the primary roles of effective management to accommodate competing, and often conflicting, uses. When tourism and leisure-based activities are major revenue sources in a region, it is important to strike the correct balance between the demands of water-based recreation and the needs of the environment. Different recreational activities can also interact in a counter-productive fashion — angling, boating, surfing, bathing and water-skiing cannot all take place on the same area of water at the same time without some form of regulation. It is essential that effective planning and consultation processes exist to ensure that appropriate management practices are implemented and monitored. Such planning may enable what is generally a limited resource to be channelled into the most appropriate activities at a particular location. Visitor pressure can be managed more effectively by such means and the health and well-being of the individual and the environment is usually best served in this way.

1.4 Good practice in monitoring

The chapters of this book each include an element of good practice in the monitoring and assessment of recreational waters. Together these elements constitute a Code of Good Practice. This Code of Good Practice comprises a series of statements of principle or objectives which, if adhered to, would lead to the design and implementation of a monitoring programme of scientific credibility. The Code applies to the monitoring of all waters used for recreational activities that involve repeated or continuous direct contact by people with the water. In many circumstances there are different approaches or methods that can be applied to achieve the objective stated in the Code. Although each approach is equally valid in isolation, adoption of diverse approaches within a single programme would not lead to the comparability of results that may be required by an inter-location study or by an enforcement programme. Where data are to be compared between laboratories or between sites, all available measures should be implemented to ensure comparability of results. These include:

- A quality assurance programme based on internal and external (inter-laboratory comparisons) controls.

- The development of criteria for dealing with participating laboratories consistently failing to comply with minimum analytical quality. This should be stated prior to data collection and it should be adhered to.
- Procedures for dealing with data and sampling anomalies and omissions, which should be agreed prior to data collection and adhered to.

In regulatory monitoring programmes, factors such as sampling frequency, analytical methods, data analysis, interpretation and reporting, sample site selection and criteria for recreational water use areas, are usually defined by a regulatory agency and should take account of the principles outlined in the Code of Good Practice.

1.5 Legislative context

Effective coastal or freshwater zone management requires an effective legislative framework to define the roles of different bodies and levels of government, as well as to provide environmental objectives. Management is not restricted to national issues; water quality, pollution control, international tourism and shipping are amongst the activities that affect the coastal zone and that also extend beyond national boundaries. No single government or agency can be responsible for the wide range of issues that need to be addressed in the coastal–freshwater zone and legislation should be considered at the international, national and local levels. In general, however, the fragmented and often duplicated responsibilities in the coastal zone are severe impediments to effective planning and management in many countries.

The structure and responsibility of local government differs throughout the world. In the UK, for example, County Councils are responsible for the strategic planning, structure plans and waste disposal, while the District Councils are responsible for housing, local planning, local plans, environmental health, coast protection, waste collection and noise control. In Australia, the Local Councils have general responsibilities for the production of coastline management plans, coastline hazard mitigation, hazard awareness and beach management, as well as specific responsibilities under the Environmental Planning and Assessment Act.

1.6 Socio-economic issues

The development of tourism may create conflict. Displacement effects, such as movement of the indigenous population out of coastal areas, banning fishing from tourist beaches, or inappropriate adaptation of cultural and historic resources, are pervasive and may lead to political and social reactions against tourism. These social pressures also tend to reinforce pressures for enclosed resorts and concentrated tourism enclaves that may increase the adverse environmental effects of tourism. Private beaches that charge entry fees provide a means of socio-economic selection. Furthermore, social problems may arise in a range of situations that include but are not limited to:

- *Conflicts between beach users.* Conflicts may develop particularly on intensively used beaches. They may occur between visitors, swimmers, surfers and boat users and may be resolved by delimiting zones of the beach and nearshore sea.
- *Complaints by people using a beach.* Noise from radios, vehicles or boats and other potentially insensitive behaviour may lead to conflict situations.
- *Camping on beaches.* Camping on beaches is tolerated in many countries, especially away from seaside resorts and other urbanised areas. However, this may be a cause of conflict between visitors and local people.
- *Visual intrusion.* Structures such as fishermen's huts and beachcombers' shacks may be considered unsightly.
- *Animals.* Fouling by dogs can be a nuisance and a health risk.
- *Alcohol.* Some authorities prohibit the drinking of alcohol on the beach, mainly because of unpleasant behaviour and the increased litter resulting from empty drinks cans and broken bottles.

Measures for coping with the adverse social and cultural impacts of tourism include studies of a social carrying capacity for tourism, public education programmes and improved security measures to address increased crime and drug problems. Observations of beach behaviour suggest that there is a tendency for beach users to segregate themselves on the basis of race and class (Bird, 1996). The benefits of tourism, including tourism in coastal areas, derive from direct and indirect income, new jobs, foreign investment, infrastructure development, increased local support for environmental amenities, and conversion to less stressful use. The main costs of beach tourism could be ascribed to the effects on water quality and availability, sewage and solid waste disposal, loss of non-renewable resources (e.g. sand mining from beaches is a major negative impact related to tourism development in the Caribbean), overharvesting of renewable resources, increased social tensions, stimulation of imports of foods and other consumables, costs of damage from natural hazards, increased densities of people and conversion to more stressful uses.

Benefits and costs can be measured in quantitative, financial terms if there have been sufficient econometric studies to determine shadow prices for known differences in environmental and social effects of tourism. National and regional development planners need studies that define the local rate of retained earnings or the local factors enhancing the effects of various types of tourism facilities. Given this basic information, local planners could relate these economic benefits to their assessment of the costs of environmental effects of alternative types of tourism development, or of specific project proposals. In the absence of these data it is not possible to make an assessment of the relative costs and benefits from the various mixtures of different tourism facilities. The valuation of beaches is a key element for developing a cost analysis of the environmental impacts of tourism (Houston, 1995).

1.7 Framework

This book concentrates on providing the practical information necessary to design and implement monitoring programmes and studies of recreational water and bathing beach quality and to create the link between the information generated and action to protect human health. The elements outlined throughout this book should be implemented flexibly according to the different objectives and priorities that exist in the area under consideration.

This book therefore comprises a series of 12 further chapters (Figure 1.2), each culminating in a section on elements of good practice as described above. The introductory comments in this section lead into Chapters 2 and 3 where guidelines on the selection of appropriate variables for the successful monitoring and assessment of recreational water and bathing beach quality are elaborated. This includes selection of suitable areas that can be designated for recreational use and the location of sample collecting stations and necessary variables at each sample station. Logistic issues in implementing the monitoring programme are also explored.

Chapter 4 provides the background for implementing a full and reliable analytical quality assurance system to ensure confidence in, and reliability of, the data gathered. Chapter 5 introduces the management of bathing beaches to maximise health protection of recreational users. A variety of methods of involving the public in the whole process are discussed in Chapter 6. Construction of a public information strategy is discussed together with involving those with a vested interest in ensuring that the most suitable strategy is adopted. Chapters 7 to 11 focus specifically on issues related to a specific type, or group of types, of hazard. These include hazards related to drowning and injury (Chapter 7).

The microbiological quality of recreational waters, its assessment in a consistent and coherent fashion and the interpretation of microbiological water quality data are the themes of Chapters 8 and 9. These chapters look at approaches to microbiological quality and sanitary assessment and the methods employed. The concept of pollution or sanitary inspection is introduced as a rapid and effective means of providing information for the monitoring process. Chapter 10 moves on to examine the potential health effects due to concentrations of cyanobacteria and microalgae in recreational waters; methods of assessing reliably the likely risks from such sources are addressed. Chapter 11 elaborates on monitoring associated with other

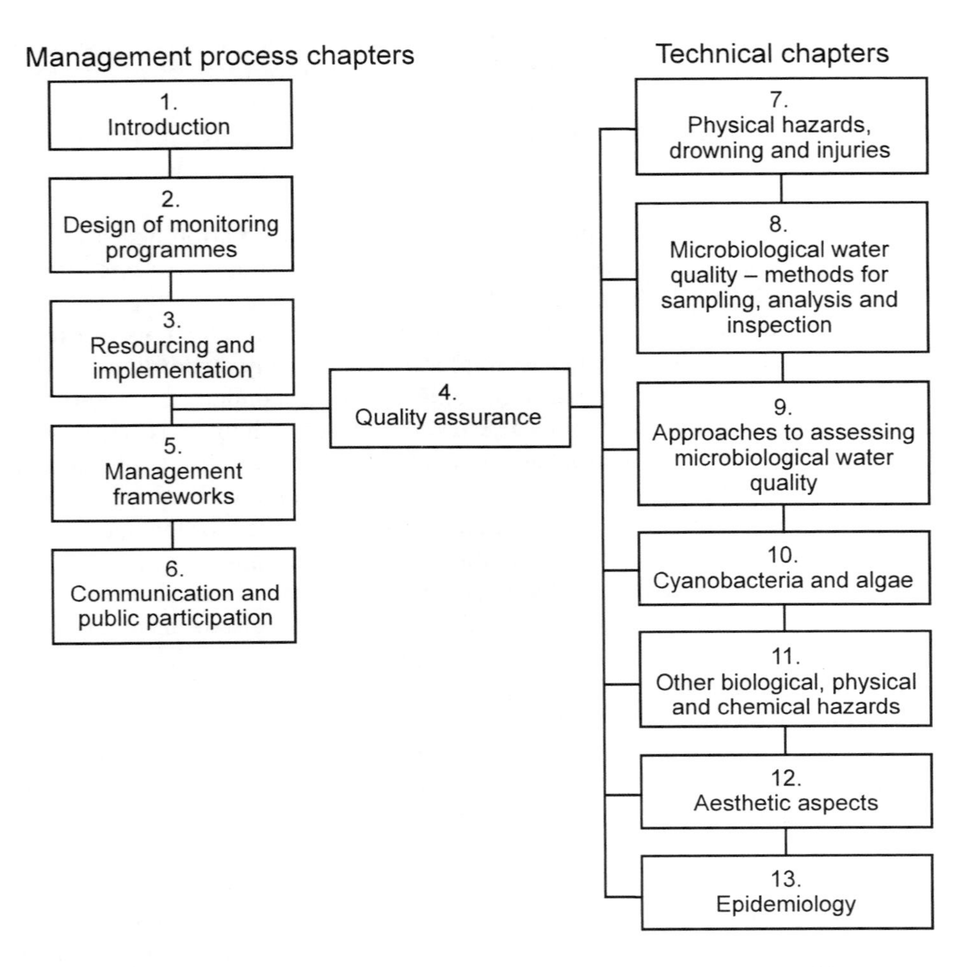

Figure 1.2 Management processes and technical aspects of monitoring bathing waters as discussed in the various chapters of this book

biological, chemical and physical hazards and Chapter 12 extends the principles of monitoring into the context of aesthetic aspects of beach quality. Issues of perception of recreational water and bathing beach quality derived from aesthetic pollution indicators are explored further. The applied and effective nature of such monitoring and assessment schemes is seen as particularly suitable to coastal and estuarine environments. Chapter 13 introduces the complex practice of epidemiological surveys and their application to recreational waters.

1.8 References

Bird, E.C.F. 1996 *Beach Management.* John Wiley and Sons, Chichester, 281 pp.

Chorus, I. and Bartram, J. [Eds] 1999 *Toxic Cyanobacteria in Water. A Guide to Public Health Significance, Monitoring and Management.* E & FN Spon, London, 416 pp.

Feder, N.M. and Blanchard, A. 1998 The deep benthos of Prince William Sound, Alaska, 16 months after the Exxon Valdez oil spill. *Marine Pollution Bulletin,* **36**(2), 116–130.

Houston, J.R. 1995 Beach replenishment. *Shore and Beach,* **63,** 2124.

WHO 1998 *Guidelines for Safe Recreational-Water Environments. Volume 1: Coastal and Fresh-Waters.* Draft for consultation. WHO/EOS/98.14, World Health Organization, Geneva, 208 pp.

Chapter 2*

DESIGN OF MONITORING PROGRAMMES

Traditionally the primary reason for the assessment of the quality of an environment has been to verify suitability for intended uses. Monitoring has also evolved to determine trends in the quality of the environment and to determine how quality is affected by anthropogenic activities, including for example waste treatment operations (the latter is known as impact monitoring). Monitoring the background quality of recreational water environments is also now widely carried out to provide a means of assessing impacts and to check whether unexpected change is occurring. In regulatory monitoring programmes, factors such as sampling frequency, analytical methods, data analysis, interpretation and reporting, sample site selection and criteria for recreational water-use areas are generally defined by the regulatory agency.

General definitions for various types of environmental observation programmes have been proposed (e.g. Chapman, 1996) which may also be modified and interpreted in relation to recreational water use, as follows:

- *Monitoring.* Long-term, standardised measurement and observation of the environment in order to define status and trends.
- *Survey.* A finite duration, intensive programme to measure and observe the quality of the environment for a specific purpose.
- *Surveillance.* Continuous, specific measurement and observation for the purpose of management and operational activities.

Each of the above activities are often not clearly distinguished one from another and all may be referred to as "monitoring", because they all involve collection of information at set locations and intervals. They do, nevertheless, differ in relation to their principal use in the recreational water quality assessment process.

2.1 Aims and objectives of monitoring

No assessment programme should be started without identifying the specific need(s) for information. Thus water quality assessment should take into account hydrological factors, water uses, economic development, policy and legislation, etc. The decisions that will result from the assessment

* *This chapter was prepared by K. Pond, W. Robertson, G. Rees and H. Salas*

Box 2.1 Setting objectives for microbiological monitoring of bathing areas

The objectives of microbiological monitoring programmes can be diverse. However, microbiological monitoring of bathing areas is, in most cases, undertaken to comply with regulations and/or to establish the degree of microbiological pollution in order to protect public health and the environment. Those macro-objectives only answer the question: *Why?* More specific objectives need to be defined which will also tackle the question of *Where?* (location of bathing area, sampling points and frequency). Some aspects are fixed by the regulations. Others, such as the location of sampling points, are only generally defined and require preliminary screening. Such screening will establish the spatial and temporal variations of microbiological water quality to select the optimal sampling points and frequency to obtain data representative of those fluctuations. Questions of *What*? (variables or indicators to be determined) and *How?* (methodology of inspections and analysis) are sometimes only partially defined by the regulations. These variables must be those that are more representative of sewage pollution as a measure of health risk. Specific comparative studies of standardised indicators and procedures are advisable at each specific geographical area.

Public information and participation may be another objective included in the regulations. Microbiological results given to the public should include visual inspections for aesthetic factors that bathers will be confronted with and should be expressed in a clearly understandable ranking system.

Other specific objectives will assess the impact on the microbiological quality of river outlets at the sea bathing area and any discharges at inland reservoirs as well as the effects of rain. Spatial and temporal variations identified before will have to be taken into account and their impact assessed over a representative period so that remedial and/or preventative measures (such as indications of risk) can be encouraged.

Source: Based on the approach used by the Unit of Microbiology, Faculty of Medicine, University Rovira i Virgili, Spain

programme determine whether emphasis should be put on concentrations or loads and on spatial or temporal distribution, as well as determining the most appropriate monitoring media. There are generally several competing beneficial uses of the recreational water-use area, and the monitoring activities should reflect the data needs of the various users involved.

The objectives of the assessments may focus activities on the spatial distribution of quality (a large number of sample stations), on trends (high sampling frequency), or on pollutants (in-depth inventories) (Box 2.1). Full coverage of all three requirements is virtually impossible and costly. Preliminary surveys are generally necessary in order to determine the appropriate

Table 2.1 Types and objectives of principal water quality assessment operations

Type of assessment	Major focus of water quality assessment
Multipurpose monitoring	Space and time distribution of water quality in general
Trend monitoring	Long-term evolution of pollution (concentrations and loads)
Basic survey	Identification and location of major problems and their spatial distribution
Operational surveillance	Water quality and related water quality descriptors (variables) for specific uses
Background monitoring	Background levels for studying natural processes; often used as reference point for pollution and impact assessments
Preliminary surveys	Inventory of pollutants and their space and time variability; usually prior to designing and establishing a routine monitoring programme
Emergency surveys	Rapid inventory and analysis of pollutants for rapid situation assessment following a catastrophic event
Impact surveys	Sampling limited in time and space, generally focusing on a few variables near pollution sources
Modelling surveys	Intensive water quality assessment limited in time, space and choice of variables to support, for example, eutrophication models or oxygen balance models
Early warning surveillance	At critical water use locations (continuous and sensitive measurements)

Source: Bartram and Ballance, 1996

focus of activities. Table 2.1 summarises the principal types of water quality operations in relation to their main objectives.

It cannot be overemphasised that the benefits of careful preliminary planning and investigation far outweigh the efforts spent during this initial phase. Mistakes and oversights during this part of the programme may lead to costly deficiencies, or overspending, during many years of routine monitoring.

2.2 Elements of recreational water quality assessment

Once objectives have been set, the scope of the monitoring programme should be defined. This includes definition of criteria for inclusion or exclusion of recreational water-use areas and the preparation of an inventory of areas included or excluded as recreational water-use areas. A review of existing data and the compilation of a catalogue of the basic characteristics of the area, supported by preliminary surveys, determines the monitoring design. The completed review and catalogue should be followed by recommendations to relevant authorities for management, pollution control and, eventually, the adjustment or modification of monitoring activities (Box 2.2).

Box 2.2 Beach monitoring programme Lima, Peru

The city of Lima, capital of Peru, is located on the coast of the Pacific Ocean and has a current population of about 8 million (1998). In spite of its tropical latitude of about 12° south, the marine waters of the area are relatively cold due to the Humboldt Current emanating from the polar ice cap waters of the South. During the summer months (December to March), the beaches within and near Lima are used extensively by the local population for recreational activities such as swimming.

Lima generates about 16.5 $m^3 s^{-1}$ of wastewater, the major part of which is discharged untreated directly or via the Rimac River to the coastal marine waters. There are no existing submarine outfalls although plans call for the construction of long sea outfalls with treatment during the next decade. Due to the arid climate of the region, the reuse of sewage for crop irrigation is practised to some extent and will increase in the future.

The water quality standards of Peru classify marine waters as safe for primary contact recreation when 80 per cent of five samples taken over one month period show less than 1,000 MPN per 100 ml of faecal coliforms and 5,000 MPN per 100 ml for total coliforms. This standard drives the frequency of measurement of the monitoring programmes described below.

Because of its proximity to populated areas as well as its accessibility by public transportation, Miraflores Bay is a very popular beach area referred to as the "Costa Verde". A major trunk sewer discharges about 6–7 $m^3 s^{-1}$ of raw sewage directly to Chira Beach (not used by the public) approximately 4 km east (upstream) of the closest popular beach of Costa Verde. Predominate currents in this area are parallel to the coast from east to west.

For years the media attributed the contamination of the "Costa Verde" area to the Chira outfall. In 1986, the Environmental Health Division of the Ministry of Health initiated a beach vigilance monitoring programme with 21 stations sampled on a weekly basis during the summer in the Costa Verde area. The data clearly demonstrated that there was gross pollution in the vicinity of the Chira discharge and that this contamination was beginning to encroach on the popular beach areas of Costa Verde. However, the pockets of contamination observed on some of the most popular beaches of Costa Verde could only be attributed to local direct discharges from sanitary sewer overflows, restaurants and other installations. Bather density could also have been a contributing factor.

The monitoring programme was expanded to 24 stations in Costa Verde that were sampled weekly during the summer bathing season and monthly during the winter from 1987 to 1989. These data confirmed that local discharges rather than the Chira outfall caused the contamination at some popular beaches in Costa Verde. Management action was taken in 1991 with the construction of a small trunk sewer and six pumping stations in Costa Verde to transfer sewage to the main sewer system with subsequent discharge via the Chira outfall.

Budget cuts reduced the monitoring programme to as little as 10 stations in 1991 at the most contaminated beaches. In 1992 and 1993 weekly monitoring was resumed at the 24 Costa Verde stations during the summer. These data clearly demonstrated the water quality improvements at some beaches due to

Continued

Box 2.2 Continued

the management action taken in 1991. The monitoring programme was even able to pick up the impact on beach water quality when the Costa Verde trunk sewer pumping stations were not operating during the energy blackouts caused by the drought of 1992.

In 1994 the monitoring programme was expanded to beaches to the north and south of Costa Verde which are also used by the Lima population. A total of 61 stations were sampled weekly in the summer and monthly in the winter.

The monitoring programme was expanded to national coverage in 1997. The programme presently takes samples at 148 stations: 77 stations at the beaches used by the Greater Lima Metropolitan Area population (these are sampled weekly in the summer and fortnightly in the winter) and 71 stations from Tumbes in the north to Tacna in the south (sampled weekly in the summer and monthly in the winter). In 1996, enterococci, *Escherichia coli* and *Vibrio cholerae* were added to the total and faecal coliform measurements at 46 stations. The programme conducts approximately 13,560 microbiological measurements per year.

The Lima monitoring programme was the key to ascertaining the real sources of the contamination of the popular Costa Verde beaches and instrumental in convincing the authorities, which in turn led to the implementation of sound management actions. The greater investment that would have been required to dispose of the sewage discharged properly via the Chira outfall would not have changed the situation at the popular Costa Verde beaches that were contaminated by small local sources. Nevertheless, if the monitoring approach described in this guidebook had been applied, action might have been taken sooner based on the information provided by the initial sanitary inspection. Furthermore, the sanitary inspection would have served to modify the monitoring programme to focus on those beaches where potential sources of pollution were identified, thus making the monitoring programme source- and use-driven as opposed to only use-driven.

Bathers on the beach at Costa Verde, Peru. Photograph courtesy of "El Comercio"

There are certain elements that are common to all water quality monitoring and assessment programmes. They are more, or less, extensively developed depending on the type of assessment required. These elements are:

- *Preliminary surveys*. Short-term, limited activities to determine the type of monitoring media and pollutants to be considered, and the technical and financial feasibility of a complete monitoring programme.
- *Monitoring design*. The selection of variables, station location, sampling frequency, sampling apparatus, etc. (Chapters 3, 8 and 12).
- *Field monitoring*. This includes *in situ* measurements, sampling of appropriate media, sample pre-treatment and conservation, identification, storage and shipment (Chapter 8).
- *Hydrological monitoring*. Measurements of water discharge, currents, tides, water levels, thermal profiles, etc. Hydrological data should always be related to the water quality assessment activities.
- *Laboratory activities*. These include concentration measurements, biological determinations, etc.
- *Data quality control*. This consists of analytical quality assurance within each laboratory and amongst all laboratories participating in the same programme (Chapter 4).
- *Data storage and treatment*. This is now widely computerised and involves the use of databases, for data storage reporting statistical analysis, trend determinations, multifactorial correlation, etc. together with presentation and dissemination of results in appropriate forms (graphs, tabulated data, data diskettes, etc.) (Chapters 3 and 8).
- *Data interpretation*. This involves the comparison of water quality data from different stations. For specific problems, and the evaluation of the environmental significance of observed changes, external expertise may be needed. Publication and dissemination of data and reports to relevant authorities, the public, and the scientific community is the necessary final stage of assessment activities (Chapter 6).
- *Water management*. Decisions will be taken at various levels involving local, national and international bodies, and by water authorities as well as by other environmental authorities. Important decisions concern the redesign of assessment operations in order to improve the monitoring programme and to make it more cost-effective (Chapter 5).

In recreational water quality investigations, the purpose of sampling is to obtain samples that are as representative as possible with respect to the microbiological, physicochemical and aesthetic properties of the area (Chapters 8, 9, 11 and 12). Sampling should be conducted during the bathing season, but is most appropriate when recreational waters are suspected of being contaminated or a source of waterborne disease. Historical data, combined with an annual environmental health assessment, may indicate that only occasional sampling is necessary. If deterioration in quality has occurred then monitoring of the area should be undertaken. Such an approach will allow health officials to concentrate their resources on beaches of questionable quality.

2.3 Data collection

2.3.1 Beach registration

In order to improve the quality of recreational water-use areas and to select beaches that can be developed as tourist areas, planners and managers may wish to keep a continuous record of selected information. The information necessary for selecting those parts of the coastline that will be used as bathing beaches now or in the near or distant future can be stored in a beach registration system. The aim of such a system is to establish a catalogue of all beaches, to use a checklist to collect the information needed to plan a monitoring programme and to decide if and how the beach will be developed in the future.

The checklist for the registration of beaches may be amended as required; for example, to assess the hazards present for swimmers in a particular area in order to develop the beach for tourism, and to prepare a beach management plan for the planning and co-ordination of all the resources related to providing a safe aquatic environment for the public. This approach could be especially useful when resources are not abundant and have to be employed as efficiently as possible (Chapter 3).

Beach registration is typically divided into four components:

- *Description of the surroundings.* The registration should include information on accessibility (roads, tracks, public transport, no access), hazard mitigating measures (information signs and information sources, lifeguards, showers, first aid posts, swimming and diving safety warnings) and facilities (restaurants, hotels, bars, toilets, drinking water, litter bins, car parks and camping grounds).
- *Description of the beach.* This should include an estimation of the area of the beach (length, width), beach material and visitors per day (estimate the peak numbers according to season, whole bathing season, main holiday period, public holidays and weekends). The number of visitors per day should be compared with the visitor capacity of the area.
- *Description of the water environment.* This includes details of the bathing zone (direction and speed of the current, slope, bottom material) and its use (fishing, jetskiing, intensive yachting, swimming, diving, etc.).
- *Counter indications.* Designated sensitive areas (resting place for water fowl, breeding place for rare birds, sanctuary, conservation area and other kinds of protected area such as military sites or other areas where public access is prohibited).

The information listed above should be collected by means of a desk survey of the existing information and during a subsequent field inspection on the beach. The baseline information should be revised annually. Ideally a map should accompany each registration and should show the extension of the beach, the accessibility, the surroundings, etc. The information may then be transferred to a computer system and amended as necessary.

In gathering data for inclusion in a beach registration system it is important to involve the local community. Often local people, local politicians, shopkeepers and, in particular, those charged with the operation of the beach (the local authority, beach operator, lifeguard) will have valuable information. Relevant non-governmental organisations (NGOs) such as nature conservation groups, angling clubs and yacht clubs, water skiing clubs and lifesaving associations, may provide useful information (Chapter 6).

The development of a registration system in the form of a database could aid coastal managers in their decision-making process through highlighting the suitability of beaches for particular uses. For example, it may become apparent that some beaches should not be promoted for recreational use because the area is ecologically sensitive or because bathing might be dangerous due to currents, bottom conditions or particular health hazards.

2.3.2 Environmental health assessment

In the past, sanitary inspections or surveys were directed primarily towards microbiological contamination of recreational waters but in recent times they have been broadened to include chemical contamination and other biological and physical hazards. The term environmental health assessment is now used to reflect this broadened scope. An environmental health assessment can be defined as a comprehensive search and evaluation of existing and potential microbiological and chemical pollution and biological and physical hazards that could affect the overall safety of a particular stretch of recreational water or bathing beach. Potential influences on water quality (such as river mouths, sewage outlets, harbour areas, other wastewater outlets) and physical hazards (rocks, open and rough water, rip-tides, shallow water, etc.) should also be considered. A comprehensive environmental health assessment consists of pre-inspection preparations, an on-site visit and the preparation of an assessment report. The environmental health assessment typically relies upon on-site inspection of hazards and mitigating factors for physical and microbiological hazards (Chapters 7 and 8), and on water quality testing, especially for microbiological quality (Chapter 8). When undertaking an assessment the sampling techniques employed are particularly important. Some guidance for obtaining statistically valid measurements is given below. Full details of methods and quality assurance procedures to be followed in sampling programmes are provided in Chapters 3, 4 and 8.

2.3.3 Quality monitoring

During routine visits, the site should be surveyed for signs of microbiological and chemical contamination. For example, visible sewage plumes, oil slicks, suspicious odours and fish or bird kills should be considered as immediate indications of unacceptable water quality. Beach monitoring for litter, tar balls, etc. should also be undertaken. The task of deciding the optimum number of samples to take and the most suitable locations in order to characterise quality in a meaningful way, and with the most economic use of resources, can be quite daunting. Statistically-based methods of sampling design can help this task and can also ensure that the data collected are appropriate for later statistical analysis and interpretation (Chapters 3 and 8). Basic sampling design naturally falls into seven aspects:

- Reasons to sample.
- What to sample.
- How to sample.
- When to sample.
- Where to sample.
- How many samples to take.
- Sampling evaluation.

The issues of what, when and how to sample are defined by the assessment programme objectives.

The results of any quality monitoring programme depend on where the pollution comes from, and therefore on where the samples have been taken. The physical factors characterising sampling stations may vary widely between stations, resulting in large differences in analytical values; for example, in bathing water monitoring stations water depth, current speed and direction, existence of haloclines and/or thermoclines, mixing processes, sampling depth and distance to sewage outlets and other pollution sources may all affect water quality. In reservoirs and lakes the phenomenon of thermal stratification is a source of complexity in sampling design because of its potential affect on vertical water quality differences. The most usual basis for design in such circumstances is stratified random sampling.

Sampling sites should be selected on the basis of information gathered during the beach registration and the first on-site inspection. Ideally, the sites chosen should be representative of the water quality or beach area throughout the whole area where users are exposed. The selection of sites should pay particular attention to site-specific conditions that may influence the concentrations and distribution of indicator organisms and pathogens.

Monitoring and surveillance programmes generally rely on observations made on discrete samples obtained within spatial and temporal constraints. An essential component of a monitoring programme is ensuring that the sample obtained is representative of the phenomenon under study. Errors introduced during sample collection and preparation are usually several orders of magnitude higher than errors due to analytical determinations.

The main aspects to be considered for obtaining a representative sample are: the adequate selection of the sampling points, sampling stations, frequency and timing of sampling; the strict adherence to proper sampling and quality assurance procedures; the complete identification of the sample; the adequate preservation of the sample; and the prompt transport of the sample to the laboratory.

The exact location of sampling points in any monitoring programme, including the distance between them, varies with each individual beach. Chapter 9 provides a sampling protocol including sampling and analysis criteria to be followed for microbiological water quality. This should be used in conjunction with the guidance provided in Chapter 3 to adapt the monitoring programme to the resources available. An environmental health assessment in the recreational area provides a good basis for establishing the location and number of sampling stations. The results of an intensive sampling programme, together with a detailed survey of water currents and water discharges, will identify any particular pattern of water quality deterioration that has to be considered when selecting sampling stations representative of the whole recreational area (Chapter 9). The intensive sampling programme should include the analysis of water samples taken at different water depths, at different hours of the day, during different tidal phases, and during any other known source of possible variation. The experience gained during the implementation of the monitoring programme should serve to modify and improve the initial sampling programme. Advanced statistical analysis may be used to identify spatial and temporal patterns amongst sampling results drawn from special surveys. These patterns may then be used as the basis for more general sampling site allocation. This is a particularly important aspect for dealing with the variability associated with reservoirs.

Details of relatively simple types of water sampling equipment are contained in Bartram and Ballance (1996) (see also Chapter 8). For surface and subsurface sampling, the containers used should be bottles of dark-coloured borosilicate glass of 200 to 300 ml capacity, with wide-mouths and ground glass stoppers. The same type of bottle may be used for subsurface sampling with the addition of an extension arm and clamp. Specific sampling procedures are contained in the recommended methods for determination of specific indicator organisms and pathogenic bacteria described in Chapters 8 and 9. For sediments, several types of bottom samplers are available commercially that can be used for collection of samples of sediments for microbiological analysis. The equipment required to monitor aesthetic aspects of recreational water-use areas is generally less than for water quality monitoring and varies depending on the method used (Chapter 12).

Sampling should be performed in a systematic manner to reduce variation between individual results. For this reason, it is necessary to keep constant as many factors as possible. These include the period of sampling (i.e. time of day) and the sampling method, as well as the location and depth (in the case of water sampling) of individual sampling points. Sampling can be considered as completed once the sample is transferred to the sterile container, whether on the beach or aboard a sampling vessel.

When sample transit does not allow the use of a central laboratory, other alternatives must be considered. These may include analysis of samples in an approved laboratory nearby, use of an approved laboratory field kit or use of a mobile laboratory. Such alternatives should undergo thorough testing and comparison before they are adopted.

2.4 Elements of good practice

- The objective(s) of a monitoring programme or study should be identified formally before designing the programme and they should be stated prior to data gathering.
- Objectives should be described in a manner that can be related to the scientific validity of the results obtained. The required quality of any data should be derived from the statement of objectives and should be stated at the outset.
- In designing and implementing monitoring programmes, all interested parties (legislators, NGOs, local communities, laboratories, etc.) should be consulted. Every attempt should be made to address all relevant disciplines and to involve relevant expertise.
- The scope of any monitoring programme or study should be defined. This would normally take the form of definition of criteria for inclusion and exclusion of recreational water-use areas and preparation of an inventory of recreational water-use areas.
- A catalogue of basic characteristics of all recreational water-use areas should be prepared and updated periodically (generally annually) (and also in response to specific incidents) in a standardised format. It should include as a minimum the extent and nature of recreational activities that take place at the recreational water-use area and the types of hazards to human health that may be present or encountered. Unless specifically excluded, the list of potential hazards to human health would normally include the microbiological quality of water, cyanobacteria or harmful algae, drowning and physical hazards. Monitoring programmes frequently also address aesthetic aspects and amenity parameters because of their importance to health and well being.
- Programme or study design should take account of information derived from the inventory of recreational water-use areas and catalogue of basic

characteristics which, in turn, may require refinement of programme objectives.

2.5 References

Bartram, J. and Ballance, R. [Eds] 1996 *Water Quality Monitoring. A Practical Guide to the Design and Implementation of Freshwater Quality Studies and Monitoring Programmes*. E & FN Spon, London, 383 pp.

Chapman, D. [Ed.] 1996 *Water Quality Assessments. A Guide to the Use of Biota, Sediments and Water in Environmental Monitoring*. Second edition. Chapman & Hall, London, 626 pp.

Chapter 3*

RESOURCING AND IMPLEMENTATION

Monitoring programmes, by necessity, must be commensurate with the socio-economic and technical and scientific development of the country where they are implemented. For example the extent of development of national legislation and co-ordinating or oversight programmes will affect activities undertaken. Similarly, more complex analytical variables require highly trained technicians and costly laboratory facilities. In general terms, it is possible to distinguish three levels into which monitoring programmes can be classified (Table 3.1). All elements of assessment, from objective setting to data interpretation, are related to these three levels. The aim is to progressively develop monitoring operations from the basic level to the more comprehensive levels. Each level is associated with increasing demands on staff (expertise and numbers), inspection and fieldwork (complexity and frequency), laboratory facilities (range of analysis, throughput) and data management and reporting capacity.

3.1 Staffing and training

The personnel in charge of sample collection, field handling and field measurements must be trained for these activities (Table 3.2). The choice of personnel for sampling depends on a number of factors, including the geographic features of the region and the systems for transportation. For example, in a small country with good transport infrastructure, sampling may be carried out by laboratory personnel going to the field to take samples, conduct field analyses and transport samples back to the laboratory. In countries of a larger size that possess a more developed monitoring system, specially-trained field personnel often conduct the sampling and inspection. In large countries that have a poor transportation system, relatively more personnel are required. In this situation specialists from decentralised facilities, such as health centres or hydrometereological and hydrological stations, may be involved in sampling, inspection and testing. Such personnel may not always possess all appropriate training.

* *This chapter was prepared by G. Rees, J. Bartram, E.B. Pike and W. Robertson*

Table 3.1 Levels of monitoring competence in relation to resource requirements

Level	Basic information/ visit rate	Accident hazards[1]
Local (no national organisation yet)	Local action comparable to basic level in some locations only	Local action comparable to basic level in some locations only
Basic (no access to equipment or staff resources at national level; limited local resources)	At least one pre-season visit; creation of a catalogue of basic characteristics; all beaches registered, but more used and higher risk beaches inspected and and monitored	Annual inspection for identification of any hazards and interventions (e.g. signs, warning systems)
Intermediate (limited access to resources both local and national level)	Comprehensive cataloguing and timetabling of visits; additional visits during peak seasons (e.g. monthly); greater proportion of beaches monitored	Periodic verification of interventions during bathing season; central capacity for incident investigation
Full (no significant resource limitations)	Additional visits during peak seasons (e.g. fortnightly or weekly); complete cataloguing, including updating for each recreational area; all beaches with significant use monitored	Central register of recorded incidents; decentralised capacity and procedure for incident investigation

Each level also demands inclusion of the requirements specified at lower levels

[1] See Chapter 7

This guidebook provides a major part of the information needed for carrying out successful and adequate fieldwork and sampling. The quality of information produced by a monitoring programme depends on the quality of the work undertaken by field and laboratory staff. The importance of appropriate training cannot be over-stressed.

Separate training packages should be developed for field staff, laboratory staff and others. It should be emphasised that training is not a "once-only" activity, but should be continuous. Supervision, as a form of training, is especially relevant to laboratory and field staff. It is vital that the training function

Table 3.1 Continued

Microbiological parameters[2]	Cyanobacteria and algae[3]	Other
Local action comparable to basic level in some locations only	Local action comparable to basic level in some locations only	Local action comparable to basic level in some locations only
Inspection for faecal pollution or sewage odour; delimitation of high risk areas; initial screening of faecal streptococci (marine or freshwaters), *E. coli* or faecal coliforms (freshwater) for primary classification; internal quality control at laboratories; at least one sample a month in season once the beach is classified	Inspection for scum, type, and transparency	Register of local special problems
Identification and cataloguing of potential sources of contamination; all beaches at primary classification; monthly sampling; resampling and investigation of unexpected peak values; reclassification scheme initiated; investigation of rain effects and design of preventative measures; internal quality control at laboratories; occasional inter-laboratory comparison studies	Phosphate analysis (freshwater) Chlorophyll *a* (freshwater)	Check on local information availability; active warning and management response
Additional microbiological parameters if necessary; possible reclassification investigated where indicated; internal and external quality controls regularly operated; convergence amongst participating laboratories	Toxicity detection and toxin analysis capacity if necessary (not routine); remote sensing methods where relevant	Chemical monitoring (for necessary parameters)

[2] See Chapters 8 and 9 [3] See Chapter 10

is flexible, responding to experience and feedback and taking account of specific needs. Training is especially important when programmes for monitoring are implemented in several countries or independently managed areas or regions, from which the results will be compared and used outside the country or region where the monitoring has taken place.

Assuming a good general education or relevant previous experience, the training period for staff responsible for fieldwork and sampling is about one week, with approximately one additional week of further training as monitoring develops through basic, intermediate to full levels. In order to maintain

Table 3.2 Principal tasks undertaken by staff type as an indication of skills and training requirements for differing levels of monitoring competence

Level	Field staff[1]	Laboratory staff
Local	Basic inventory for beach registration Collection of water samples for microbiological and cyanobacteria analysis Transparency measurement (Secchi disc) Cyanobacterial scum recognition Basic sanitary survey	Data analysis and management Analysis of faecal indicator bacteria (according to indicator and method available)
Basic	More complex sanitary survey Selection of sampling sites Intensive microbiological sampling for primary classification Observational verification of effectiveness of interventions	Phosphate analysis Chlorophyll *a* analysis Organisation and implementation of necessary quality assurance
Intermediate	Local follow-up of unexpected peak microbiological values (including liaison with local authorities)	Participation in occasional, informal interlaboratory comparisons (probably "round-robin")
Full	Participation in accident investigation	Cyanobacterial toxicity detection and toxin analysis, where relevant Participation in regular, formal inter-laboratory comparisons

For a description of the different levels of monitoring competence, see Table 3.1

[1] The balance between field and laboratory staff is determined by local factors

motivation, it is recommended, that short follow-up events are provided. If the staff are less experienced more training will be necessary.

Although not exhaustive, training for field staff should include the objectives of the water quality monitoring programme and its local, national and international significance. The training should stress the importance of samples being of good quality and representative of the water body from which they are taken and it should give guidance on how to ensure that the samples meet those requirements. The training should also include planning of field sampling and map reading, as well as how to make field notes describing the sampling site and station and how to undertake on-site inspections. Safety aspects of field sampling are an important component of the training programme.

Staffing requirements for servicing a monitoring programme may vary widely and it is not possible to make general statements about the number of staff needed for fieldwork. Estimating staffing requirements includes allowing for travel time or distance between beaches and laboratory, the choice of laboratory infrastructure (e.g. centralised or decentralised) and co-ordination between participating institutions.

The head of the laboratory and/or the programme manager are generally responsible for:

- Laboratory management.
- Determining and procuring the equipment and supplies that will be required.
- Ensuring that Standard Operating Procedures (SOPs) are being followed (Chapter 4).
- Ensuring that adequate quality control procedures are being followed.
- Enforcing safety procedures.

Laboratory technicians must have suitable training. They will generally be responsible for:

- Maintenance of the laboratory including, cleanliness and safe storage of all equipment, glassware and other reusables.
- Storage and preparation of reagents and media.
- Checking the accuracy of field equipment.
- Training of junior staff.
- Performing the tests and recording the results of field analyses.

3.2 Laboratory and analytical facilities

The choice of laboratories to be involved in the monitoring programme can have a major influence on the time required for collecting the samples. There may be reasons such as ensuring staff training, equipment repair and analytical quality control, that suggests using a central laboratory is more appropriate. Nevertheless, when total sample numbers are high and samples are transported over long distances, it may be preferable to use more than one laboratory. Under these circumstances it is normal practice for one of the laboratories to act as the central or co-ordinating laboratory. Some countries have developed mobile laboratories (typically in small vans or minibuses) as an alternative solution to the problems of sample transport. Where long distances are encountered this has often been found to be an effective approach. The need for co-ordination between institutions, for example when sampling is undertaken by one agency and analysis by another, may reduce the necessary workload but may lead to inefficiency if the co-ordination is not effective.

In many countries, monitoring laboratories are organised on two tiers: regional laboratories (lower level) to conduct basic determinations not requiring very complex equipment, and central laboratories (higher level) to conduct more complex analyses requiring elaborate equipment and well trained personnel. In addition, the central laboratories often provide the regional laboratories with methodologies and analytical data quality control.

During the initial stages of development of a recreational water monitoring system, it is reasonable to focus on the basic variables that, as a rule, do not require expensive and sophisticated equipment. Gradually, the number of variables measured can be increased in relation to the financial resources of the monitoring agency. Even in fully-developed monitoring programmes, the elaborate equipment and technical skills necessary for the measurement of complex variables are not needed in every laboratory.

Laboratories must be selected or set up to meet the objectives of each assessment programme. Attention should be paid to the choice of analytical methods. The range of concentrations measured by the chosen methods must correspond to the concentrations of the variable in a water body and to the concentrations set by any applicable water quality standards. Ideally, a laboratory should consist of four sections:

- Reception and registration of samples.
- Production of analytical media.
- Sample analysis.
- Washing up and autoclaving equipment.

However, it is possible for a laboratory to function properly in fewer sections, depending on the number of analyses to be undertaken.

3.3 Transport and scheduling

Resource elements in fieldwork include transportation, personnel requirements and equipment. Problems of transport are largely related to distance and accessibility of sampling sites. Where access to sites is known to be problematic, reliance has been placed traditionally on four-wheel drive vehicles, but these are often expensive to purchase and to operate. It may generally be assumed that the main part of the beaches included in the routine monitoring programme can be reached using ordinary vehicles because beaches difficult to access have most probably been excluded when selecting the participating beaches. Some beaches may be located such that using motorcycles or public transport is a possibility; however, most frequently, ordinary cars equipped for transportation of the samples will be required. Sampling transport equipment can consist of an insulated box with melting ice, or a refrigerator installed in the vehicles, to ensure that the samples are kept cool.

Travel time to and from sampling sites is a major constraint for staff undertaking fieldwork. Realistic estimates of travel time to each sampling site should be made as early in the programme as possible. An inventory of sampling stations should be developed, including actual travelling time, in order to facilitate programme planning. When planning the route for visiting the sites, the demand of having the samples analysed at the laboratory within the appropriate time frame must also be taken into account (see Chapter 8 for microbiological analyses and Chapter 10 for cyanobacteria).

3.4 Inspection forms and programmes

Public health authorities should have, at least, a basic inspection programme in place for all recreational-water sites within their jurisdiction. The primary purpose of the inspection programme is to minimise the risk of illness or accident to bathers. The programme should be based on an SOP (refer to Chapter 4). This will ensure that the on-site inspection, laboratory analyses and interpretation of data are carried out in an objective and uniform manner. An on-site survey form should be prepared as a guide for inspectors to make certain that all aspects of the site receive adequate review and evaluation.

In most programmes at least two forms will be used. The first "basic registration" form collects the minimum background information to construct a register. The beach register is a high priority during the basic level of monitoring. The principal uses of the beach register are:

- To determine which beaches should be considered by the programme, based upon criteria such as extent of use, degree of development (or plans to develop) and already-recognised hazards.
- To provide data (such as transport options and travel times) to assist in programme planning.

Details of the information to be included in a beach registration form are provided in Chapter 2.

Once a beach enters into a monitoring programme it will be subject to periodic, usually annual, inspections. These inspections are principally orientated towards the identification of hazards that might lead to physical injury or contribute to drowning; towards the adequacy of measures (signs, lifeguards, communications) in place to reduce these risks; and towards sources of microbiological pollution such as sewage outfalls, combined sewer overflows, rivers and storm drains. At freshwater sites, the inspection may also be concerned with the likelihood of cyanobacterial blooms (see Chapter 10). An example of a sanitary survey form is included in Chapter 8 (see Box 8.1) and the components of an on-site inspection are discussed further in Chapter 7 (physical hazards), Chapters 8 and 9 (microbiological aspects) and Chapter 12 (aesthetic aspects). Other inspections may be required sporadically depending on local circumstances, for example to assess bather load (Chapter 8), cyanobacterial hazards (Chapter 10) or to verify the effectiveness of interventions to control microbiological quality (Chapter 9).

A key requirement of any inspection programme is the availability of qualified personnel. Ideally, inspectors should have a basic knowledge of public health microbiology, environmental chemistry, limnology, oceanography, estuaries and meteorology. Training workshops, based on SOPs and actual case studies, should be held during programme development and implementation. Additional workshops should be held periodically to review inspection procedures.

The co-operation of all involved or interested individuals or organisations is essential if assessments and monitoring programmes are to result in improved water quality and reduced health risks. Representatives from user groups, tourist associations, beach and resort owners, industries and sewage treatment facilities and public health laboratories should be aware of, and invited to participate in, the programme. The complexity and completeness of surveys are likely to increase as programmes develop from local through basic and intermediate to full levels (see Table 3.1). At the most advanced level, for each site, three to five days may be required to prepare for the survey, conduct an inspection and prepare the report, although subsequent re-verification visits may require less time. From these findings public health authorities will be able to classify the beaches within their jurisdiction with ratings ranging from excellent to very poor and to relate these ratings to requirements for monitoring and management. Chapter 9 discusses in depth the development of classification schemes for microbiological quality. In this way, resources can be directed to those beaches that present the greatest risks to public health.

3.5 Data processing and interpretation

If data have been collected in a careful manner in a properly designed study, they will be representative and unbiased and suitable for analysis, interpretation and reporting. Processing, management and storage of data can be collectively described by the term "data handling". This activity has a central position in any study as shown in Figure 3.1. The data handlers receive data from samplers and interviewers in the field, as well as the results of analyses from laboratories. They then assemble and archive the data in central files and analyse the data as instructed, so that they can be interpreted by those responsible for reporting the results. It is most desirable to prevent bias, from the introduction of personal or political perceptions, by arranging for separate groups of people to carry out the collection, handling and interpretation of data. Data handlers must be appropriately trained and qualified for their duties. It is also desirable that the data handlers do not have any particular interest in the results of the study, apart from carrying out their work with dedication and accuracy. The whole process of the study should be under the control of a study director, who will receive instructions on the objectives and on the programme to be carried out as well as statistical and other scientific advice from appropriate experts. In this way, the requirements for effective data management (Box 3.1) will be fulfilled. The period over which the data may be needed and the purpose for which they could be used should be

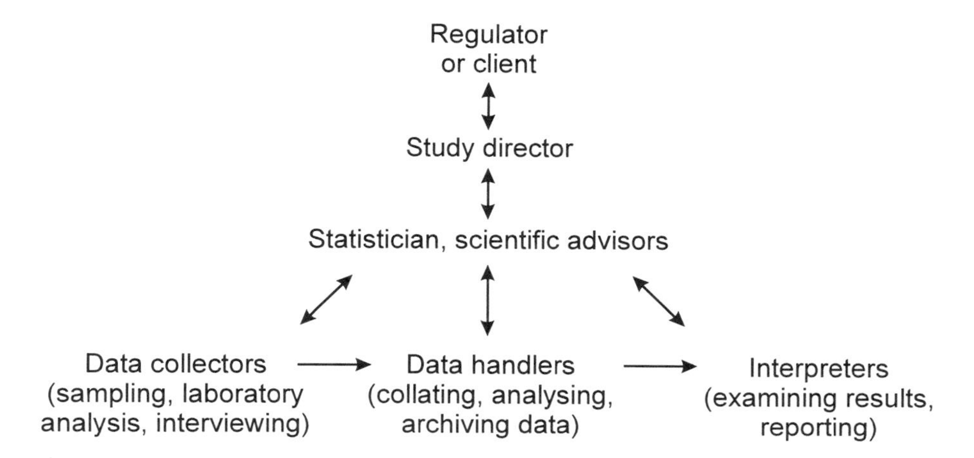

Figure 3.1 The central relationship of data handling in the conduct of monitoring programmes and scientific studies

considered at the outset, particularly for large or continuing studies. Computer hardware and software systems inevitably change over time and future compatibility could be a problem. For example, it is doubtful whether data stored on punched cards or paper tape some 30 years ago could now be retrieved. Moreover, changes in methods of analysis will reduce the compatibility of data from older studies with those of recent studies, thus rendering the raw data (but not necessarily the conclusions) less useful. Consideration should be given to storing data in several places, and in more than one format, in order to forestall the risk of loss and damage.

For the long term, it is vital to ensure that published reports carry adequate details of the methods of surveys and that the key data and basic statistics, such as means and standard deviations of measurements, are recorded as appendices to the final reports. The basic test of the adequacy of survey recording is whether an independent investigator could understand what was done sufficiently to be able to repeat the work and to be able to test the results and conclusions from the data presented.

The greatest cost of any study is the data collection. These data will be unique and there will usually be no opportunity of repeating a missing or erroneously recorded observation. The raw data sheets should be examined critically by the person in charge of collecting the data as soon as possible after recording, so that discrepancies can be detected and corrected while the events can still be remembered. It may be possible to check suspect meteorological and tidal records with those obtained at nearby, official recording stations. Different sets of data (e.g. from different sites, sampling runs or

Box 3.1 Requirements for effective data management

The requirements of the monitoring programme must be defined. This is the responsibility of the study director, instructed by the regulator or client and as advised by the statistician and scientific experts. In the case of a scientific survey or epidemiological study, the effects under study are framed as conceptual models, requiring the testing of null hypotheses by statistical methods. The data required are determined by the monitoring schedule imposed by the regulatory authority or, in the case of a survey, by the desired statistical power.

Appropriate forms or other recording instruments are produced for collection of the raw data required to meet the above requirement. These must be approved by the collectors and handlers of the data, so that all suppliers of data use a standard, approved format. It is essential that the raw data are in a form that can be transcribed easily and with minimal risk of error to a permanent file, suitable for processing and archiving.

The data are analysed by appropriate methods to produce the desired output. The appropriateness of the methods must be decided earlier in fulfilment of the first requirements. Suitable analytical quality control procedures must be used for input and analysis of data.

days) must not be pooled, unless it has been shown that there is no significant statistical difference between the data sets.

It must be remembered that data are collected for a specific purpose, and thus they may not be useful for meeting subsequent objectives. Most data from routine, regulatory monitoring fall into the "data-rich, information-poor" category because other information, necessary for explaining the circumstances or trends, was not recorded at the time. The use of statistical methods to test *a priori* null hypotheses is an obligatory part of scientific method. However, the temptation to carry out further analysis of the complete data set at a later date, in order to detect statistically significant associations or correlations between variables, is strictly invalid unless used solely as a method of suggesting hypotheses for further study. Such "data dredging" is analogous to fitting targets to holes (Jolley, 1993).

The effects studied in behavioural and epidemiological research of water recreation are typically small and are difficult to separate from confounding factors (i.e. factors that affect the responses of the exposed and control groups differently). Potentially confounding factors must be identified at the planning stages of a study and suitable measures should be introduced into the design of the study (e.g. by the matching of exposed and non-exposed subjects, or by random assignment) and in the analysis of the data to detect or nullify them. Confounding can be detected or corrected for in analysis by various methods, such as stratification of the exposed and non-exposed groups (e.g. by common potential confounders such as age, sex or socio-economic class) or by including potential confounders as variables in multi-variate analyses, such as logistic regression analyses (for a description, see McCullagh and Nelder, 1989). It is not possible to eliminate the effects of confounding in the data, if it has not been considered, measured and examined in the design of the study (see Chapter 13) (Datta, 1993; Leon, 1993; WHO, 1998).

3.5.1 Database construction

The construction of the database will depend on:

- The level of monitoring as defined in section 3.2 and the intended use of the data, e.g. for compliance monitoring, risk assessment, baseline data, acquisition, new scheme design, epidemiological investigations and/or post-audit evaluations or remediation efficacy.
- The technology available.
- The requirement for data transfer to regulators or to other agencies.
- The requirements for "data audit" and "chain of evidence" procedures.

There are many data storage systems available that have been developed by laboratories and software engineers. Perhaps the most stringent and sophisticated database would be characterised by a data storage system able to accommodate the information suitable for a full "chain of evidence" assessment. This type of database is being developed in drinking water surveillance monitoring. The reason for the high level of laboratory audit stems from the potential litigation that could derive from contaminated waters and associated disease outbreaks. This type of database is emerging in drinking water assessment in the UK following a consultation document produced by the Drinking Water Inspectorate (DWI, 1998). The requirements of such a data storage programme are that the data could be used in a court of law to achieve a criminal prosecution. To achieve this the prosecution is generally required to demonstrate that the data are accurate beyond reasonable doubt, i.e. the analyst can prove that there are systems in place to prevent tendentious or accidental misreporting through laboratory analytical error or database construction mistakes (such as recording a result from the wrong sample).

Laboratory audit control is now routine in laboratories undertaking compliance assessment programmes for recreational and drinking waters in some countries. Here, the recording laboratory must be in a position to prove that the recorded value is correct and that laboratory procedures have been followed. The level of internal control is less than for "chain of evidence" assessments, i.e. the laboratory would have to prove that all reasonable measures had been taken but it would not have to demonstrate that there was no error "beyond reasonable doubt".

Any environmental sampling programme should implement appropriate intralaboratory and interlaboratory analytical quality control (AQC) procedures. These would normally involve the collection of field and laboratory blank (e.g. sterile) samples to investigate the integrity of aseptic techniques and field duplicate samples to investigate reproducibility. Generally, it is preferable to use split samples from the same bottle in the case of duplicate enumerations. In inter-laboratory trials duplicate samples are split, or spiked samples are prepared and delivered to participating laboratories for analysis. Analytical quality control is described in full in Chapter 4.

There are many commercial data storage software systems available. The developing international standard is a spreadsheet that facilitates storage, rudimentary statistical analyses, graphical representation and data export for external communication and reporting. Spreadsheet packages are commonly used for the storage and recording of recreational water quality data worldwide. Such systems are appropriate for the user familiar with computation methods and standard package use. They may require adaptation for the lay user or to make them effective tools for clerical staff with a low level of information technology (IT) skills. In such circumstances, the spreadsheet can be modified with a Graphical User Interface (GUI). This facilitates rigid, but user friendly, data access and export, appropriate and tightly controlled analysis and, where appropriate, data cleaning and security. Perhaps the most widely used programming language in the production of a GUI is Visual Basic. This language can be used to design forms and screen menus to control data input, data analysis and reporting, as well as to provide security through the use of passwords. Visual Basic programming requires a competent programmer but the advantages in its use for data security and quality are very significant.

The data storage system should be appropriate to the intended purpose of the storage exercise. If the objective is to demonstrate chain of evidence or a future audit of the data for compliance, a bespoke system will probably need to be constructed to facilitate clerical input, data checking by non-technical staff, data cleaning after extreme values have been flagged, and clearly defined data reporting as defined by legislation. Here, linking a GUI with a spreadsheet through Visual Basic is probably the most appropriate route. Where the data are simply being stored for scientific and/or baseline definition purposes, and no immediate audit beyond normal AQC is required, the spreadsheet alone can be a suitable vehicle for data storage.

3.5.2 Preliminary examination of data

It is important, in the interests of consistency, to agree at the outset the procedure to be used for dealing with missing, indeterminate and outlying values, and to adhere to these procedures consistently. Sets of data frequently contain missing values. These values can often be estimated from trends in the data set (where the subsequent analysis requires a complete set) but it must be realised that such "patching" reduces the number of degrees of freedom on which to base statistical decisions. Missing data cannot be estimated or reported when monitoring to assess compliance with a standard for water quality.

Indeterminate values are frequently found, implying that the volume of sample examined, or the concentration of the determinand, was too small or too large to be within the limits of the method. The analysts should be instructed to record the facts, i.e. "less than" or "greater than" the analytical limit, together with the volume examined. This, at least, enables a rank order of values to be established. Procedures for estimating a mean from data with indeterminate values are described below. The particular difficulties encountered in microbiological analyses are explained in Chapter 8.

Detailed examination of data often reveals values that lie outside the normal range of values or trends in the data and which therefore seem improbable (known as "outliers" or doubtful values). Some computer programmes are able to identify outliers, according to defined criteria. Such doubtful values must be investigated by going back to the original records and, if at all possible, to all laboratory personnel and samplers who were responsible for obtaining the value recorded. Only if there are strong technical reasons (e.g. contamination of a batch of culture medium, or a fault in a recording instrument) should such values be deleted from the data set. There are no valid statistical reasons for excluding outliers from sets of data. Their occurrence provides a strong case for investigation of sampling and laboratory procedures. They also indicate the value of carrying out simple checks for the consistency of data in the laboratory at the time when they are first recorded. If no technical problem is found, such values must be accepted. These values could be the first indication of a change in water quality or they may be a random, infrequent event.

Due regard must be taken of the underlying nature of the probability distribution within the data collected, because this will determine the most appropriate way of expressing the central tendency and dispersion of the data. This should be taken into account when standards are derived, because standards will invariably specify their requirements and limiting values in terms of statistics, such as the average, median or geometric mean, or an upper percentile value.

Table 3.3 Examples of the different probability distributions which may be encountered in surveys of recreational water use areas

Distribution	Properties and examples
Discrete (whole-number) random variables	
Binomial	Results of a sequence of independent trials, specified in advance and with two possible outcomes ("yes", "no") and constant probability of success from trial to trial.
Poisson	Describes occurrence of random events in a continuum (e.g. annual deaths by drowning in a region or counts of randomly distributed particles in independent, identically sized samples); variance and mean are identical.
Continuous random variables	
Normal	Conforms to the normal probability density function, distributed symmetrically about the arithmetical mean (average); mean and median are identical. Distribution is described by a mean and the standard deviation. Heights and weights of individuals, errors of analysis and data for many physical and chemical measurements usually approximate to normality.
Log-normal	The logarithms of the values are normally distributed. Distribution is described by the geometric mean, which is equivalent to the median, and by the standard deviation of log values (the log standard deviation). Generated by random variation of the rate constant in natural processes subject to exponential decay or growth; sets of microbiological counts or chemicals diffusing in water, or particle-size analyses of sediments are therefore typically log-normally distributed.
Ordered, categorical variables	
	Data can be arranged and ranked in order of size or categorised, but do not have discrete or continuous values. For example, water users can be arbitrarily ordered according to their observed degree of contact with water (i.e. none, wading and paddling or head immersion). In other cases, the categories do not have a size or implied order (e.g. water-contact recreation may be swimming, surfing, rafting or diving). Data can be tested by appropriate non-parametric methods.

Parametric tests for statistical significance assume that the data conform to a particular model of distribution, such as the normal distribution, and thus are only valid when applied to data that are known to conform approximately to that distribution, or can be transformed appropriately so that they do so. This is a problem where statistical advice must be sought before attempting analysis of data. However, many non-parametric tests of significance, such as those using ranked values instead of actual numbers, are inherently distribution-free, but they lack the statistical power of the parametric tests.

Table 3.3 lists examples of frequency distributions that are commonly encountered in the data collected in surveys of recreational water-use areas, together with their properties. Invariably, there will be data sets that do not conform to the frequency distributions listed in Table 3.3 and, at best, only an approximate fit to the frequency distributions will be possible. A full treatment of probability distributions and their properties can be obtained from textbooks of probability and statistics (e.g. Devore, 1991).

The use of the mean and the standard deviation to describe central tendency and variability in data is appropriate if the data are distributed normally or approximately normally. This is usually the case with physical and chemical data. Microbiological, virological and biological data are almost invariably found to be skewed and distributed log-normally or approximately log-normally. This is thought to occur because of growth, decay and dispersion processes in natural waters, which tend to follow exponential (first-order) reactions, in which the rate constants are subjected to random environmental fluctuations. Skewness caused by log-normality can be detected in several ways. It should be suspected if the data contain many relatively small values and relatively few very large values, or if the average (arithmetical mean) is much larger than the median.

Tests for skewness are given in many statistical textbooks. Log-normally distributed data can be made to conform to the normal distribution before analysis if they are first transformed to logarithms. The geometric mean (i.e. the antilog of the average of the log values) and the standard deviation of the logarithms are the appropriate statistics for such data. The geometric mean cannot be calculated if any of the values are zero (e.g. "undetectable in the volume examined" or "less than the limit of detection"). In such cases, the median can be recorded as an equivalent to the geometric mean, with an explanatory note, or a log $(x+1)$ transformation can be used, by adding 1 to all the values of x before taking log values. The reverse transformation is used in expressing the geometric mean.

Other transformations are sometimes used. Whole number counts of random events, such as deaths by drowning, usually conform to the Poisson distribution. This should be suspected if the variance (the square of the standard deviation) of the data is similar numerically to the mean. The appropriate transformation is to take square roots of the values, if they lie in the range 10–100, or $\sqrt{(x + 0.5)}$ if the values are less than 10. If most of the numbers are greater than 100, transformation is not needed. In the case of values which are rates or ratios, such as velocities (m s^{-1}), reciprocals should be used and the correct mean of the n values of x is the harmonic mean:

$$\frac{1}{\sum\left(\frac{1}{x}\right)/n}$$

Skewness and the effects of transformation can be detected by constructing frequency distributions. This is most readily done by graphical analysis on the computer, but may also be done with small data sets by constructing cumulative frequency plots on normal probability graph paper. Such plots will approximate to straight lines, with slopes proportional to the standard deviation if the data are normally distributed, or if the transformation is appropriate.

3.5.3 Internal data check mechanisms

Mechanisms for checking the quality of data require links between sampling and AQC (which is achieved through rigorous programme design) and appropriate data checking mechanisms. For purposely built systems, the links can be built into the GUI or Spreadsheet or Graphics system. For example, the implementation of field and laboratory blank (sterile) and duplicate samples, together with participation in inter-laboratory AQC programmes, should ensure the numerical integrity of the data acquired, provided appropriate dilutions are employed. Thus, data reporting unexpected low or high values should provide a true representation of indicator bacterial concentration or physicochemical parameters. However, the entry of the data to the database offers a further opportunity for automated data checking, principally in the form identifying outliers in the data set. The definition of an outlier requires a historical data set. Identification of a specific data item as an outlier value is based on a knowledge of the statistical distribution of the environmental determinand. In the case of microbiological data, it is generally accepted that environmental data measured at recreational waters follow a $\log_{10}$-normal probability density function. Thus an outlier could be defined as a data item where the $\log_{10}$ value was greater than, for example, two standard deviations from the historical $\log_{10}$ mean value.

This above approach is perhaps the most scientifically valid, although, it presupposes historical data and has a significant problem which derives from the choice of a cut-off at which the data item is considered an outlier. It is certainly the case that microbiological concentrations in recreational waters can commonly increase by several orders of magnitude (i.e. 3–4) following rainfall events. Thus, the definition of a numerically low cut-point tied to an automatic data cleaning system designed into a GUI and operated by clerical staff with little scientific insight into the acquired data, might result in very significant data loss from precisely the most high-risk periods against which the system was seeking to provide protection. For this reason, it is essential that data cleaning is closely supervised by competent scientific staff and that any automated systems simply define the items on which a scientific decision is required, rather than being allowed to carry out any automated change to the raw data matrix by deletion of apparent outlier values.

3.5.4 Determining compliance with a standard

Many monitoring programmes are designed to ascertain compliance with a standard or other objective for water quality. The standard should be carefully checked to determine the method that will be needed to assess compliance. Any deviation from the specified method will lead to doubts about comparability of the conclusions between parallel programmes in different regions, nationally or internationally. The compliance method will have three basic components: the design of the sampling scheme, (including the number of samples, their frequency and the period of sampling); the description of water quality (such as the chemical, physical or microbiological variable), the units of measurement and the descriptive statistics used to describe the level attained in the set of observations (such as the range of values, the average, median, or geometric mean, the standard deviation, or a given percentile value); and the criterion for judging passing or failing of the standard, or for classifying the quality at the water recreation area.

Water quality at a particular site is, essentially, a continuous population of measurements of quality, from which a sampling programme can only provide a limited number of discrete measurements. Because of the errors of measurement that occur when a small number of samples are taken, the sampling programme only gives an approximate estimate of the conditions existing over the whole period of the programme. These errors may, or may not, be regarded as important, although they will have most significance when assessing compliance of waters which are of borderline quality in terms of the standard. Once this problem is recognised, it is important to consider the burden of proof required to assess compliance. It is most usual to take results at their face value, without taking account of sampling error. This is justifiable, but is an empirical decision which will always carry a risk of misclassifying waters, particularly if small numbers of samples are taken. Alternatively, allowance may be made for sampling errors by adopting a "benefit of the doubt" or a fail-safe approach, so that when water quality is borderline the risk of incorrectly failing or passing the standard, respectively, is acceptably small and defined in statistical terms. A general description of these problems in the design of water quality monitoring programmes has been given by Ellis (1989).

Most standards specify a limiting value, which may be either a measure of central tendency, such as a mean or median value obtained over a specified period, or an upper limit which must not be exceeded. An upper limit may be absolute, or it may allow for the natural variability that occurs with time and which is caused by such factors as weather and tidal conditions. Such allowance can be made by defining an upper confidence limit, defined statistically by the observed variability or by the percentage of samples (percentile) which must not exceed the upper limit.

3.5.5 Data presentation

In all except the smallest surveys, the study director and data handlers will wish to use electronic computation for speed and accuracy. They should, however, use reliable and proven statistical packages and assure themselves that the calculation routines give the correct output. Manual calculation may be the only method available to small teams, handling small data sets and lacking computation facilities, or for field use. It can also be invaluable as a means of checking the data as they are produced, or for checking the output from computation. Desk calculators that provide a printed output of the operations and calculations are recommended because the printout provides a means of checking errors of input.

Graphical methods can be used manually when the data set is small. Linear probability graph paper, which has ordinates ruled in equal divisions and abscissae ruled proportionally to the percentage points of the standard normal density function, can be used to check that the data (transformed if necessary) conform approximately to the normal distribution. If so, the graph paper can then be used to estimate the median, other percentile points and the standard deviation of a set of data, with two-figure accuracy.

Much data processing is associated with assessments of microbiological quality and this is described in Chapter 8. Water quality data points may be plotted sequentially on a chart to indicate changes of quality with date of sampling. The occurrence of high values may be used to initiate investigation. In these circumstances the plot becomes a control chart. The choice of a limit value and appropriate action, for when the limit is exceeded, will be chosen to suit the need. The limit value can be set to coincide with the value given in a standard. More conventionally, two upper limit values are set on a control chart: at the mean plus twice and at the mean plus three times the sample standard deviation. These represent values that would be expected to be exceeded only once in 20 or 100 samples and which indicate, respectively, a warning and the need for remedial action.

Use of a computer is obligatory for analysing large sets of data. Even when it is possible to use a hand-held calculator or even to use graphical methods, the computer is able to produce results free from error and, in many cases, to reduce the total time expended in data analysis. There are many statistical packages available that can take raw data from a spreadsheet and subject them to statistical analyses and in many cases use them to produce excellent graphics. Such packages are available to suit different needs. For general use, typical packages that are commercially available are MINITAB Statistical

Software, SAS (Statistical Analysis System), SPSS (Statistical Package for the Social Sciences) and BMD (Bio-Medical Data programs). Specialised computer programs include GENSTAT (a comprehensive collection of statistical programs available from Numerical Algorithms Group Ltd) and EpiInfo (public domain software for epidemiological investigations). Each of these has graphical capabilities adequate for most users, although they can also produce output appropriate for more sophisticated and internationally available graphics and publishing software.

3.5.6 Data interpretation and communication

Interpretation of data and communication of results are the final two steps in an assessment programme. Correctly interpreted data will not be of much use if they are not disseminated to relevant authorities, to scientists and the public, in a form that is readily understandable by, and acceptable to, the target audience. The form and level of data presentation is, therefore, crucial. Often, it is advisable to produce two types of reports: a comprehensive, detailed report containing all relevant data and associated interpretation and an executive summary (in an illustrated and simple form) which highlights the major findings. Usually, the interpretation of data is undertaken by specialised professionals, such as the relevant scientists, (e.g. microbiologists, chemists or epidemiologists) the data treatment team and professionals from other organisations such as environmental protection agencies, health authorities, national resource agencies. As a courtesy, the results and recommendations should be discussed with all interested groups and individuals before reports are formally released. A contingency plan should also be developed with the assistance of all those with a vested interest to investigate and respond to cases of adverse health effects or to any unforeseen event or conditions that could lead to a deterioration in water quality and possibly increase the risk of illness to bathers.

A very important part of any sampling exercise is to review the extent to which the desired objectives have been achieved. There are many reasons for periodic adjustments to any assessment or monitoring programme for recreational water quality. The initial objectives may have been achieved and the programme may need reorientating from a baseline study to routine monitoring, with the establishment of new objectives and possibly the addition of supplementary monitoring variables or the substitution of existing activities with new activities. Once the samples have been taken, contemporary information is available on distributions and variability. If the required precision has not been achieved, these new data may be used to establish how many extra samples are required, and how the sampling strategy may be further optimised. If necessary, any extra sampling may then be carried out immediately.

3.6 Elements of good practice

The logistical planning of any monitoring programme or study should take account of socio-economic, technical or scientific and institutional capacities, staffing, equipment availability, consumable demands, travel and safety requirements and sample numbers, without compromising achievement of the objectives or scientific validity of the programme or study.

- The hierarchy of authority, responsibility and actions within a programme or study should be defined. All persons taking part in the programme or study should be aware of their roles and interrelationships.
- Staff should be trained adequately and be appropriately qualified, including in respect of health and safety aspects.
- Collection of data and information should use the most effective combination of methods of investigation, including observation, water quality sampling and analysis, interview of appropriate persons and review of published and unpublished literature.
- Frequency and timing of sampling and selection of sampling sites should reflect beach types, use types and density of use, as well as temporal and spatial variations in the recreational water-use area that may arise from seasonality, tidal cycles, rainfall, discharge and abstraction patterns, beach types and usage.
- Sampling should provide a data set amenable to statistical analysis.
- Data handling and interpretation of results should be done objectively without personal or political interference.
- The need for transformation of raw data, before analysis, to meet the conditions for statistical analysis should be agreed with a statistical expert before commencing analysis.
- Data handlers and collectors should agree on a common format for recording results of analyses and surveys and should be aware of the ultimate size of the data matrix. Forms and survey instruments should be compatible with this format. Likewise, data handlers should agree on a format for the output of results with those responsible for interpreting and presenting the data.
- Procedures for dealing with inconsistencies, such as omissions in records, indeterminate results (e.g. indecipherable characters, results outside the limits of the analytical methods) and obvious errors should be agreed in advance of data collection. On receipt from the data collectors, record forms should be examined and the agreed procedure followed. Discrepancies should be referred immediately to the data collector for correction or amendment. Where re-sampling is impossible, estimates are preferable to leaving gaps in the data record, (estimates should always be recorded as such) although they will reduce the statistical degrees of freedom.
- Ideally, arrangements should be made to store data in more than one location and format, to avoid the hazards of loss and obsolescence. Data should

be transcribed accurately, handled appropriately and analysed to prevent errors and bias in the reporting.

- The statistical routine should be selected by a statistical expert.
- Data should be handled and stored in such a way to ensure that the results are available in the future for further study and for assessing temporal trends.
- Data should be interpreted and assessed by experts with relevant recommendations for management actions prior to submission to decision makers. Interpretations should always refer to the objectives and should also propose improvements, including simplifications, in the monitoring activities — stressing the needs for future research and guidelines for environmental planning.
- Interpretation of results should take account of all available sources of information, including those derived from inventory, catalogue of basic characteristics, sanitary and hazard inspection, water quality sampling and analysis, and interview, including any historical records available.

3.7 References

Datta, M. 1993 You cannot exclude the explanation you have not considered. *The Lancet*, **342,** 345–347.

Devore, J.L. 1991 *Probability and Statistics for Engineering and the Sciences,* 3rd Edition. Duxbury Press, Belmont, California.

DWI 1998 *Standard Operating Protocol for the Sampling of* Cryptosporidium *Oocysts in Treated Drinking Water.* Drinking Water Inspectorate, London.

Ellis, J.C. 1989 *Handbook on the Design and Interpretation of Monitoring Programmes.* WRc Publication NS 29, WRc plc, Medmenham, Marlow, Bucks, UK.

Jolley, D. 1993 The glitter of the table. *The Lancet,* **342,** 27–29.

Leon, D.A. 1993 Failed or misleading adjustment for confounding. *The Lancet,* **342,** 479–481.

McCullagh, P. and Nelder, J.A. 1989 *Generalised Linear Models*, 2nd Edition, Chapman & Hall, London.

WHO 1998 *Guidelines for Safe Recreational Water Environments.* Draft for consultation. World Health Organization, Geneva.

Chapter 4*

QUALITY ASSURANCE

Quality Assurance (QA) is a management method that is defined as "all those planned and systematic actions needed to provide adequate confidence that a product, service or result will satisfy given requirements for quality and be fit for use". A Quality Assurance programme is defined as "the sum total of the activities aimed at achieving that required standard" (ISO, 1994).

Any monitoring programme or assessment must aim to produce information that is accurate, reliable and adequate for the intended purpose. This means that a clear idea of the type and specifications of the information sought must be known before the project starts, i.e. there must be a data quality objective. Data quality objectives are qualitative and quantitative specifications that are used to design the system that will limit the uncertainty to an acceptable level within the constraints allowed. These objectives are often set by the end users of the data (usually those funding the project) in conjunction with the technical experts concerned.

Quality Assurance for a recreational water monitoring programme will, apart from helping to ensure that the results obtained are correct, increase the confidence of funding bodies and the public. Quality Assurance extends to all aspects of data collection from sanitary surveys to laboratory procedures. Unless the data can be checked they should not be included in any assessment; unconfirmed observations have little value and can result in misclassification.

4.1 Components of Quality Assurance

The components of a QA programme are often grouped into three levels, variously labelled: the strategic or organisational level (dealing with the quality policy, objectives and management and usually produced as the Quality Manual); the tactical or functional level (dealing with general practices such as training, facilities, operation of QA); and the operational level (dealing with the Standard Operating Procedures (SOPs) worksheets and other aspects of day to day operations).

* *This chapter was prepared by A. Storey, R. Briggs, H. Jones and R. Russell*

4.1.1 Setting up the system

There is no single method for establishing a QA system. Each organisation has its own problems that will require special consideration and planning. However, once the decision to implement a QA system has been taken and the necessary funds and facilities have been made available, then a plan must be drawn up. For a new project the QA system can be drawn up before the start but if the project is already established then a QA system can be retrofitted. In the latter situation, existing practices must be evaluated with respect to QA needs and any QA checks and procedures that are already in place. It is better to build on procedures already in place and only to remove them if they are clearly unsatisfactory. If too many changes are imposed too quickly, especially where they are seen to increase work load, they are unlikely to be met with a favourable response and implementation will be poor. The QA programme must be seen to be practical and realistic and not to include trivial or unnecessarily time-consuming or difficult tasks (WHO/UNEP/VKI, 1997).

4.1.2 The Quality Manual

The Quality Manual is composed of the management documents needed to implement the QA programme and includes (ISO, 1990):

- A quality policy statement, including objectives and commitments.
- The organisation and management structure of the project, its place in any parent organisation and relevant organisational charts.
- The relationship between management, technical operations, support services and the quality system.
- Procedures for control and maintenance of documentation.
- Job descriptions for key staff and reference to the job descriptions of other staff.
- Identification of approved signatories.
- Procedures for ensuring traceability of all paperwork, data and reports.
- The laboratory's scope for calibrations and tests.
- Arrangements for ensuring that all new projects are reviewed to ensure that there are adequate resources to manage them properly.
- Reference to the calibration, verification and testing procedures used.
- Procedures for handling calibration and test items.
- Reference to the major equipment and reference measurement standards used.
- Reference to procedures for calibration, verification and maintenance of equipment.
- Reference to verification practices including inter-laboratory comparisons, proficiency testing programmes, use of reference materials and internal quality control schemes.
- Procedures to be followed for feedback and corrective actions whenever testing discrepancies or departure from documented procedures are detected.

- Procedures to be followed for feedback and corrective actions whenever testing discrepancies or departure from documented procedures are detected.
- Complaints procedure.
- Procedures for protecting confidentiality and property rights.
- Procedures for audit and review.

4.1.3 Training

The development of the programme must include all staff. Typically, the management commit resources, establish policy and standards, approve plans, assign responsibilities and maintain accountability. The supervisory staff take responsibility for the development and implementation of the programme and operating personnel provide technical expertise and advice. At all stages, the operating personnel must be consulted about the practicalities of any proposed changes. In turn, they must notify management of any problems or changes that may affect the programme.

4.1.4 Standard Operating Procedures

Standard Operating Procedures (SOPs) are the documents detailing all specific operations and methods, including sampling, transportation, analysis, use of and calibration of equipment, production of reports and interpretation of data. They are the internal reference manual for the particular procedure and should detail every relevant step. Anybody of the appropriate training level should be able to follow the SOP. They should, where necessary, cross-reference other SOPs and refer to them by number. Method SOPs may originate from organisations such as the International Organization for Standardization (ISO), British Standards Institute (BSI), American Standard Technical Method (ASTM) or from the instructions that come with the test kit where a commercially produced method is used. Such SOPs have the advantage of not requiring verification and save time in writing "in-house" SOPs. However, if they are used they must be used without modification. If any modification at all takes place, the alterations must be documented. Sometimes "in house" methods are preferred, and it is vital that such methods are properly verified. This may be done by reference to scientific literature and by "in house" validation.

The procedure should be written in short, clear sentences. Equipment SOPs should include methods and frequency of maintenance, cleaning, calibration and servicing. Method SOPs should include all the information necessary to carry out the procedure without reference to other documents with the exception of fully documented SOPs. Any statements regarding ranges for measurement variables such as temperature, weights, etc. should be within the scope of the facility, i.e. not so wide that they affect the result but not so narrow that they are not practically achievable or necessary.

Calculations should include any equations and demonstration of statistical control. Where applicable, criteria for the acceptance of data should be stated and acceptable ranges quoted. Disposal methods for reagents, test materials and other consumables should also be stated.

Some SOPs, such as those for office procedures, will be customised. The person most technically competent to carry out the procedure described should write the SOP. An SOP should have a descriptive title and also have a unique reference and version number. The purpose of the SOP should be stated alongside the variables measured, the expected range of values, the limitations of the method and the expected precision and accuracy. Any documents regarding the source of the method should be stated. Safety notes should include any foreseeable risks involved in the procedure, alongside procedures to minimise risk and procedures in case of an accident. Any special training required for the operator, and special apparatus required for the procedure (including all reagents and materials required) should be stated along with such information as the grade, reference number, size and company of origin. The storage, handling, recording and subsequent disposal of the sample should also be covered in the SOP, including storage temperatures, sample splitting, traceability, and any other issues. The style and format of the final data report should be given where applicable and reporting procedures and archiving requirements should also be included.

4.1.5 The Quality Assurance manager

For larger projects, proper management of QA will require the appointment of a QA manager to liaise with staff, to manage data archives, to conduct regular audits and reviews and to report on any QA issues. The manager is responsible for inspecting all aspects of the system regularly to ensure compliance, for reporting on such inspections and audits to management and for recommending improvements. These activities involve inspecting facilities and procedures regularly, tracing samples and documents back through the system and ensuring that all appropriate records have been kept.

Where QA is the responsibility of a separate section within an organisation many of the management difficulties are minimised. Appointment of a full time QA manager is difficult in a small organisation and in these cases the responsibility for QA should be assigned on a part-time basis, to a suitable member of staff.

4.1.6 Auditing and checking compliance

When all the documentation for the QA system is in place, it should be piloted. During this time, the QA manager should conduct a series of audits covering all aspects of the system. Traceability of data is a key component which can be checked by picking data at random and tracing them back through all relevant paperwork to the sampling procedure. A review of the system with positive and negative areas clearly defined should be written at the end of the pilot phase.

One method of implementation is to apply for accreditation from a recognised QA system. The ISO standard, ISO 9000, is suitable for the monitoring programme as a whole and is available in many countries. These systems are expensive but do allow the QA programme to be assessed independently against an agreed standard. Sometimes formal accreditation is required by regulatory and commercial bodies.

4.1.7 Maintaining Quality Assurance

In order to maintain the QA system, it is necessary to check periodically each area of the system for compliance. This involves auditing the component parts to assess whether they continue to meet the original criteria. This procedure should be formerly documented. Reports on all audits should be made available to management and to the persons responsible for the work concerned. Deviations from required standards must be corrected as soon as possible. The audit must be independent, and should be thorough and unannounced.

4.2 Equipment maintenance and calibration

All equipment, whether site, office or laboratory, must be maintained on a regular basis as documented in the relevant SOPs, codes of practice and manufacturer's guidelines. Laboratories must apply standards within the limits established for the care of a particular piece of equipment. This applies to general equipment, such as glassware, as well as to sophisticated analytical instruments and vehicles. It especially applies to field equipment.

The care and cleaning of equipment is very important to ensure analytical quality. Regular internal and external calibration checks must be performed on equipment such as balances, pipettes and pH meters. The frequency of these checks depends on the stability of the equipment in question but should be based on established practice. The form and frequency of these checks should be documented in the relevant SOPs. Calibration and maintenance records should be kept for all equipment, thus allowing the repair status to be monitored.

4.3 Sampling

Any analysis can only be as good as the sample taken. Variations in sampling procedures can have a marked effect on the results of analysis. It is very difficult to quantify these effects and therefore procedures for sampling operations should be documented carefully so that all relevant information is recorded at the time of sampling by the field worker.

4.3.1 The sampling plan

For any sampling programme, a sampling plan must be prepared to allow full control of the sampling process so that any change seen between two sampling rounds can be attributed to changes in environmental conditions and not to changes in procedure. Items to be considered in preparing a sampling plan include planning issues, fieldwork procedures and field safety issues.

Planning issues

Planning issues include identification of the objective of sampling (e.g. to test compliance with a bathing water regulation), choice of site (location, type of water body), the type and number of samples to be collected (sample types, e.g. water, sediment, the number of samples, appropriate equipment) and timing of sampling (considering the state of tides).

Fieldwork procedures

Consideration must be given to sampling SOPs (for equipment, sampling method, storage, etc.), as well as size of sample and sample containers. Preservation must be decided in consultation with staff from the analysing laboratory, who will advise clients on the volume and type of sample and who will usually provide sampling containers and preservatives where necessary. Ensuring field quality control includes the use of blanks, duplicate samples, replicate samples and spiked samples. Storage and holding time (conditions for storage, such as in an ice box, maximum time before analysis for unstable parameters, etc.) must also be considered.

The laboratory staff must be made aware when samples are due to arrive so that they can make the appropriate arrangements. When choosing an analytical laboratory it is important to be aware of the location of the laboratory in relation to the sampling site, as well as the latest time of day that they are prepared to accept samples.

Other factors include deciding where to carry out analysis, i.e. in the laboratory or on site. Some analyses may be better performed on site, such as dissolved oxygen measurements, calibration of field measuring equipment, flow pumps and thermometers, etc. and sample treatments such as filtration. Some samples need to be split or subsampled. Where this is done, great care needs to be taken because samples are frequently very variable. Contingency plans need to be prepared for situations such as bad weather and vehicle breakdown. Field sampling sheets also need to be prepared. These can be filled in manually on paper forms or on a portable computer providing that the software has been properly validated. When designing field report forms it is important that the place, time and date of sampling, sampling conditions, any field measured variables, equipment used (with an inventory number), any necessary sample preparation and the name of the operator are included in the form. Practical difficulties, such as how many samples the field worker will need to carry, parking and access to the site also need to be considered.

Field safety issues
Field safety can have a bearing on the quality of data generated where field operators may be inclined to use a less than optimum procedure in order to protect themselves. This must be taken into account when writing the sampling procedure. For example, insisting on sampling water at chest height may deter some operators if the conditions in the water to be sampled are rough. Sampling from boats can be especially hazardous in rough weather. Even the 30 cm depth stipulation of the European Union's Bathing Water Directive can be difficult to comply with. When devising a plan, areas of risk may have to be borne in mind, including water depth and sampling conditions, currents, wildlife, traffic and weather. Staff must always be provided with the appropriate protective equipment and SOPs should be developed with the safety of operators of paramount concern.

4.3.2 Field quality assurance

In spite of the difficulties involved in site work, QA is critical at this point. If a good, practical, field QA programme is put into operation, confidence in the data collected should be ensured (WHO/UNEP/VKI, 1997). All equipment must be kept clean and in good order, and records should be kept of all maintenance and of any irregularities that may affect the results. Conditions in the working area should not expose the operator to undue risk of any type.

Standardised and approved methodologies must be used at all times. If a method proves unworkable on site, then an alternative must be found quickly and agreed by all those involved. Operators must not change procedures without referral to the management procedure. Where unavoidable changes are made, for example, in bad weather, they must be fully documented. Nevertheless, a good sampling plan should make provisions for bad weather.

Prevention of sample contamination and losses
It is important that samples are protected from contamination and deterioration before their arrival in the laboratory. This can be ensured by using only recommended sample containers. Where reusable containers are used, it is essential that they have been cleaned properly and, if necessary, sterilised before use. Containers that have been sterilised must remain sterile until the sample is collected. The inner portion of the sample container should not be touched by the operator. If the seal on the bottle is broken (in the case of a commercially purchased microbiological sample bottle), or if the protective paper or foil has been lost from the top (home-made sampling containers), the bottle should be discarded.

Recommended preservation methods must be used. Where this involves chemical preservatives, the chemicals must be of analytical grade, and provided and tested for efficacy by the analytical laboratory.

Field measurements, such as pH and temperature, must be made on a separate subsample which is then discarded in order not to contaminate samples for interlaboratory analysis. Conductivity measurements should not be made with a sample that has been used previously for measuring pH, because potassium chloride from the pH probe may affect the conductivity reading.

All sample containers should be kept in a clean environment, away from dust, dirt and fumes. Petroleum products and fumes may contaminate samples with heavy metals and hydrocarbons. This can be a major problem on boats, where leaks and seepage of petroleum products are common. Samples must be stored in a cool box or portable refrigerator and transported to the laboratory as soon as possible. Cool boxes are more efficient if they contain some water.

Field Quality Control

Quality Control (QC) is an essential part of the field QA programme. It requires the collection of replicate samples to check the repeatability of sampling (see section 4.5.1), and the submission of field blanks and duplicates to check for contamination, handling and storage problems and other errors that may affect the results from the time of sampling to the time of analysis. The timing and frequency of these samples should be documented in the sampling plan.

4.4 Laboratory facilities

Except for any on-site analysis, analysis is usually performed in a laboratory. It is essential that any facilities are adequately equipped to deal with the analyses required and are convenient for the delivery of samples. This should have been ascertained before the start of the monitoring programme (see Chapter 2).

Small-scale organisations responsible for monitoring may find it more convenient to use outside facilities for analysis and sometimes for sampling. In these cases, the use of a laboratory belonging to an accreditation scheme is advisable and, moreover the laboratory should be inspected for compliance by an experienced member of the monitoring programme. An inspection should take into account the following features (ISO, 1984):

- Lines of communication between staff and management.
- Staff training and qualifications.
- Resources.
- Equipment maintenance and calibration.
- Standard Operating Procedures.

- Traceability of results.
- Sample handling and storage.

Where in-house facilities are used, it is essential that the monitoring work does not overload the laboratory. Resources (staff, space, equipment and supplies) must be sufficient for the planned workload. The laboratory must be well managed and must conform to all relevant health and safety guidelines. All analyses performed must be within the remit and expertise of the facility and SOPs must be in operation for all analyses (see Chapter 2).

4.4.1 Sample receipt and storage

Procedures for sample handling, transport and storage prior to analysis should ensure that the quality of the sample is not compromised. The condition of each sample and its storage location should be recorded along with its proposed analyses. If the sample is split, this must also be recorded. All samples must be identified uniquely with a number or code. It is important to ensure that the passage of a sample, and any associated paperwork, through the laboratory is fully documented and, therefore, traceable.

4.4.2 Reporting

The efforts of QA are directed ultimately towards ensuring that any data produced are suitable for their intended use; this applies to the results and any interpretations. The first stage in the reporting process is the examination of the results to see if they are fit to report (although raw data should have been checked prior to this stage). Results must be reported accurately and in a way that aids interpretation. To facilitate this, information may need to be included that has a bearing on interpretation, such as sampling conditions or the method of analysis. All data included must be checked by an experienced analyst with reference to site reports, calibration and QC data. Many laboratories have a system which requires the checking and countersigning of analytical reports (usually by the laboratory manager) to act as a safeguard against erroneous or misleading data leaving the laboratory. This type of system is only effective when conscientiously applied.

4.5 Analytical Quality Control

Analytical Quality Control consists of two elements: internal quality control (IQC) and external quality control (EQC). External quality control or inter-laboratory control is carried out periodically and checked by the laboratory responsible for the monitoring system. Internal quality control consists of the operational techniques used by the laboratory staff for continuous assessment of the quality of the results of individual analytical procedures. The focus is principally on monitoring precision, although accuracy is not ignored. It is necessarily part of the wider QA programme, but differs from it by the emphasis placed on quantifying precision and accuracy. Whereas QA strives to achieve quality by regulating procedures using management techniques, IQC focuses on the individual method and tests its performance against mathematically-derived quality criteria.

4.5.1 Internal quality control in the chemical laboratory

Internal quality control within the chemical laboratory comprises a variety of activities, some of which are described below (Briggs, 1996).

Choice of analytical method

A variety of different analytical methods are usually available for determining the concentration of any variable in a water sample. The choice of method is critical for ensuring that the results of the analysis meet the laboratory's requirements, because different methods have different precisions and sensitivities and are subject to different potential interferences. Consideration must be given to these parameters before a method is chosen. A number of standard methods are available for most of the analytical determinations involved in water quality monitoring, and in some cases the method is named in the regulations. These standard methods frequently include extensive validation data that allow the method to be evaluated easily. In addition, many methods are sanctioned by appropriate international or national organisations. It is important that any method selected meets the individual programme requirements. The performance of a method can be affected unpredictably by many factors.

Before any analytical method is put into routine use it is essential that it is properly validated. A minimum programme of validation includes a number of elements. One of these elements is the determination of linearity — the calibration point should be determined and if possible a linear response curve should be demonstrated. In addition, the limit of detection (the lowest concentration of the variable that can be distinguished from zero with 95 per cent confidence) should be determined. Within-and between-day coefficients of variation should be performed at three concentration levels to determine precision. Analysis of reference materials with known concentrations of the variable, or comparison analysis with existing methods in other laboratories, should be performed where possible.

Validity checking

After a method has been validated, found to be suitable and introduced into routine use in the laboratory, it is necessary to ensure that it continues to produce satisfactory results. Validity checks should be made on every batch of samples, or at frequent, regular intervals if batches are large or if testing is continuous. Validity checking is an extension of the checks carried out before the method was selected and is intended to confirm regularly the conclusions reached at that time.

Calibration check

If a calibration curve is being used, standard solutions should be analysed from time to time within the required range of concentration. The ideal calibration curve is linear within its most useful range, with a regression coefficient of 0.99 or greater. The response of the measuring equipment to the concentration of the variable in a standard solution (in terms of absorbance or some other parameter) should be recorded when it is expected that this parameter will be comparable from assay to assay. In addition, the deviation of individual calibration points from the line of best fit can be used to assess the precision of the calibration, which should be within the mean precision limits for the method.

Use of blanks

Method blanks and, where possible, field blanks should be analysed with each batch of samples. A method blank consists of reagent water, usually double-distilled water. A field blank is reagent water that has been bottled in the laboratory, shipped with sample bottles to the sampling site, processed and preserved as a routine sample and returned with the routine samples to the laboratory for analysis. The analysis of a blank should not yield a value higher than that allowed by the acceptance criteria. This procedure checks interference and the limit of detection of the assay.

Recovery checking

A specimen spiked with a known amount of the variable should be tested in each batch and the closeness of fit to the expected value calculated. In most cases this procedure provides a check on accuracy but, in assays where a variable is extracted from the original matrix (such as in many sample cleanup procedures used prior to chromatographic analysis), it can be used to monitor the extraction step. It is important that the matrix of the spiked specimen matches the real sample matrix as closely as possible. Many laboratories use real samples with low natural values of the variable for this purpose, spiking them with known amounts of the variable and including both the spiked and natural samples in the same assay batch.

Precision and accuracy checks

Precision and accuracy checks are an extension of the validity checking described above. These checks allow the quality of the assay to be monitored over time using techniques such as control charting. The validity checks described above only allow acceptance or rejection of the assay data.

Precision and accuracy checking should allow slow deterioration of data quality to be identified and corrected before data have to be rejected. This results in increased efficiency and reduced costs for the laboratory.

Control by duplicate analysis
Use of duplicate analysis as a method of precision checking has two distinct advantages: quality control materials are matrix-matched and the materials are readily available at no extra cost. Because the samples are analysed using the same method, equipment and reagents, the same bias will affect all results. Consequently, duplicate analyses are only useful for checking precision; they provide no indication of the accuracy of the analyses. Results from duplicate analyses can be used to calculate a relative range value, R, by using the equation:

$$R = \frac{(X1 - X2)}{(X1 + X2)/2}$$

where $X1$ and $X2$ are the duplicate results from an individual sample and $X1–X2$ is the absolute difference between $X1$ and $X2$. These values are then compared with the mean relative range values previously calculated for the assay during validation. The simplest method of assessment is to use the Upper Concentration Limit (UCL), where UCL = 3.27 × mean R value. When any value is greater than the UCL, the analytical procedure is out of control. This method, although statistically valid, provides no indication of deteriorating precision.

Precision control using pooled reference material
A more sophisticated approach is to use acceptance criteria based on warning and action limits. This method has the advantage of providing some monitoring of accuracy but is a viable control only if the material to be used will be stable in storage for sufficient time. The reference material is normally prepared by taking previously analysed samples with known concentrations of the variable under investigation, mixing them and aliquoting the resultant pool. The aliquots are then stored in readiness for analysis. A small sample of the aliquots is analysed to determine the mean concentration of the variable, and the standard deviation and the coefficient of variance at that concentration level. Data may be used only if they come from analysis that are in control. This approach requires that the new pool materials must be prepared before the old ones are finished.

A typical precision control exercise would involve the analysis of four aliquots from each pool in each of five assays, thus obtaining 20 results. The material from the pool should be analysed at several different times with different batches, because between batch variance is always slightly greater than within batch variance. Once 20 or more analyses have been made on this pool of material, the mean and standard deviations of the results are calculated. Any result that is more than three standard deviations from the mean is discarded and both of the statistics are recalculated. The mean is the "target" value and ideally, will be a close approximation of the true concentration of the variable in the reference material. The mean and standard deviation become the basis of the acceptance criteria for the assay method and may be used to draw up control charts.

At least three separate reference materials with different mean values of variable concentration should be in use at any one time in order to provide control of the analytical method across a range of concentrations. If precision is checked at only one concentration of the variable, it is impossible to detect whether precision is deteriorating at other concentrations. Use of several reference materials also allows their preparation to be staggered so that they become exhausted at different times. This assures greater continuity of control, because two or more old pools will still be in use during the first few assays of a new reference material.

Although the monitoring of accuracy by assessing deviation from the reference material mean (target value) is possible, care must be taken because the target value is only an approximation of the true value. As reference materials become exhausted and new ones are made, there will be a slow deterioration in accuracy. Accuracy can be safeguarded by regular participation in EQC exercises (see section 4.5.5) and by the use of certified reference materials.

Certified reference materials

Certified reference materials (CRMs) are matrix-matched materials with assigned target values and ranges for each variable, reliably determined from data produced by repeated analysis. Target and range values may be generated from data produced by several laboratories using different analytical methods or calculated from data obtained by the use of one analytical method (usually a reference method). Consequently, there may be bias in the target value. The target values assigned to each variable in the matrix in certified reference materials are generally very close to the true value. For some variables, however, there is an appreciable difference in bias between different analytical methods and this may lead to wide assigned ranges. When a laboratory is not using one of the reference methods the "all method" range may be so wide that it is practically meaningless. Certified reference materials are also only practical for variables that are stable in long-term storage.

Certified reference materials are prepared and checked under carefully controlled conditions and, as a result, they are costly to produce, correspondingly expensive to purchase and they may be difficult to obtain in some countries. Some authorities advocate the routine use of CRMs as precision control materials, but it is more cost effective to use them for the periodic checking of accuracy, in combination with a rigorous IQC programme.

Use of control charts

The principle of control charts is that IQC data can be graphically plotted so that they can be readily used and interpreted. Consequently, a control chart must be easy to use, easy to understand and easy to act upon. The Shewhart chart is the most widely used control chart (Shewhart, 1986). It is a graph with time (or assay batch) on the x-axis and the concentration of the variable in the reference material on the y-axis. Target, warning and action lines are marked parallel to the x-axis. Data obtained from precision control using reference materials (as described above) are usually plotted on a Shewhart chart. In this application, the target line is at the mean concentration of the variable for that specific pool of material and warning lines are placed at two standard deviations to either side of the target line. Provided the distribution is normal, 95 per cent of results from assays in control will fall between the two warning lines. Action lines are normally placed at three standard deviations to either side of the target line and 99 per cent of normally distributed results should be between the action lines.

In the regular use of a Shewhart chart, an aliquot from an appropriate reference material is analysed with every batch of samples and the measured concentration of the variable in the aliquot is plotted on the chart. Normally, no more than 1 in 20 consecutive results should fall outside the warning lines. If this frequency is exceeded, or if a result falls outside the action lines, the method is out of control.

The scatter of the assay results for the reference material around the target line provides an indication of the precision of the method, while the mean of the assay results relative to the target value indicates whether there is any bias (consistent deviation) in the results. If the analysis on one or more of the control specimens yields a result that it is outside the warning or action lines on the chart, the following action should be taken:

- A single result outside the warning lines should lead to careful review of data from that analytical batch and two or three subsequent batches.
- Results outside the warning lines more frequently than once every 20 consecutive analyses of control specimens should prompt detailed checking of the analytical method and rejection of the assay data.
- A result outside the action limits should prompt detailed checking of the analytical method and rejection of the assay data.

4.5.2 Internal quality control in the microbiology laboratory

Internal quality control in microbiology laboratories poses special problems of reproducibility due to the naturally wide variation in the number of organisms found between subsamples (see Chapter 8). Apart from method and field blanks (where the method blank should be sterile distilled water and the field blank should be a natural sample either guaranteed free of the test organisms or sterilised natural water), control samples should be analysed which are known to contain appropriate numbers of the micro-organisms that are normally sought. It is possible to purchase sets of freeze dried wild-type bacterial reference cultures for quality control and accreditation requirements. These cultures should be reconstituted and diluted with quarter strength Ringer's solution to give a suitable number of organisms similar to that which would normally be seen in the natural samples. These cultures are expensive and therefore it is not feasible to use a new culture for every batch of samples or media. However, frequent subculture of reference strains is to be discouraged due to problems with contamination and mutation. This is a special problem with coliphage analysis, where mutation of the host species can prevent the detection of viral plaques. This problem can be solved by freezing down the cultures in glycerol broth and either storing in liquid nitrogen or a –70 °C freezer or, more conveniently, on commercially available plastic storage beads and freezing at –20 °C. Alternatively, some media companies supply standardised cultures in an easy to use form. These cultures may be qualitative or quantitative and have the advantage of eliminating the trial and error diluting of suspensions to achieve the desired count.

Shewhart charts can be used in water microbiology despite the problems of natural random variation. However, this means that wide control limits are necessary. For example, if the count reported for the first half of a duplicate sample is 11, then the 95 per cent confidence interval (CI) for the count of the second sample will be 3–23. Tables giving the CIs of counts are available in reference works on water analyses (HMSO, 1994). The microbiology laboratory should carry out several duplicate analyses regularly and plot the results on a control chart. Each half should be treated as a separate sample and analysed routinely. They should be inserted in the sample run in a random fashion without the knowledge of the analyst (if at all possible) and all results should be read by the same person. The first count should be recorded on the control sheet, and the corresponding CI for the second count entered. The second count is then recorded along these figures. If this count falls outside the CI, this fact should be highlighted. If a Shewhart chart is made up of these results, then any trend can be identified. If, over a period of time, the second count falls outside the CI for more than 95 per cent of the time, the reason should be investigated. As with the use of duplicate samples in chemistry, this approach keeps a check on precision and not on accuracy.

In addition to blanks and a manufactured control, the use of a known wild positive can be included. This can be chosen from the last batch of samples run. However, there can be problems with this, due to alteration in bacterial numbers over time and storage.

All prepared media should be checked for performance and sterility and identified by batch reference number. Both negative and positive control strains of bacteria should be included. Manufacturers of dried media will usually recommend control strains if requested. Where a medium is meant to inhibit the growth of a particular organism, this should also be tested.

4.5.3 Summary of an internal quality control programme

A summary of the IQC programme recommended by the GEMS/Water programme is given below. This programme offers a simple but effective introduction to IQC and is described in more detail in the *GEMS/Water Operational Guide* (WHO, 1992). For each variable the following should be applied:

- For chemical variables, analyse five standard solutions at six different known concentrations covering the working range to develop a calibration curve or, when a calibration curve already exists, analyse two standard solutions at different known concentrations covering the working range to validate the existing calibration curve.
- Analyse one method blank per set of 20 samples.
- Analyse one field blank per set of samples.
- Analyse one duplicate of a sample chosen at random from each set of up to 20 samples.
- Analyse one specimen that has been spiked with a known amount of the variable as a recovery check. This specimen should have a matrix similar to those of the samples being processed.

4.5.4 Remedial action

If any of the QC procedures indicate that a method is out of control or that a problem exists, corrective action must be taken. The main checks to make are calculations and records, standard solutions, reagents, equipment and QC materials (Table 4.1).

4.5.5 External quality control

External quality control is a way of establishing the accuracy of analytical methods and procedures by comparing the results of analyses made in one laboratory with the results obtained by others conducting the same analysis on the same material. This is usually accomplished by one laboratory sending out sets of samples, with known and unknown concentrations of variables, to all of the specified laboratories. Each participant analyses the samples for the specified variables and reports the results to the reference laboratory (Box 4.1). The results from all participating laboratories are collated by the

Table 4.1 Checks to be carried out when a problem is detected with an analytical method

Problem area	Checks
Calculations and records	Check calculations for transposition of digits or arithmetic errors; confirm that results have been recorded in the proper units and that any transfer of data has been made correctly
Standard solutions	Check the standard solutions that are used for calibrating equipment; check their storage conditions and shelf-life (an old solution may have deteriorated or a new one made up incorrectly)
Reagents and media	Check for deterioration of old products; check QC records to see if new reagents performed correctly and if they were properly prepared; check their storage conditions and shelf-life
Equipment	Check calibration and maintenance records for all relevant dispensers and measuring equipment where a method is out of control; items such as automatic pipettes, balances and spectrophotometers should be checked regularly and recalibrated as necessary
	Ascertain that equipment is being properly used; check that any QC material has not deteriorated and is properly stored; run analyses on several aliquots to determine whether the concentration of the variable remains within the allowed deviation from the target value and close to the mean of the last 20 determinations

Source: Briggs, 1996

organisers of the EQC programme and then subjected to detailed statistical analysis. A report to each laboratory is generated, giving a target value for the reference sample or samples (usually consensus mean or median), a histogram illustrating distribution of results for each material, and an individual performance score relating the individual laboratory results to the target value. The calculations for performance indicators are often quite complex because multiple specimens have to be considered and the method variance varies with the concentration of the variable. However, the general principle of providing a method of performance comparison remains the same in all EQC exercises.

External quality control reports should indicate clearly whether performance is satisfactory or not. If it is not satisfactory, two general actions must be taken. First, the analysis at fault must be examined to determine the cause of poor performance. Secondly, the IQC programme that allowed the deterioration to progress unchecked must be closely examined to establish where inadequacies exist. Both must be corrected.

Box 4.1 External Quality Assurance: the experience of a microbiology laboratory

The Unit of Microbiology, Faculty of Medicine, University Rovira i Virgili (Reus, Spain) participates in an EQA on microbial recovery where each participating laboratory analyses an external sample and the results from all participating laboratories are analysed statistically.

A critical element of the EQA is that the test sample should be processed in an identical manner to routine samples. If an analyst is aware that an external sample is to be processed the exercise is viewed as a test of their competence and modifications may be made to routine procedures in order to enhance recovery. Analysts may also use the exercise to test potential new methodologies, but this is not the purpose of EQA and statistically unreliable results may be obtained if new methods are deliberately applied.

Although relatively expensive, properly operated EQA exercises can be of great benefit to participating laboratories because they can identify failures in internal quality control and, if undertaken over a period of time, laboratories can use them to evaluate regularly the performance of their methods. Corrective measures can then be applied whenever the methods are found to be producing poor results.

The key issue identified in participating in such EQA exercises was that they must be carried out anonymously so that the samples are dealt with in exactly the same manner as routine samples.

The general objective of EQC is to assess the accuracy of analytical results measured in participating laboratories and to improve interlaboratory comparability. Wherever possible, laboratories should participate in EQC programmes for each variable that is analysed routinely. This is only worthwhile where IQC is also part of a laboratory's normal procedures. Participation in relevant EQC programmes, and maintenance of adequate performance in those programmes, is often a requirement for laboratory accreditation.

The organisation of an EQC exercise is expensive. Large quantities of stable reference materials must be prepared, these materials must be transported to the participating laboratories, data must be analysed and detailed reports on performance must be prepared. Participating laboratories are usually charged for the service provided.

4.6 Elements of good practice

- Monitoring programmes should include appropriate QA which does not infringe on health and safety and which covers the integrity of all observation, interviews, field sampling and water quality analyses as well as data input, analysis and reporting.

- A QA manager should be appointed who audits all aspects of the operation regularly with special regard to procedures, traceability of the data and reporting.
- Essential elements of QA programmes include:
  - The writing and implementation of a Quality Manual and SOPs. All SOPs should be overhauled regularly and updated as necessary, and any deficiencies should be reported and appropriate remedial action taken.
  - SOPs should include maintenance and updating of inventories and catalogues; methodologies for all major equipment, all sampling and analytical procedures; sample receipt, screening and storage; and reporting.
- Samples should be registered on arrival at the laboratory. The applied laboratory procedures should conform to the SOPs defined at the laboratory. Where possible, all analytical procedures should follow defined ISO or American Public Health Association (APHA) protocols. All equipment should be calibrated regularly and the operational procedures submitted to quality control staff in order to guarantee traceability of the data.
- The programme should be evaluated periodically, as well as whenever the general situation or any particular influence on the environment is changed.

4.7 References

Briggs, R. 1996 *Analytical Quality Assurance in Water Quality Monitoring.* World Health Organization, Geneva.

HMSO 1994 *The Microbiology of Drinking Waters.* Report 71. Her Majesty's Stationery Office, London.

ISO 1984 *Development and Operations of Laboratory Proficiency Testing Guide 43(E)*, International Organization for Standardization, Geneva.

ISO 1994 *Quality Management and Quality Assurance — a Vocabulary.* International Organization for Standardization, Geneva.

ISO 1990 *General Requirements for the Competence of Calibration and Testing Laboratories.* Guide 25. International Organization for Standardization, Geneva.

Shewhart, W.A. 1986 *Statistical Method from the Viewpoint of Quality Control.* Dover Publications, New York.

WHO 1992 *GEMS/WATER Operational Guide.* Third Edition, World Health Organization, Geneva.

WHO/UNEP/VKI 1997 *Analytical Quality Assurance and Control.* World Health Organization, Geneva.

Chapter 5*

MANAGEMENT FRAMEWORKS

"Beach management seeks to maintain or improve a beach as a recreational resource and a means of coast protection, while providing facilities that meet the needs and aspirations of those who use the beach. It includes the framing and policing of any necessary regulations, and decisions on the design and location of any structures needed to facilitate the use and enjoyment of the beach environment" (Bird, 1996).

In most countries, a single "beach manager" who undertakes all the activities such as monitoring, planning and decision-making does not exist. These activities are generally devolved among different persons and authorities at various levels (national, federal, regional, provincial, local). In order to achieve effective management of recreational water use areas, managers should have background knowledge on a range of aspects concerned with the coastal or freshwater area, such as inflows and outflows, water quality, physical aspects of the water use area and potential health hazards. Beach managers should, therefore, be aware of the social and economic dimension embedded in their decision-making. Of importance to beach managers and decision-makers are:

- The techniques available to measure the impact of tourism.
- Notions and principles of sustainability.
- Local strategies towards sustainability.
- Sustainability indicators and issues related to water quality management.
- Water analysis and water safety.

5.1 Management concerns and approaches

The coastal zone and freshwater bodies are important areas for human habitation, industry and recreation. There are thus competing uses, not only of water for bathing, surfing, sailing, scuba diving, aquaculture and other maritime industries, but also of land use, such as for residential developments,

* *This chapter was prepared by K. Pond, M. Cavalieri and B. Van-Maele*

harbours, ports, marinas and tourism industries. Offshore activities such as oil and gas exploration, disposal of sewage or radioactive waste and shipping, are also responsible for the release of contaminants into the aquatic environments. Many of the pressures mentioned above are common throughout the world and many of the threats facing the quality of recreational water bodies have arisen as a direct or indirect consequence of human activities. Such conflict of uses makes management of recreational water use areas a particular challenge. Water bodies are often used as a repository for waste and the relatively concentrated period of tourism activity in specific areas increases the environmental pressures on such water bodies. Recreational use of water and tourism activities depend highly on the quality of the natural environment for their continued success. Lack of effective management can lead to the loss of habitats, over exploitation of resources and an associated loss of income.

5.1.1 Tourism

The coastline is a major element in the geographic, recreational, commercial and ecological fabric of many countries and provides major destinations for local, national and international tourists. Freshwater areas, such as large lakes, are popular for recreation and in many cases are being developed for tourism. Associated villages and towns grow and develop their economy in accordance with the prevailing and seasonal tourism needs (main commercial streets, hotels, restaurants, clubs, shops and related activities; recreational activities on the beach or on lakes; transportation facilities, etc.). Socio-economic problems usually derive from the seasonality of the pressures on tourism-related facilities. There are a few data available on the contribution of coastal zone tourism to national economies (Grenon and Batisse, 1989) but for smaller coastal countries and island countries, especially those without industrial or agricultural outputs, tourism can be a substantial part of the economy.

There are concerns associated with the development of touristic ventures that can apply to both marine and freshwater areas. These concerns include tourism-associated aquatic transportation and the resulting pollution from vessels (oil, tar or litter on the adjacent area), as well as stress on populations and the environment where tourism is the major factor in the economy (Bird and Nurse, 1988). The development of tourist facilities may have a particular effect in developing countries where food security problems, pressures from recreational lobbies and public opinion may threaten alternative sources of income, such as the local fishery economy. Mariculture and the recovery of oil and gas can compete for the same space as that desired by recreational users. Tourists themselves can contribute to the waste and pollution of the host area, with a degrading influence on the quality of the recreational water use area arising from noise (primarily from transportation), recreational activities such as boating, and from solid and liquid wastes. Biodiversity reduction, resource depletion and human health problems may result from the accumulated environmental effects of tourism, including direct human impact (such as trampling).

There are a number of issues that may deter tourists from a recreational water use area. These include aesthetic and health problems arising from domestic waste disposal into marine and freshwaters that jeopardises the quality of food and the possibility of recreational activities on beaches in developing and developed nations. Eutrophication can degrade beaches and adjacent waters aesthetically through the accumulation of rotting marine plants; this can lead to a significant loss in tourist revenue. Aesthetic factors, such as litter, have a high deterrent value on visitors to bathing water areas (see Chapter 12).

5.1.2 Integrated management

Integrated coastal zone management (ICZM) is understood most simply as management of the coastal zone as a whole in relation to local, regional, national and international goals. It implies a focus on the interactions between the various activities and resource demands that occur within the coastal zone, and between coastal zone activities and activities in other regions. This might mean, for example, the incorporation of coastal environment protection goals into economic and technical decision-making processes or the co-ordination of tourism policies with nature conservation policies. Although ICZM has been promoted widely in recent years, it has not always been implemented successfully. This was mainly because of a lack of understanding of underlying coastal processes and it has only become apparent relatively recently that multidisciplinary land-use planning in the coastal zone is essential. Financial constraints have also been a contributory factor.

The same principles of ICZM can be applied to freshwater management and therefore marine and freshwater zones will be treated as synonymous in this chapter. In general, recreational water use areas, whether fresh or coastal waters, require similar management actions. However, lakes and other freshwater recreational water use areas are generally smaller bodies of water and are, by nature, more fragile than seas and oceans. The impact of human activities are apparent more quickly and failure to ensure adequate management will accentuate any degradation (Box 5.1).

Integrated coastal zone management programmes must address a range of issues, including habitat (loss of habitat or degradation of coral reefs, seagrass beds, wetlands, beaches and dunes, lagoons and estuaries), water quality (sources and nature of pollution, reduction and flow rates); management of natural hazards, and degradation of cultural resources and

Box 5.1 Problems of management of a freshwater recreational water-use area: Lake Geneva

Lake Geneva provides a unique example of a freshwater recreational water use area. It lies on the border of Switzerland and France and thus requires the integrated management of the two area authorities. A total of 41 beach resorts cover 4.5 per cent of the lake shore. These are artificial beaches with access to the lake, natural beaches and artificial beaches with a swimming pool. Boating and windsurfing are popular with visitors to the Lake and about 35,000 boats are registered on the Lake. To accommodate these, several yachting harbours and boat yards have been constructed. The main activities in the Lake Geneva basin are trade, tourism, banking and insurance and wine growing.

The bacteriological quality of the water in the beach resorts may vary considerably. There are a number of local pollution "blackspots" and restrictions in many areas prohibit swimming. On the Swiss side of the Lake the water is monitored according to the EU Directive on the quality of bathing waters (CEC, 1976) and also according to the procedure described in 'Examen et évaluation de la qualité hygiénique des bains de lac et de rivière' (Eschmann and Lüönd, 1965). Monitoring is only undertaken for Salmonellae and *E. coli*. On the French side, the monitoring is undertaken by the Ministry of Health, in compliance with the EU Directive (CEC, 1976). Domestic and industrial sewage systems have been installed and storm drains are being phased out. The quality of the water is therefore expected to improve. Human activities around the Lake are generally in conflict with the natural environment. There is intensive development around the shores of large private properties and woodland estates, that is having direct physical and aesthetic effects on the landscape.

In terms of management of the shores of Lake Geneva there is very little co-ordination between France and Switzerland despite certain provisions, such as France's Coastal Law and Switzerland's Cantonal Masterplans, because priority is given to the economy. There is, however, more co-ordination between the Swiss and French laboratories in relation to monitoring water quality in the Lake.

There is a real need for integrated lakeside management in order to consider all activities in the lakeside area. Development should be restricted to suit the capacity of the natural environment and specific lakeside provisions need to be drawn up and enforced. A trans-border structure for collaboration and co-ordination is also required which would promote an integrated approach to management. Economic instruments, both as incentives and disincentives, should be developed to integrate the environment into lakeside management. It is of paramount importance to promote environmental awareness and education throughout society if integrated and environmentally sound lakeside management is to be achieved.

Source: Adapted from OECD, 1993

management of developments (mariculture, extractive industries, tourism, shore front development, major facility siting). Coastal managers must also consider any decline in fisheries, public access, biodiversity protection, sea level rise and degradation of scenic quality.

Integrated management ensures that priorities are given to all users of the water zone. Through policy supported by legislation and regulations, the most appropriate activity or activities can be given preference and investments in the area can concentrate selectively on these activities. Funding is a common problem in environmental management and for this reason some form of classification scheme combining priority for action and type of action required is especially useful. The classification then becomes an important tool in assisting planners in developing a strategy for improving the quality of the bathing water and the beaches. When the problem can be rectified through local efforts (such as beach cleaning), the management process should seek appropriate action from the municipality and reclassify the beach accordingly. Where the cause requires major investments or decisions on regional or perhaps national level, the authority should ensure that health concerns are represented adequately.

It is important to emphasise that improving the bathing water and beach quality strictly with the purpose of increasing the amount of tourists visiting a region or a country, can conflict with interests in protecting ecologically important areas or designated sensitive areas, etc. It must be an important issue for planners at regional, as well as national level to develop a plan for selecting and protecting these areas that are not to be transformed into large intensive tourist resorts. This is a premise for working with a classification scheme (see also Chapter 9).

5.2 Management framework

The WHO *Guidelines for Safe Recreational-water Environments* (WHO, 1998) present a management framework within which different levels of health risk and associated interventions are ordered in four major fields under the umbrella of integrated management (see Figure 1.1). The major fields of intervention are clustered as regulatory compliance, public awareness and information, control and abatement technology, and public health advice and intervention.

5.2.1 Legislation and regulation

Effective coastal or freshwater zone management requires an effective legislative framework to define the roles of different bodies and levels of government, as well as to provide environmental objectives. Management is not restricted to national issues — water quality, pollution control, international tourism and shipping are amongst the activities that affect the coastal zone and that extend beyond the national boundary. It is therefore obvious that a single government or agency cannot be responsible for the wide range of issues that need to be addressed in the coastal or freshwater zone. Effective legislation must provide a framework within which the roles and responsibilities of different organisations or interested groups are defined and must accommodate capacity to act at the international, national and local levels.

At an international level, legislation of particular relevance often relates to the management of international or transboundary waters. Whilst the legislation itself may be "hard law" or "soft law", it may provide for harmonisation or standardisation in data generation and exchange, and create obligations to notify other concerned parties regarding hazards and quality changes.

At a national level, regulatory measures are often considered to be inflexible but they are easy to operate and provide a clear and common framework for all parties concerned. In general, some form of basic water law provides a framework within which specified agencies are empowered to regulate. In the field of recreational water use, the national level is particularly important in establishing common ground for the assessment and reporting of safety and thereby supporting "informed personal choice". However, laws and regulations of relevance to safe beach management may derive from diverse influences, such as public health, social integration and rights of the disabled people, navigation for pleasure purposes, aquatic sports, fishing activities (in the sea and on the shore, e.g. bivalves), relevant flying activities, trade activities on public areas and the concession of State lands.

The diversity of national regulatory structures requires diverse approaches and solutions but, in general, managers concerned with recreational water use areas should consider both common law and statutory law. In most countries under common law, liability and negligence arise from the breach of duty known as "duty of care". This applies to members of the public as well as to operators. The duty specified is to take reasonable care. In the case of the safe management of the beach, the responsibility for taking adequate precautions rests with the operator. Of particular importance to those concerned with beach operation is the standard of care arising from their activities, i.e. that of an ordinary skilled person exercising or professing to have a particular skill. This is of particular relevance to lifeguards, for example, who are expected to conduct themselves as one would expect of the competent qualified lifeguard.

Those who employ staff on a beach may have specific duties to those staff under statutory law. In general these duties cover premises, written operating procedures, general working conditions, training, appropriate health and safety policy statements, consultation with safety representatives, safety procedures, the free provision of appropriate uniform, protective clothing and personal safety equipment and the provision of adequate first aid facilities for employees. For the employers there is an obligation to conduct operations in such a way as to ensure that members of the public are not exposed to risks to their health and safety.

The occupier of premises has a duty of care to any visitor using the premises for the purpose for which he is invited or permitted to be there. In general the operator of a natural beach will not be exposed to liability, although only one attraction, such as a diving platform, would expose the operator to liability if the duty of care is breached. The same applies to any operator deriving income from the provision of services for visiting swimmers. The operator may be exonerated from liability if a danger is brought to the visitors' attention and the operator takes appropriate measures.

5.2.2 Public awareness and information

The concept of reciprocal rights and responsibilities (as implicit in the concept of duty of care) highlights the importance of the capacity of the individual to make healthy or safe choices. In order to participate successfully in healthy recreation, members of the public require awareness (i.e. in this context knowledge regarding hazards and safe behaviours) and access to information to enable them to make informed choices. However, informed personal choice contributes not only to the protection of the individual but creates an incentive for improvement in the quality and safety of recreational water use areas — as users demand safer locations, the economic incentive to provide safe and attractive facilities increases.

Education and awareness

A basic appreciation of the health hazards that may be encountered during recreational water use and regarding safe behaviours is a prerequisite to health protection through the exercise of improved personal choice. A variety of special interest groups, such as lifeguard organisations, are instrumental in promoting education and public awareness activities. Watersport clubs, such as sailing, scuba diving, canoeing and swimming clubs, teach members basic first aid, safety procedures and a respect for the water environment. All watersports are potentially hazardous and participants should be made aware of the particular hazards associated with their sport. A variety of formal courses, usually culminating in an examination, are in existence to ensure that such activities are undertaken in a safe and responsible manner.

Beach classification

One tool to support informed personal choice that has received great attention in recent years is that of beach classification. In order to be effective a beach classification scheme needs to be based upon health and safety and must be of interest to users. It must also be based on reliable comparable data and overseen by a credible and impartial agency. Whilst classification schemes (e.g. Chapter 9) are often designed specifically to support and encourage informed personal choice, other classification schemes may be used for purely management information purposes (see Chapter 6), such as for determining beaches inappropriate for tourism development, or for identifying those suitable for the encouragement of tourism and perhaps those eligible for certain forms of aid or awards from a national or regional authority.

In order to decide to which class a beach belongs, certain criteria for each class need to be defined. The issue of beach classification systems and associated management response is discussed fully in Chapter 9. A classification system allows differences between classes to express differences in the problems to be solved. This means that different programmes or plans are needed depending on what class a specific recreational water use area belongs to. It also implies that the classification of beaches and water is a continuous process. A certain beach or water may belong to one class in one year, but as different measures are taken and problems are solved, the beach or water may change class next year.

On-site and local information

Users of recreational water use areas rely on information about safety, hazards to health and facilities, that comes from the news media, local authority notice boards and signs, environmental groups and tourist publicity, as well as relying on their own perceptions. Users can only control risks actively by acting on knowledge provided to them. Public awareness measures at the local level, in combination with national policies, are thus essential in order to achieve effective management of the recreational water use area and to reduce risks to users (see Chapter 7).

5.2.3 Control and abatement technology

Not all hazards encountered during recreational use of the water environment may be addressed effectively, or their associated adverse health effects averted, through informed personal choice. In some instances removal of the hazard or preventing access to the hazard is the preferred management option. In precluding access to a hazard high intervention approaches (such as fencing) or low intervention such as making access difficult (no development of car parks or public transport access), can be used. Such measures are relevant to a range of hazards, such as areas with strong currents, rocky environments, poor water quality or areas subject to toxic cyanobacterial blooms. To achieve long-term improvements in the quality of recreational waters investment must be made in pollution abatement technologies (see Chapter 9).

5.2.4 Public health intervention

Despite concern for the aesthetic aspects of recreational water use areas, the driving force behind much activity is public health and safety. In many cases circumstances may lead to situations that present an unfamiliar or unacceptable risk to public health. Such circumstances may relate to a breakdown in sewage treatment and disposal infrastructure, to toxic cyanobacterial blooms or to new or transient water uses that are incompatible with existing patterns of use. Under such circumstances the authorities responsible for public health are generally required to take a lead role in determining what actions should be taken and for what period. Such decisions are, in practice, often made under pressure of time and with inadequate information, but may be assisted by the existence of a national point of reference where experience and information on such incidents is maintained.

In addition to emergency response, some countries have, in recent years, instigated some form of accident emergency plans. These deal with, for example, major oil accidents at sea, or a chemical industry or nuclear accident. A structure for alert systems or notification relays, including home numbers for authority staff, may already exist. Unfortunately, the more common pollution accidents, such as failure of a sewage treatment plant or unusual wind direction forcing polluted waters onto a beach which is normally clean, or an algal bloom causing skin irritation, are often not included in this system. They do, however, affect the public more directly. It is therefore preferable to establish warning systems for these kind of events.

Beaches with a full lifeguard or warden service have most of an alert system in place, i.e. a dedicated person with warning signals on site during the bathing season. However, lifeguards are often only responsible for alerting the public to specific hazards, such as high winds, and not to pollution incidents, although in reality they, along with other coastal workers (e.g. rangers, wardens or coastguards) would alert the public to such incidents.

5.3 The role of organisations or individuals with a vested interest

The large number of interested organisations at all levels involved in the coastal and freshwater zones requires particular co-ordination and co-operation (see Figure 1.1). Central and local governments have a particular responsibility to establish standards and regulations to limit the health hazards to users of recreational water environments. Of particular importance is the layout of a management strategy for the achievement of integrated management. This requires addressing the issues of resources, economic development activities and societal needs in recreational water use areas. National government is also instrumental in directing, promoting and co-ordinating all activities relating to the application of laws concerning the coastal zone, including defining general criteria and methodologies for the monitoring of recreational water environments and activities, mode and frequency of sampling and analytical techniques. Organisations such as the World Health Organization Advisory Committee on the Protection of the Sea and other such bodies can provide advice and guidance. Research institutions, universities, non-governmental organisations (NGOs), special interest groups and the tourism industry can aid in the technical assessment of hazards and in monitoring changes. Industry, in particular the tourism industry, is increasingly adopting a more proactive role in monitoring the environment. Local interest groups, NGOs, local authorities, the tourism industry and the media are involved in raising awareness of users to some of the hazards associated with recreational activities (see Chapter 6). Citizens are often instrumental in contributing to remedial measures and are increasingly involved in public participation activities, such as beach cleaning, riverine fly tipping area cleanups and monitoring (see Chapters 6 and 12).

5.3.1 Municipalities

Local authorities are frequently the legal agency of the government. They often take a key role in bringing interested organisations together and gaining their collaboration and co-operation in decision-making, and participation in the implementation of decisions. Local authorities also contribute to the development and enforcement of standards and regulations. In general, public health laws and acts state that a local authority may make bylaws with respect to public bathing and beach management, including public bathing and coastal zone management. Municipalities may therefore be responsible for regulation of the areas and the hours when bathing will be permitted; they may also be responsible for requiring the persons providing accommodation for bathing to provide and maintain lifesaving appliances and lifeguards, as well as being responsible for the regulations for preventing danger to bathers. Municipalities may also enforce regulations regarding the navigation and speed of vehicles for pleasure purposes within any area allotted for public bathing. This is of particular importance because it permits the zoning of pleasure vehicles in relation to bathers. Municipalities may also be responsible for all the inland and adjacent areas above the low water mark where these bylaws have affect; protection of public health and safety, including monitoring of water and adjacent land. Protection, preservation, restoration and enhancement of coastal natural resources (including beaches, floodplains and dunes) as well as the use of beachfront property in a manner compatible with preserving public property may be the responsibility of the municipality. Municipalities may be required to produce coastline management plans as part of their normal planning responsibilities and any local authority with land subject to coastline hazards should plan and manage that land in accordance with its hazard susceptibility. Local authorities are therefore responsible for the investigation, design, construction and maintenance of works and measures to mitigate coastline hazards and also for promoting hazard awareness in their community in an attempt to reduce the social disruption and damage caused by coastline hazards. The latter can be done by supplying information and advice to property owners, residents, visitors, potential purchasers and investors (see Chapter 6). It is also a local authority responsibility to improve and maintain beaches and their amenity.

The structure and responsibilities of local governments differ throughout the world. In the UK, for example, County Councils are responsible for strategic planning, structure plans and waste disposal and the District Councils are responsible for housing, local planning, environmental health, coast protection, waste collection and noise control. In Australia, the Local Councils have general responsibilities for the production of coastline management plans, coastline hazard mitigation, hazard awareness and beach management, as well as specific responsibilities under the Environmental Planning and Assessment Act. In general, however, the fragmented and often duplicated responsibilities in the coastal zone are identified as severe impediments to effective planning and management.

Specific regulations for a beach are disseminated at the local level on the basis of the physical, environmental and social characteristics of the area. Regulations for the management and operation of a beach and for water activities are usually promulgated by the City Council or (as in Italy) by the nearest Harbour Maritime Authority and are addressed to the concessionaires or managers of the maritime State land.

5.3.2 Facility operators

Once a beach manager has a complete picture of the beach characteristics (beach registration and classification) all decision elements are available for the daily operation of the beach and for (mid- to long-term) management plans. It is suggested that the competent authority should designate an operator, or another responsible person, to be on duty when a beach is open for visitors. This operator should take decisions relating to the beach and should take appropriate action when requested by the authority when accidents or spills take place leading to beach contamination or when water quality becomes unacceptable or for safety reasons (such as weather, inadequate lifeguards or safety equipment).

There are also a number of considerations to take into account when designing the facilities to support a public beach. Resources become an important issue and therefore the provision of facilities should be prioritised according to the needs and uses of the area under question. Monitoring for potential health hazards and associated management actions should always be considered a priority over the provision of shops and refreshment kiosks when developing recreational water use areas. However, it is acknowledged that this approach would not always attract tourists and the associated finance to an area. Research has shown that visitors to the coastline place more value on the cleanliness of an area and on the provision of facilities than on unseen human health hazards such as microbiological parameters (Oldridge, 1992; Morgan *et al.,* 1995) (see also Chapter 12). Education and public awareness must, therefore, become an essential part of integrated management, especially where resources are low and prohibit the provision of facilities (see Chapter 6). Consultation between those with a vested interest and local communities is essential if the various conflicts of use are to be resolved.

5.4 Management options

The various different attitudes of visitors to the recreational water use area also determine the necessary level of facilities that are desired. Different cultural contrasts exist in the use of beaches; in many tropical areas the sea is used as a cleaning place or for trade amongst fishermen, whereas throughout Europe beaches are generally used as places of passive activity. Despite many people looking for seclusion at a beach, the pressures to develop recreational water use areas to support growing populations and increases in tourism are so great that it is becoming very difficult to find underdeveloped beaches, particularly in countries with a warm climate.

Management options and preferences vary according to the level of development of an area and the preferences of visitors to that area. Management actions need to take into account local economic needs as well as the desires of the users. In general, two broad categories of user can be identified: those seeking resort areas with facilities, entertainment and easy access, and those seeking secluded or rural areas.

5.4.1 Resort beaches

The following guidelines are provided for management of resort beaches where tourism is a priority. Where resources are a restraining factor careful prioritisation should occur that will minimise public health risks. Topography, including slope and bottom material, need to be considered. Beach cleaning should also address the removal of litter and debris from the lake or seabed where they present a hazard to bathers. Specific regulations may be adopted for the prohibition of potentially hazardous items, such as glass, on the beaches. In addition, adequate litter bins should be provided. Clearly visible depth markers should be provided at the points of maximum depth of all designated areas and at diving boards, platforms and similar facilities. Zoning is an important measure in minimising risk where different user groups coexist within a confined area (i.e. dog-free zones, conservation areas and naturist zones, zones for swimming, sailboarding, powerboating, etc.). Swimming may be limited to a specific area, i.e. the least hazardous, which also facilitates supervision and segregation of incompatible activities (see Chapter 7). Wastewater from toilets and showers should be discharged to the local municipal sewage system. If that is not possible an alternative treatment should be established that is acceptable to local or national standards. Where possible, toilet facilities and showers should be provided in adequate number. To prevent cars and vehicles driving on the beach, access facilities should be provided to beach parking areas. Access to beach areas for emergency vehicles should be provided and appropriately signposted. Easily read and understood information boards should be used to display beach regulations, general information on beach and water quality and facilities. The signs should be located so that they will be seen at the access points before entering the beach, the resort or the swimming area. Where appropriate, more formal regulations, in the form of bylaws, may be adopted to control activities at the coastal zone, particularly noise, fires, dog fouling and litter.

Safety aspects are of particular importance to coastal managers of tourist beaches (see Chapter 7). Clearly identified warning signs should be provided where appropriate indicating, for example, when the beach is closed for swimming, the times when lifeguards are on duty, danger of swimming during heavy storms or after sunset or in dangerous currents.

5.4.2 Rural beaches

Ideally, as with resort beaches, rural beaches should be monitored for potential health risks as a priority. Rural beaches are generally popular with walkers, naturalists, and fishermen and for other kinds of casual enjoyment. Such beaches should be cleaned "as needed" but at least four times a year. Beaches that are particularly frequently used shall be cleaned at least once a week during the summer and each month during the winter.

On rural beaches, safety boards should be displayed at all principal access points to beaches, in car parks (if present) and at particular hazard spots. The hazards of the particular beach should be clearly indicated, together with the times of high and low water, the distance of the nearest telephone and some useful telephone numbers, and the location of the nearest first aid facilities. It is suggested that public rescue equipment should also be in place in the more frequented rural beaches.

5.5 Elements of good practice

- The management framework developed for a bathing water area must take into account the impact of various competing activities, sustainable management processes, water quality issues and associated safety issues.

- Such a framework must reconcile development pressures with socio-economic, cultural and environmental criteria.
- The full range of legislative and regulatory controls that interplay with coastal or freshwater recreational water management must be incorporated into the management framework including duty of care, health and safety legislation, water quality regulations, pollution control and international articles governing international tourism and shipping. Such measures will vary from local bylaws through national to international law.
- The development of an integrated management framework must include a range of issues including nature conservation, water quality, management of coastal development access and environmental degradation.
- The role of a local municipality is central to an effective coastal management framework. Their activities must be co-ordinated within a coherent national context.
- A beach classification scheme can be constructed which provides a discreet hierarchy of categories and concomitant management activities.
- On completion of a full catalogue of the characteristics of a particular recreational water area, the beach manager has the framework from within which to establish the operational activity.

5.6 References

Bird, J.B. and Nurse, L. 1988 The Coasts of Barbados: an economic resource under stress. In: K.Ruddle, W.B. Morgan and J.R. Pfufflin [Eds] *The Coastal Zone: Man's Response to Change.* Hardwood Academic Publishers, Chur Switzerland, 197–211.

Bird, E.F. 1996 Causes of beach erosion. In: E.F. Bird [Ed.] *Beach Management.* John & Sons, Chichester, West Sussex, 73–110.

CEC (Commission of the European Communities) 1976 Council Directive of 8 December 1975 concerning the quality of bathing water (76/160/EEC). *Official Journal,* **31**, 5.2.1976.

Eschmann, K.H. and Lüönd, H. 1965 *Examen et appréciation de la qualité hygiénique des eaux de baignade: proposition de la Commission d'hygiène et de bactériologie des chimistes.* Cantonaux et Municipaux de Swisse, Geneva.

Grenon, M. and Batisse, M. 1989 *Futures for the Mediterranean basin. The Blue Plan.* Oxford University Press. XVIII, 279.

Morgan, R., Bursahoglu, B., Hapoglu-Balas, L., Jones, T.C., Ozhan, E. and Williams, A.T. 1995 Beach user opinions and beach ratings: A pilot study on the Turkish Aegaen coast. In: E. Ozhan [Ed.] *Proceedings of the Second International Conference on the Mediterranean Coastal Environment. MedCoast '9,* MEDCOAST, Middle East Technical University, Ankara, Turkey, 373–383.

OECD 1993 *Coastal Zone Management: Selected Case Studies*. Organisation for Economic Co-operation and Development, Paris.

Oldridge, S. 1992 Bathing water quality: a local authority perspective. In: D. Kay [Ed.] *Recreational Water Quality Management, Vol 1., Coastal Waters*. Ellis Horwood Ltd., Chichester, 33–37.

WHO 1998 *Guidelines for Safe Recreational Water Environments: Coastal and Fresh-waters.* Draft for consultation. World Health Organization, Geneva.

Chapter 6*

PUBLIC PARTICIPATION AND COMMUNICATION

Successful beach management requires an understanding of the nature and dynamics of a beach system, i.e. the physical, chemical and biological interactions that take place on and around the beach, the requirements and perceptions of the beach users, economic and tourism interests and environmental protection measures. Inevitably, there are conflicts between these elements, although many of these conflicts can be resolved through effective communication at an early stage, through information and, above all, active participation of all parties, particularly the public.

Large differences exist between the capacity and mechanisms for communication in resort beaches near big urban areas and in rural beaches used only by a limited number of people. Nevertheless, beach managers should consult with, and inform, beach users at all appropriate stages. The success of beach management depends very much on the active participation and involvement of the local population and of beach visitors (Camhis and Coccossis, 1982; Gubbay, 1994). The underlying principle is that the public has a right to know, a right to be heard and a right of co-decision. In keeping with the principles of Agenda 21 (UNCED, 1992), the public should be involved in information gathering and management of recreational water use areas. In resort areas, the management tasks are usually predominantly in the hands of the local government or health authority. Progressively more responsibility lies with the local community or individual user as beaches become more rural.

The public can take an active role in a variety of practical activities concerned with beach management. The participation of the public in monitoring helps to raise awareness of the condition of the recreational water use area and provides a cost-effective method of gathering large amounts of data which can then be acted upon by beach managers (see Chapter 12). Involvement of the public in special interest groups, such as voluntary lifeguard organisations, helps to educate the public for self protection. Beach managers have a responsibility to educate the public about hazards related to the recreational water use area, to provide warnings to the public and to provide other information. There are a variety of methods for communicating with the

* *This chapter was prepared by B. Van Maele, K. Pond, A.T. Williams and K. Dubsky*

public, such as flags, signs, literature or beach awards. Whichever method is used, it is imperative that the public understands clearly the message being conveyed.

6.1 Public participation schemes

A number of community participation schemes have been developed world-wide. An example is the "Officer Snook Program" which was initiated in 1992 at Miami Beach and was sponsored by the United States Coast Guard. This scheme includes videos, slide shows, competitions, cleanups and recycling programmes involving 25,000 elementary schools (Sevin and Sevin, 1995). In Glacier Bay, Alaska the prevention of marine debris is an integral part of the visitor management and education programme (Synder, 1995). Some schemes are aimed at specific types of marine debris. In Tasmania, Australia the use of television advertisements was an integral part of a community awareness programme initiated in response to the growing entanglement of marine mammals and seabirds in marine debris (Slater, 1995). In southern Africa, the Dolphin Action and Protection Group launched a national campaign in 1987 entitled "Save our Sealife: Prevent Plastic Pollution". The scheme targeted shipping and fishing companies, industry, schools and the general public, and involved the distribution of pamphlets, the initiation of beach-cleans and the raising of the issue in Parliament. The scheme has now been extended to Antarctica, Namibia and islands in the Southern Atlantic and Indian Oceans (Rice, 1995).

Education and public awareness are key elements in the reduction of marine debris. Public involvement in beach litter management takes two forms: direct action such as beach cleanups and monitoring; and indirect action, such as education, award schemes and legislation. The involvement of the public in beach monitoring and cleanup programmes has dual advantages in that it allows a large sample size to be achieved, and raises awareness among society which will then translate into effective individual action to reduce litter at source. Involvement of the public in such campaigns has been achieved world-wide. Coastwatch Europe, for example, involves tens of thousands of volunteers each year in monitoring marine debris (Dubsky, 1995).

The largest network organising beach clean events is the Center for Marine Conservation (CMC) based in Washington DC, USA. This centre organises annual beach cleans during "Coastweek" at the end of September and beginning of October. Volunteers use standard recording cards which divide debris into eight major categories according to the fabrication material and a further 65 categories according to type of item. Guidance notes are provided, including an identification guide and information on how the data is used. Volunteers are asked to record the location of the beach and the nearest city, the estimated distance covered and the number of bags filled with debris (Bierce and O'Hara, 1992, 1993).

6.2 Local communication

Chapters 2 and 9 provide examples of schemes involving registration and classification designed to help managers identify the characteristics of their recreational water use area and its different uses. It is extremely important at the stage of beach classification to involve all interested parties. The more people feel they are involved, the easier it is to get active and constructive participation and support for monitoring programmes, development plans and environmental protection measures. When beach management plans are operational, further information updates are important. This information needs to involve aspects of beach safety and water quality.

Baseline surveys, eco-audits, or shore and hinterland surveys are an excellent way to gather data about the bathing water itself, the surrounding aquatic environment and the hinterland. For resort beaches, the baseline survey should be carried out during the main bathing season, when caravan parks are full, local restaurants are running at peak capacity and facilities such as public toilets and showers are being well used. If a survey is undertaken out of season unexpected events, such as seepage from overloaded septic tanks and storm drains, are likely to be missed. It is essential that the baseline survey is augmented with background information concerning seasonal changes. Such information can be gained from local people, such as year-round swimmers or non-governmental organisations.

Local public participation should be part of the whole exercise from survey to subsequent action. However, it must be borne in mind that involving local public participation during baseline survey data gathering might occasionally invite bias into reporting unless care is taken by the survey team. The baseline survey will provide a variety of information that can be used in plans for informing the beach users and visitors on safety and health risks.

6.3 Types of information

6.3.1 Beach safety

Unless users are aware of the hazards and regulations applicable to particular areas, they are unable to make an informed choice about their destination or to react appropriately to management strategies. While there may be resort areas with an abundance of public information and controls, it is not economically viable, nor necessarily desirable, to extend such infrastructure to rural bathing places, used by only a few people. However, it is in everybody's interest that bathing is as safe as possible even in these isolated areas.

Strategies for accident prevention should first address the removal of hazards. If this is not possible, steps should be taken to reduce the level of risk. Information is particularly important where less can be done physically to reduce the risk. In this regard, all available techniques should be used to convey safety messages, such as the provision of safety signs and notices, flags and brochures (see Chapter 7).

6.3.2 Water quality

The primary reason for monitoring bathing water quality and for informing the public is to protect public health. Members of the public are unlikely to want to know the details of sample treatment in the laboratory, although they would need to know whether the water quality is safe. It is essential that information provided to the public is presented in a clear, unambiguous and easily understood way. Some of the cheapest and quickest approaches to assembling and presenting summary data are often the most effective.

6.4 Award schemes

Award schemes are often used as an incentive programme to involve all parties concerned in participating in optimising beach safety, water quality and education activities. Awards are generally the recognition of effort, or of standard achieved. Most award schemes look at only a few of the parameters associated with beach classification. They often fail to take account of the beach user's perception of the environment. The ideal scheme should consider physical, biological and human parameters. The first two are relatively easy to measure (see Chapters 8, 10 and 11), the latter is more difficult to assess.

Beach award and evaluation systems are valuable tools for the promotion and management of beaches and tourism. Annual and systematic surveys of a variety of parameters, including beach litter, have been undertaken for a number of award schemes. Beach awards can be important agents for change, integrating a variety of factors, including water quality, safety, litter, and beach management practice in general. Resorts, in particular, want these awards and manage their beaches to ensure that they comply with the requirements of the award.

6.4.1 Blue Flag

Probably the most widely known beach award within the European and Mediterranean context is the Blue Flag Award. The Blue Flag scheme is organised by the Federation of Environmental Education in Europe (FEEE) (FEEE, 1998). The Blue Flag Campaign was started in 1987 as one of the many activities of the European Year of the Environment. It is a Europe-wide initiative involving more than 1,000 beaches and 500 marinas in 19 European countries. Within the European Union, only "identified" bathing waters within the terms of the Bathing Waters Directive 76/160/EEC are eligible for the award. Outside the European Union, almost any beach could apply for the award via the national operator. Qualification is based on a wide variety of criteria (some of which are obligatory, others are guideline criteria) divided into four groups: environmental education and information, environmental management, water quality, and safety and services. In terms of the environmental education criteria "the aim of the campaign is to increase the public's environmental awareness and to create a platform for active participation in the protection of the environment" (FEEE, 1998). Co-operation between FEEE and the United Nations Environment Programme, Industry and Environment (UNEP IE) office resulted in a pilot project for implementing the Blue Flag concept in non-European regions.

To combine monitoring of Blue Flag holders with gathering extra information, an "In Season Beach Award" was run in Ireland in 1992 for 100 beaches. Points were allocated for a range of criteria and each beach was visited and checked thoroughly by one of a volunteer team (see Box 6.1).

6.4.2 Costa Rica

Chaverri (1989) devised a rating system to identify beaches suitable for governmental and private tourist development in Costa Rica under the authority of the Marine and Terrestrial Act (Ley Maritimo Terrestre). Up to 113 factors, classed as either positive and negative, were given a score between zero and four, with the final rating score for the beach obtained by subtracting the sum of the "negative" scores from the sum of the "positive" scores. The factors comprised six groups. Some selected factors were water, beach, sand, rock, general beach environment and the surrounding area.

No attempt was made to attribute quantitative values to scores for any of the factors, so that the beach score for any factor was based purely on the subjective judgement of the particular assessor. In addition, no attempt was made to assess the importance attached by beach users to any of the factors in the checklist, to assess which factors were of importance for various types of beaches (apart from a differentiation between sand and rock areas), or to attach weightings to the various factors. Even the rigid division of factors into "positive" and "negative" categories could be considered to be subjective.

6.4.3 Black Sea Environment Programme

The Black Sea Environment Programme aims to strengthen and create regional capacities for managing the Black Sea, in particular by developing policies and legislative frameworks relating to pollution, health, biodiversity, and to attract investment. The programme emphasises the importance of harmonisation of methodologies and standards for evaluation of bathing beaches and beach quality. It provides guidelines for assessment of bathing

Box 6.1 The In-Season Beach Award

The presence of a Blue Flag indicates that the visitor should find dependable water quality, cleaning, toilets and other facilities on a managed beach. It does not relate to wind shelter, diving facilities, beautiful scenery, etc., nor does it allow the visitor to predict when swimming might be safe or unsafe and on which days lifeguards are supposed to be on duty. Nevertheless, such questions would be asked by more concerned tourists before booking their holiday or before heading off to any particular beach from several possible beaches that are at an equal distance away.

In order to combine the checking of Blue Flag winners with the gathering of extra information, an in-season beach award was designed and run in Ireland in 1992 for the 100 top beaches. A national weekly newspaper sponsored the award. A list of the top 100 beaches was prepared from Blue Flag entries, augmented by further beaches known from local community notes. A detailed questionnaire was designed and tested on different beaches before being adapted. A volunteer team with a good environmental background was established and trained together. Each beach was visited in the peak July and August bathing season and checked thoroughly by a member of the team. Where possible, local people were interviewed. Photographs and sketches augmented the reports filed by the team.

Points were allocated, on a predetermined scale, for natural assets and facilities provided, with an option of bonus points for special quality. In the health category, for example, a stream was considered something positive as a natural asset and allocated a point. Sixty-seven beaches had such an asset. Unfortunately, the majority of the streams turned out to be polluted when checked for faecal streptococci. A clean stream with good invertebrate diversity was thus a rare quality, and was awarded an extra six points. The display of water quality results and minimal frequency monitoring was also awarded a point. Moreover, when members of the public questioned at random found the information clear and understandable, an extra three points were awarded. Winners were announced in each of the following sections: water quality, other facilities, natural beauty and wildlife value.

The scheme received very good publicity and initiated a lot of local activity to remove accumulations of litter. Those beaches shortlisted as final award winners were revisited over a two-day period by the sponsor's helicopter. While a beach with undependable water quality could not become a winner, it could get a very high number of points for other assets. The results could then be used to argue for improvement of the weakest feature, e.g. water quality.

beaches and bathing water quality, and on how to implement assessment programmes and to evaluate the results.

The programme suggests a questionnaire for registering beach quality that takes into account details concerning beach facilities, physical characteristics of the beach, usage, accessibility, water quality and designation. It does not involve "scoring" the beaches. The final classification is based on the following definition: *"a good beach is a safe beach as well as a beach with good water and beach quality"* (WHO, 1995). The beach is classified according to any problems discovered and, using this classification, an action programme can be identified. The objective of the programme is then to encourage the use of data to refine the action programmes to solve problems that have been highlighted through the monitoring programmes.

6.4.4 Schemes developed for Turkey

Morgan *et al.* (1995) used a questionnaire based on beach users preferences and priorities linked with a 47 factor checklist for five Turkish beaches; Oludeniz beach scored the highest with 87 per cent. Additionally, Morgan *et al.* (1996) carried out further studies on Turkish, Spanish and Maltese beaches by testing beach user perception for 50 beach aspects. Williams and Morgan (1995) have also assessed 28 Turkish beaches in terms of 50 physical, biological and human parameters based on the views of a range of international coastal experts; Dalaman beach rated the highest at 93 per cent. Beaches were scored for each parameter on a scale of one (poor) to five (good). Williams *et al.* (1993b) and Leatherman (1997) have used a similar scale for 182 beaches in the south west peninsula of the UK and 650 beaches in the USA respectively. These checklists could be readily improved because many aspects of the beach environment were classified as good or bad without regard to the varying preferences of different types of users, and various uses, of the beach environment. Many factors were judged on a subjective basis with no weightings attached. In addition no attempt was made to resolve the problem of different views and preferences of visitors to different types of beach.

6.4.5 Local quality schemes

Various local schemes exist to assess beach and water quality, such as the Solent Water Quality Awards, which were established in 1992 and are administered through the Solent Water Quality Conference, a consortium of local authorities and interest groups in Hampshire, UK. All bathing waters in the Solent region (identified and non-identified) that are used regularly for bathing can enter the scheme. The criteria for achieving an award are:

- At least one representative sampling point must be selected for each beach.
- Imperative standards of the EU bathing water directive (CEC, 1976) must be met.
- The water must not contain any gross pollution by faeces or other sewage-related debris, or suffer from persistent occurrence of oil, tar or a significant smell.
- Supporting information, such as water quality results from the previous years must be given.

The main criticism of these awards is that they do not consider the beach itself and are restricted to the water quality.

6.4.6 Other schemes

Recent studies suggest strongly that people with different personalities and demographic variables have different requirements for the beach environment and prefer to visit different types of beaches (Morgan *et al.,* 1993; Williams *et al.,* 1993b; Williams and Morgan, 1995). This poses a problem for beach ratings, but it can be overcome by dividing the beaches into a number of categories on the basis of degree of commercialisation, i.e. presence or absence of particular facilities. For example, Williams *et al.* (1993a,b) and Morgan *et al.* (1993, 1995) used questionnaire surveys as a basis for establishing preferences and priorities of beach users at various beach types, and to weight the various factors in a beach quality rating scale. The scheme was carried out in two main stages. Firstly, an assessment was made of the preferences for various beach features (such as pocket, log spiral or linear beaches) and facilities (such as toilets) and the attributes of the visitors to different types of beaches. This enabled the various factors in the beach quality rating scale to be optimised and correctly weighted. This was followed by the introduction of a checklist for the beach quality rating scale containing classifications and categories of 48 beach aspects closely matched to those in the questionnaire. As many beach aspects as were reasonably possible were assigned classifications based on quantifiable values. Weighting and scoring of the various beach aspects on the checklist was generated by analysis of questionnaire responses.

6.4.7 Standardising grades and categories

The standardisation of grades and categories for describing and informing the public of the quality of recreational water use areas is complicated by the variety of aims that exist amongst different schemes. Recently, Earll *et al.* (1997) put forward the idea of a standardised litter pollution category, i.e. the "ABCD" grading system used in the "Code of Practice on Litter and Refuse" developed by the 1990 Environmental Protection Act (DoE, 1991) and the Thames Clean Project (Lloyd, 1996). Litter categories suggested by Earll *et al.* (1997) are:

Grade A Absent, no evidence of litter anywhere.
Grade B Trace, small items only.
Grade C Unacceptable, widespread distribution with minor accumulations.
Grade D Objectionable amount, area heavily littered with accumulations along the boundaries.

Litter categories of concern to the general public include sewage-related debris, litter accumulations and harmful litter such as medical waste. The

Table 6.1 Proposed classification scheme for the assessment of aesthetic quality of coastal and bathing beaches

Category/type	A	B	C	D
Sewage-related debris				
General	0	0	1–5	6+
Cotton buds	0	1–9	10–49	50+
Litter				
Gross	0	1–5	10–24	25+
General	0–49	50–99	100–999	1,000+
Harmful	0	0	1–3	4+
Accumulations				
Number	0	0	1–3	4+
Total items	0	1–5	4–49	50+
Oil	Absent	Trace	Some	Objectionable
Faeces	0	1	2–9	10+

Source: Earll *et al.*, 1997

number of items listed in Table 6.1 relates to a 100 m stretch of beach at the high water strand line. A recreational water use area would receive a grading based on one of the categories falling into the worst grade, i.e. if one of the categories scores a "D" then the beach is graded a "D" beach. The actual numbers proposed in Table 6.1 are subject to further research. A constant strand line length of 100 m has been advocated but this could cause problems for small pocket beaches of less than 100 m in length.

6.5 Education

Awareness on water safety may be achieved through community education. This can be by means of talks to groups and schools, information sheets and posters, videos or practical activities. Public participation and education can be promoted through government advisory committees, citizens advisory committees, interest group representatives, public hearings, broad dissemination, information gatherings, community meetings, media campaigns, brochures, newsletters, school programmes, community exhibitions and user group training.

6.5.1 School education

School education differs greatly between countries and also between regions and school types. Most students never see a County Council or Parliament debate and have never asked their local representative to pose a written question for them (such as why the local beach is not designated). In addition, environmental law is rarely taught in schools. Although water quality experiments might be carried out in chemistry and biology classes, and fieldwork might be undertaken, the results are rarely compared with real data generated by official monitoring programmes.

The involvement of school groups in awareness campaigns such as Coastwatch Europe (Dubsky, 1995) (see Box 6.2), in beach cleans such as those organised by the CMC in the USA (Bierce and O'Hara, 1993) and in other community participation programmes (Box 6.3) (see also Chapter 12) is becoming more widespread. Understanding provides the ability to make informed decisions. Bathing is practised so widely as a form of recreation that information relating to its safe enjoyment should be widely disseminated beyond swimming classes. A basic understanding of water pollution, water quality and dangers on the shore may be taught beneficially in school such that the knowledge gained is applied early.

6.5.2 Special interest groups

Swimming, lifesaving and other local interest groups play an important part in the education and awareness of the public towards recreational water-use quality and safety (Box 6.3). Recognition of beach hazards has led to the introduction of various beach safety regulations and the establishment of lifesaving clubs at many resorts, particularly in the USA, Australia and New Zealand. Surf Life Saving, Australia, for example, is a national organisation co-ordinating 255 Life Saving Clubs and professional lifeguards who patrol 300 beaches and make over 10,000 rescues each year. This organisation has also sponsored the Beach Safety Management Programme, documenting coastal hazards and their impacts on public safety on more than 7,000 Australian ocean beaches. It has developed a database for every beach, showing location, access, nature, physical characteristics, facilities, use, and beach and surf conditions, together with an assessment of risk levels (on a scale of 1 to 10) and a prediction of the cost required to maintain adequate levels of public safety on each beach (Short *et al.,* 1993).

6.6 Elements of good practice

- The findings of any monitoring programmes should be discussed with the appropriate local, regional and/or national authorities and others involved in management (including integrated water resource management), such as the industrial development or national planning boards.
- The results of monitoring programmes should be reported to all concerned parties, including the public, legislators and planners. Any information relating to the quality of recreational water use areas should be clear, concise and should integrate microbiological, aesthetic and safety aspects.
- In issuing information to interested parties (the public, regulators, NGOs, legislators, etc.), it is essential that their concerns are kept in mind.

Box 6.2 Coastwatch Europe survey

Coastwatch Europe is an international network of universities and environmental groups co-operating on coastal zone management issues, as well as public information, participation and training schemes. The core Coastwatch project, shared by 23 participating countries, is the Coastwatch Europe survey. A single set of questions is agreed internationally by all co-ordinators in order to give baseline information about all sections of the coast. The questionnaire is translated into national languages, and may be augmented by extra national questions and, where financially possible, with water quality testing. Local baseline surveys are undertaken by local volunteers on 500 m stretches of shore from the water's edge, covering the splash zone and immediate hinterland. The volunteers are recruited through newspaper publication of the questionnaire or through associations (schools, scouts, ladies clubs, divers or sea anglers). Since 1989 there have been over 10,000 sites surveyed every autumn, making it the largest volunteer data set for the coast of Europe.

In many countries the scheme does not just involve environmental groups, universities and local volunteers, but also local authorities. Before the survey starts, surveyors are provided with a local contact and are equipped with coded maps of their area, questionnaires, survey notes and test kits. The survey often leads to follow-up actions, such as experienced in County Louth (Ireland). The County Council asked surveyors to return questionnaires to the authority before submitting them to Coastwatch, with the pledge that officials would look through the data and act on broken pipes, illegal dumping, etc. within weeks of receiving the information. As promised, within a month of receiving the data a big coastal clean up was started by the Council, which invited local people to join in. Such co-operation in management builds good will and translates into better coastal quality.

In running the survey and various forms of follow-up action, Coastwatch co-ordinators have found that Europe-wide, specific volunteer subsets, such as fishermen and yachtsmen, have excellent knowledge based on their experiences of living and working in the locations. In most cases, simply raising the polluters' awareness of the consequences of their actions and bringing people together in the common cause of making their local water safe, brings about the required change. Sometimes, lack of finance is clearly the limiting factor.

Increasingly, it is the local people who ask for guidelines to gather baseline information and draw up a management plan for their area. The survey has often resulted in cases of co-operation between local public and officials for common aims and quality control, such as the joint management of litter, introduction of recycling campaigns or nature trails. If sewage treatment is inadequate, for example, a combined effort in lobbying the government to supply the necessary funding can be much more effective than either local people or a local authority asking alone. In cases where officials cannot be persuaded to join in, scientifically qualified environmental groups can be an alternative.

Box 6.3 Community participation schemes

Negril Coral Reef Preservation Society (NCRPS) based in Jamaica is a non-profit non-governmental organisation that was formed by a group of diving operators in 1990 because of concern over the state of the reefs. At the time of its inception, the main goal was to install reef mooring buoys on frequently dived reefs in a growing tourist town that was once a fishing village. Thirty-five state-of-the-art reef mooring buoys were installed in 1991 with the help of "REEF RELIEF", a partner organisation in Key West, Florida. Although the reef mooring buoys prevented over 20,000 anchorages annually, it was decided that the project should be expanded.

Deteriorating water quality was identified as the biggest threat to Negril's (and Jamaica's) reefs. Lack of proper sewage treatment, deforestation, poor agricultural and solid waste management practices allowed nutrient-laden effluents to enter coastal waters. The nutrients were stimulating the growth of nuisance algae, which were smothering the reefs. As a result, the coastal waters of Negril are now in the advanced stages of eutrophication and live coral coverage is less than 10 per cent, while algae dominate more than 65 per cent of the reef. The NCRPS has as one of its primary concerns the restoration of water quality so that coral reefs can, hopefully, someday return to their previous state, or at least become recovering reefs. In 1997, a small water quality monitoring laboratory was established at the NCRPS Headquarters. A water quality monitoring programme was initiated, measuring nutrient levels in rivers, streams, ground, and coastal waters throughout the Negril Environmental Protection Area and National Marine Park. Monthly samples were collected by the NCRPS rangers and analysed in the local laboratory, while some samples were sent to outside laboratories for analysis.

An aggressive public education campaign targeting schools, communities and the hospitality industry involved raising awareness of water quality issues. Annual workshops entitled "Protecting Jamaica's Coral Reef Ecosystem" allow open discussion and participatory planning of management initiatives. A Junior Ranger training programme, involving hundreds of children between the ages of 10 and 17 years, gets students and teachers within the local schools involved in learning about water quality issues and taking part in the monitoring programme. In the context of establishing a management structure for a Marine Park, the water quality initiatives are included in an overall coral reef monitoring programme. In partnership with the Jamaican government, through locally established "Resort Boards", NCRPS has also designed a watersports and recreational zoning programme. Demarcation buoys set 300 feet from shore mark a safe swimming zone, and there are plans to expand this programme by adding additional buoys for demarcation of non-motorised craft and environmental zones. The Society is responsible for the installation and maintenance of these demarcation buoys, and the rangers patrol them together with the police, to ensure that rules and regulations are adhered to.

Source: Negril Coral Reef Preservation Society, Pers. Comm.

- Reports addressing the quality of recreational water use areas should be accompanied by references to local and visitor perceptions of the aesthetic quality and risks to human health.
- The deleterious impacts of human health hazards and aesthetic pollution, and of control measures to avoid or reduce such impacts, should be introduced into environmental health education programmes in both formal and informal educational establishments.
- The usefulness of the information obtained from monitoring is severely limited unless an administrative and legal framework (together with an institutional and financial commitment to appropriate follow-up action) exists at local, regional and international level.

6.7 References

Bierce, R. and O'Hara, K. [Eds] 1992 *1991 International Coastal Cleanup Results*. Center for Marine Conservation, Florida, 217 pp.

Bierce, R. and O'Hara, K. [Eds] 1993 *1993 International Coastal Cleanup Results.* Center for Marine Conservation, Florida, 275 pp.

Camhis, M. and Coccossis, H. 1982 The national coastal management program of Greece. *Ekistics,* **293,** 131–137.

CEC 1976 Directive on bathing water quality, 8 December 1975, (76/160/EEC). *Official Journal of the European Communities*, **L31**/1, 5/2/76.

Chaverri, R. 1989 Coastal management; the Costa Rica experience. In: O. T. Magoon [Ed.] *Coastal Zone '87,* Volume 5, American Society of Civil Engineering, New York, 1112–1124.

DoE Department of the Environment 1991 *Environmental Protection Act 1990. Code of practice on litter and refuse.* Her Majesty's Stationery Office, London.

Dubsky, K. [Ed.] 1995 *Coastwatch Europe. Results of the International 1994 Autumn Survey.* Coastwatch Europe, Dublin, 65 pp.

Earll, R., Williams, A.T. and Simmons, S. 1997 Aquatic litter, management and prevention - the role of measurement. In: E. Ozhan [Ed.] *MedCoast '97. Proceedings of the fifth international conference on the Mediterranean coastal environment* Medcoast Secretariat, Middle East Technical University, Ankara, 383–396.

FEEE 1998 *The Blue Flag 1998.* Federation of Environmental Education in Europe, European Office, Copenhagen.

Gubbay, S. 1994 The future of coastal management - challenges and opportunities. In: *Proceedings of the Coastline Conference - public participation in coastal management, 31 May 1994.* Farnborough College of Technology, Hampshire, UK, 6–7.

Leatherman, S.P. 1997 Beach rating: a methodological approach. *Journal of Coastal Research*, **13**(1), 253–258.

Lloyd, M. 1996 Thames Clean methodology - strengths and weaknesses. In: R. Earll, M. Everard, N. Lowe, C. Pattinson and A.T. Williams [Eds] *A guide to methods - a working document, version 1.* National Aquatic Litter Group, Gloucester, UK.

Morgan, R., Bursahoglu, B., Hapoglu-Balas, L., Jones, T.C., Ozhan, E. and Williams, A.T. 1995 Beach user opinions and beach ratings: A pilot study on the Turkish Aegean coast. In: E. Ozhan [Ed.] *MedCoast '95, Proceedings of the third international conference on the Mediterranean coastal environment.* Medcoast Secretariat, Middle East Technical University, Ankara, 373–383.

Morgan, R., Gatell, E., Junyant, R., Micallef, A., Ozhan, E. and Williams, A.T. 1996 Pilot studies of Mediterranean beach user perceptions. In: E. Ozhan [Ed.] *ICZM in the Mediterranean and Black Sea: Immediate Needs for Research.* Middle East Technical University, Ankara, 99–110.

Morgan, R., Jones, T.C. and Williams, A.T. 1993 Opinions and perceptions of England and Wales Heritage Coast beach users: some management implications for the Glamorgan Heritage Coast, Wales. *Journal of Coastal Research,* **9**(4), 1083–1093.

Rice, N. 1995 Marine debris education in Southern Africa. In: J. Faris and K. Hart [Eds] *Seas of debris: A summary of the third international conference on marine debris.* North Carolina Sea Grant, UNC-SG-95-01, Florida, 54 pp.

Sevin, J. and Sevin, E. 1995 Officer Snook Program. In: J. Faris and K. Hart [Eds] *Seas of debris: A summary of the third international conference on marine debris.* North Carolina Sea Grant, UNC-SG-95-01, Florida, 54 pp.

Short, A.D., Williamson, B. and Hogan, C.L. 1993 The Australian beach safety and management program - Surf Life Saving Australia's approach to beach safety and coastal planning. *Proceedings of the 11th Australian conference on coastal and ocean engineering.* Townsville, Institute of Engineers, Australia, 113–118.

Slater, J. 1995 Marine debris of Tasmania, Australia: sources and solutions. In: J. Faris and K. Hart [Eds] *Seas of debris: A summary of the third international conference on marine debris.* North Carolina Sea Grant, UNC-SG-95-01, Florida, 54 pp.

Synder, J. 1995 Techniques for prevention of marine debris at Glacier Bay, Alaska. In: J. Faris and K. Hart [Eds] *Seas of debris: A summary of the third international conference on marine debris.* North Carolina Sea Grant, UNC-SG-95-01, Florida, 54 pp.

UNCED 1992 United Nations Conference on Environment and Development, *Agenda 21: Programme of Action for Sustainable Development,* UN Department of Public Information, New York, USA.

WHO 1995 Manual for recreational water and beach quality monitoring and assessment. Prepared in support of the Black Sea Environment Programme.

Draft, April 1995, World Health Organization Regional Office for Europe, Copenhagen.

Williams, A.T., Gardner, W., Jones, T.C., Morgan, R. and Ozhan, E. 1993a A psychological approach to attitudes and perceptions of beach users: Implications for coastal zone management. In: E. Ozhan [Ed.] *MedCoast 93, Proceedings of the first international conference on the Mediterranean coastal environment.* Medcoast Secretariat, Middle East Technical University, Ankara, 218–228.

Williams, A.T., Leatherman, S. P. and Simmons, S. L. 1993b Beach aesthetics: South West Peninsula, UK. In: H. Sterr, J. Hofstide and H.P. Plag [Eds] *Interdisciplinary discussions of coastal research and coastal management issues and problems*. Peter Lang, Frankfurt, 251–262.

Williams, A.T. and Morgan, R. 1995 Beach awards and rating systems. *Shore and Beach*, **63**(4), 29–33.

Chapter 7[*]

PHYSICAL HAZARDS, DROWNING AND INJURIES

Physical hazards are generally perceptible and discernible. Physical hazards, unlike many microbiological, biological and chemical hazards do not require laboratory analysis for their recognition or description. The hazards that can lead to drowning and injury may be natural or artificial. By definition a hazard is a set of circumstances that may lead to injury or death. The term "risk" is used to describe the probability that a given exposure to a hazard will lead to a certain (adverse) health outcome.

In the context of this chapter, hazards are best viewed as both the potential causes of ill health and the absence of measures to prevent exposure or mitigate against more severe adverse outcomes. Thus, an area of dangerous rocks against which swimmers may be drawn by prevailing currents or wind, the absence of local warnings, the absence of general public awareness of the types of hazards encountered in the recreational water environment, and the absence of local capacity to recognise and respond to a person in danger, may all be readily conceived as part of the hazard. The number of injuries can be reduced by elimination of the actual hazard, by restricting access to the hazard, by members of the public recognising and responding appropriately to the hazard, and by ensuring deployment of effective management actions.

The severity of the adverse health outcomes considered in this chapter differs markedly from that described elsewhere in this book. The severity of the outcomes varies widely but includes death (for example through drowning) and lifelong disability through quadriplegia, as well as blindness arising from retinal displacement. It also addresses less severe health outcomes such as cuts and lacerations that are nevertheless important in determining the pleasure derived from recreational use of water environments. Whilst the overall frequency of severe outcomes may be low they are of considerable importance for public health.

Despite the importance of the health outcomes addressed in this chapter, methods for assessment of the associated hazards and mitigating factors are relatively poorly developed and have attracted limited research when

* *This chapter was prepared by A. Mittlestaedt, J. Bartram, A. Wooler, K. Pond and E. Mood*

compared with, for example microbiological pollution of bathing waters (see Chapters 8 and 9). Nevertheless, assessment may be rapid and simple and may be readily and rapidly associated with short-and medium-term actions of immediate relevance to the protection of public health. This chapter draws heavily on the corresponding chapter of the *Guidelines for Safe Recreational Water Environments* (WHO, 1998) in which the issue of physical hazards and drowning is also discussed. This chapter summarises the key components of that chapter and provides a practitioners guide to the various issues.

7.1 Health outcomes

The most prominent health outcomes resulting from recreational use of water are:

- Drowning and near-drowning.
- Major impact injuries, especially spinal injuries, resulting in quadriplegia and less frequently, paraplegia, as well as head injuries.
- Slip, trip and fall injuries (including bone fracture and breaks).
- Cuts, lesions and punctures.
- Retinal dislocation resulting in near blindness or blindness.

7.1.1 Drowning and near drowning

Drowning and near drowning are important health issues and merit special consideration in the development and management of water recreational facilities. Informal peer supervision in more densely-used areas may contribute significantly to the prevention of drowning and, conversely, the desire for greater seclusion may be a significant contributory factor. Private pools (including ornamental, swimming and paddling pools) contribute significantly to drowning statistics, but are not addressed in this volume.

Males are more likely to drown than females (WHO, 1998) and this is, in part, associated with higher exposure to the aquatic environment (through occupational and recreational uses). In many countries, alcohol consumption is one of the most frequently reported contributory factors associated with drownings. Amongst children, lapses in parental supervision are the most frequently cited contributory factor in drownings and near drownings. Drowning and near drowning may often be associated with recreational water uses with low water contact, such as use of water craft (yachts, boats, canoes) and fishing (from water craft and from the waters edge or solid structures). Where these recreational water uses occur during cold weather, immersion cooling may be a significant contributory factor (Keatinge, 1979; Poyner, 1979). However, non-use of life jackets, even when they are readily available, is a significant contributory factor in all cases. The availability of cardiopulmonary resuscitation (CPR) and rescue skills have been reported to be important in determining the outcome of accidental immersions. However, attempted rescue represents a significant risk to the rescuer.

Most drownings occur in non-swimmers and the value of swimming lessons as a preventative measure appears logical. However, there is significant debate regarding the age at which swimming skills may be acquired safely, and the role of swimming skills in preventing drowning and near drowning is unclear. Whilst evidence does not suggest that water safety instruction increases the risk of young children drowning, their increased skills do not decrease the need for adult supervision; the impact of training on decreasing parental vigilance has not been assessed (Asher *et al.,* 1995).

Pre-existing diseases are associated risk factors and higher rates of drowning are reported amongst those with seizure disorders and paediatric seizures. Further documented contributory factors in drownings include water depth and poor water clarity (Quan *et al.,* 1989). Studies of "near drowning" show that the prognosis depends more on the effectiveness of the initial rescue and resuscitation than on the quality of subsequent hospital care. The principal contributory factors and preventative and management actions for drowning and near drowning are similar and are summarised in Table 7.1.

7.1.2 Spinal injury

Diving accidents have been found to be responsible for a variable percentage of traumatic spinal cord injuries. However, in diving accidents of all types, injuries are almost exclusively located in the cervical vertebrae and typically cause quadriplegia or, less commonly, paraplegia. In Australia, for example, diving accidents account for approximately 20 per cent of all cases of quadriplegia (Hill, 1984). The financial cost of these injuries to society is high, because those affected are frequently healthy young persons, principally males under 25 years of age (Blanksby *et al.*, 1997).

Data from the USA suggested that diving into a wave at a beach and striking the bottom was the most common cause of spinal injury, and 10 per cent of spinal injuries occurred when the person dived into water of known or unknown depth, particularly from high platforms, including trees, balconies and other structures (CDC, 1982). As with drowning, alcohol consumption may contribute significantly to the frequency of injury. Special dives, such as the swan or swallow dive are particularly dangerous because the arms are not outstretched above the head but to the side.

The role of water depth in determining the outcome of diving injuries has not been ascertained conclusively and the minimum depths for safe diving are often greater than expected. Technique and education appear to be important in preventing injury and inexperienced or unskilled swimmers require greater depths for safe diving. Most diving injuries occur in relatively shallow water

Table 7.1 Drownings[1] and near-drownings — contributory factors and principal preventative and management actions

Contributory factors	Principal preventative and management actions
Alcohol consumption	Continual adult supervision (infants)
Cold	Provision of lifeguards
Ice cover	Availability of resuscitation skills/facilities
Waves (coastal, boat, chop)	Wearing of lifejackets when boating
Underwater entanglement	Provision of rescue services (lifeboats)
Pre-existing disease	Local hazard warning notices
Sea current (including tides, undertow and rate of flow)	Development of rescue and resuscitation skills amongst general public and user groups
Offshore winds (especially with flotation devices)	Development of general public (user) awareness of hazards and safe behaviours
Bottom surface gradient and stability	Access to emergency response (e.g. telephones with emergency numbers)
Impeded visibility (including coastal configuration, structures and overcrowding)	Co-ordination with user group associations concerning hazard awareness and safe behaviours
Water transparency	

[1] In most countries males and infants constitute a disproportionate number of drownings

Source: WHO, 1998

(1.5 m or less) and a few in very shallow water (e.g. less than 0.6 m) where the hazard may be more obvious. The typical injurious dive occurs into a water body known to the individual.

The principal contributory factors and preventative and management actions for spinal cord injury are summarised in Table 7.2. Evidence suggests that preventative education and awareness-raising offer most potential for diving injury prevention, partly because people have been found to take little notice of signs and regulations. However, because of the young age of many injured persons, awareness raising and education about safe behaviour is required early in life.

7.1.3 Impact, slip, trip and fall injuries

Accidents involving limb fractures or breaks of different types have many causes and may occur in a variety of settings in or around water. The principal contributory factors and preventative and management actions are summarised in Table 7.3.

Table 7.2 Major impact injuries — contributory factors and principal preventative and management actions

Contributory factors	Principal preventative and management actions
Poor underwater visibility	Access to emergency services
Conflicting uses in one area	Use separation/segregation
Bottom surface type	Provision of lifeguards
Water depth	Local hazard warnings
Diving into a wave or into water of unknown depth	Development of general public (user) awareness of hazards and safe behaviours
Jumping into water from trees, balconies or other structures	Early education in diving hazards and safe behaviours

Source: WHO, 1998

Table 7.3 Slip, trip, fall and minor impact injuries — contributory factors and principal preventative and management actions

Contributory factors	Principal preventative and management actions
Diving into shallow water	Selection of appropriate surface type
Underwater objects (e.g. walls, piers)	Use of adjacent fencing (e.g. around docks and piers)
Adjacent surface type (e.g. water fronts, jetties)	Development of general public (user) awareness of hazards and safe behaviours
Poor underwater visibility	

Source: WHO, 1998

7.1.4 Cuts, lesions and punctures

There are many reports of injuries sustained as a result of stepping on glass, broken bottles and cans. Discarded syringes and hypodermic needles may present more serious risks and may attract greater public outcry. Simple measures, such as the use of footwear on beaches, as well as adequate litter bins and cleaning operations may contribute significantly to prevention, as may educational policies to encourage users to take their litter home. The principal contributory factors and preventative and management actions are summarised in Table 7.4.

7.1.5 Retinal dislocation

Impact to the head, resulting from diving and jumping into the water from height have been known to cause detachment of the retina in the eye. The

Table 7.4 Cuts, lesions and punctures — contributory factors and principal preventative and management actions

Contributory factors	Principal preventative and management actions
Presence of broken glass, bottles, cans, and medical wastes	Development of general public (user) awareness of hazards and safe behaviours
Walking and entering water barefoot	Development of general public (user) awareness regarding litter control
	Local availability of first aid
	Provision of litter bins
	Beach cleaning
	Adequate solid waste management

Source: WHO, 1998

Table 7.5 Retinal dislocation — contributory factors and principal preventative and management actions

Contributory factors	Principal preventative and management actions
"Bombing" (jumping onto other water users)	Development of general public (user) awareness of hazards and safe behaviours
Diving into water	
Jumping into water from height	

Source: WHO, 1998

principal contributory factors and preventative and management actions are summarised in Table 7.5.

7.2 Interventions and control measures

Control of physical hazards may involve their removal or reduction, if possible, or measures to prevent or reduce human exposure or to minimise the adverse effects of exposure. As described at the beginning of this chapter the term hazard is generally used in relation to the capacity of a substance or event to affect human health adversely. However, in the context of this chapter, the absence of appropriate control measures may be treated as a component of the chain of causation. For example, the lack of guards, rescue equipment, signs and other remedial actions can contribute to a variety of health outcomes.

The roles of various interventions and control measures in preventing human injury are discussed in the *Guidelines for Safe Recreational Water Environments* (WHO, 1998). The principal measures include public warnings and information (signs, flags, public information), lifeguarding, use separation (zoning, lines, designated areas), and infrastructure and planning, such as for emergency communication, rescue and resuscitation and emergency vehicle access. Whilst the requirement for each of these measures is largely determined locally by a variety of factors, it is important to note that most measures may be more or less effective; their effectiveness may decline after periods of limited or non-use and all are amenable to simple inspection. Importantly, for many measures, replacement or improvement may be within the capacity (financial or practical) of local authorities and, in some circumstances, user groups.

7.3 Monitoring and assessment

The assessment of hazards at a beach or water is critical to ensuring safety. The assessment should take into account several key considerations, which include:

- The presence and nature of natural or artificial hazards.
- The severity of the hazard in relation to health outcomes.
- The availability and applicability of remedial actions.
- The frequency, density and type of use of the area.
- The level of development.

The investigation of hazards in or near present or potential recreation areas, including land and water (natural and artificially constructed) results from a visual inspection procedure. The investigation of physical hazards involves an understanding of the process of causation leading to injury. Because of the importance of individual behaviours in causation, and of awareness in prevention, the involvement of the public, and of interest and user groups in particular (see Chapter 6), is especially important.

The assessment of hazards should take into account the severity and likelihood of health outcomes and the extent and density of use of the recreational area. Health risks that might be acceptable for a recreational area that is used infrequently and is undeveloped may result in immediate remedial measures at other areas that are widely used or highly developed.

Physical hazards vary greatly between sites. Monitoring of a site for existing and new hazards should be undertaken on a regular basis. The inspection and further investigation of hazards requires an understanding of the elements involved in such a programme. The identification of physical hazards, and the subsequent monitoring of any changes to the hazards depend upon potential and present water recreation areas and the hazards encountered. The purposes of inspection and investigation are to provide a routine, systematic, periodic and relevant verification of events, structures, conditions or other situations that represent hazards, whether "theoretical" or "actual" and under "real" conditions.

The following steps have been identified to evaluate an inspection process for hazards in recreational areas:

1. Determine what is to be inspected and how frequently.
2. Monitor changing conditions and use patterns regularly.
3. Establish a regular pattern of inspection.
4. Develop a series of checklists suitable for easy application throughout the system. Checklists should reflect national and local standards where they exist.
5. Establish a method for reporting faulty equipment and maintenance problems.
6. Develop a reporting and monitoring system that will allow easy access to statistics that record "when", "where", "why" and "how".
7. Investigate the frequency of positive and negative results of inspections.
8. Motivate and inform participants in the inspection process through in-service training.
9. Use outside experts to review critically the scope, adequacy and methods of the inspection programme.

7.3.1 Inspection forms and checklists

Because hazards vary greatly and because of the importance of social and behavioural factors (in causation and in prevention), it is important that checklists and inspection forms are developed, tested and refined according to local priorities and experience. Based upon Tables 7.1 to 7.5 some of the factors that may be included in an inspection protocol are described in Table 7.6. Many factors of importance described above are not included in this list because they are not amenable to an inspection-based approach.

7.3.2 Timetabling of inspections

The frequency of inspection will vary according to the intensity of use of the area and the speed with which the hazards encountered and the remedial actions in place change at a specific location. The timing of inspections should take account of periods of maximum use (e.g. inspection in time to take remedial action before major holiday periods) and periods of increased risk. The frequency of inspection therefore has to be predicted based on the size of the facility, the number of features in the facility, and the extent of past incidents or injuries. The criteria for inspections and investigations may vary from country to country. There might be legal requirements and/or voluntary standard-setting organisations.

7.3.3 Reporting and notifications

The importance of co-ordination and participation of all interested individuals or organisations is emphasised in Chapters 5 and 6. Except where minimum legal requirements are specified, action to address deficiencies

Table 7.6 Factors to consider when designing an inspection programme relevant to physical hazards and drownings in recreational waters

Hazard	Factors
Drownings and near-drownings	Sea current (including tides, undertow and rate of flow)
	Offshore winds (especially with flotation devices)
	Possibility of underwater entanglement
	Bottom surface gradient and stability
	Waves (coastal, boat, chop)
	Water transparency
	Impeded visibility (including coastal configuration, structures and overcrowding)
	Lifeguard provision
	Provision of rescue services (e.g. lifeboats)
	Access to emergency response services (e.g. telephones with emergency numbers)
	Local hazard warning notices
	Availability of resuscitation skills and facilities
	Rescue and resuscitation skills amongst user groups
	Co-ordination with user group associations concerning hazard awareness and safe behaviours
	Wearing of lifejackets when boating
Cuts and lacerations	Presence of broken glass, bottles, cans and medical wastes
	Frequency of beach cleaning
	Solid waste management
	Provision of litter bins
	Local availability of first aid
Spinal injuries	Bottom surface type
	Water depth
	Conflicting uses in one area
	Jumping into water from trees, balconies or other structures
	Underwater visibility
	Local hazard warnings
	General public (user) awareness of hazards and safe behaviours
	Early education in diving hazards and safe behaviours
	Level of separation/segregation
	Lifeguard supervision
	Access to emergency services
Slip, trip and fall accidents	Underwater objects (e.g. walls, piers)
	Underwater visibility
	Adjacent surface type (e.g. water fronts, jetties)
	Surface type selection
	Adjacent fencing (e.g. around docks and piers)

identified in inspections depends upon the goodwill of local authorities (local government, user groups and other interested parties). Maintaining co-ordination with such persons and authorities contributes greatly to the overall success of a monitoring programme in containing hazards and preventing adverse health effects.

Whilst much reporting is necessarily of a local nature, it is worthwhile to interpret and report findings at regional or national levels (where this is possible). Moreover, some approval schemes (see Chapter 5) stipulate either general requirements that management plans should be developed and implemented or that specific safety-related requirements should be met. Safety-related data may, therefore, contribute to informed personal choice (and thereby assist individuals in contributing directly to the protection of their own health) and also encourage local authorities to support safety-related improvements.

In addition to the benefits of reporting mentioned above, the availability of information concerning the existence of hazards and the deployment of remedial or preventative, measures may help to generate new insight into the effectiveness of those, and other, measures. Information on this aspect is limited at present.

7.4 Elements of good practice

- The nature and extent of any risk to, or potential hazard to, human health or well-being must be identified and characterised fully. The individual hazards must be related to a likely adverse health outcome.
- To assess the extent of risk, a suitable inspection protocol must be adopted. Such a protocol must define the components of a bathing area that may pose a risk to human health, cataloguing the water conditions, substratum, effects of climatic factors, infrastructure, management and regulatory regime, etc.
- The end-use of the bathing area, including carrying capacity and density of bathers, influences the outcome of the risk assessment.
- On completion of the initial assessment, appropriate control measures including management responses, must be defined, such as zoning, warning mechanisms and public information schemes, lifeguard provision and bathing area infrastructure.
- All situations that may give rise to adverse health outcomes at a bathing area should be reported in a consistent format and stored in an incident database that can be used to inform the level and nature of future management procedures.

7.5 References

Asher, K.N., Rivara, F.P., Felis, D., Vance, L. and Dunne, R. 1995 Water safety training as a potential means of reducing risk of young children's drowning. *Injury Prevention*, **1**, 228–233.

Blankesby, B.A., Wearne, F.K., Elliott, B.C. and Blitvitch, J.D. 1997 Aetiology and occurrence of diving injuries: a review of diving safety. *Sports Medicine*, **23**(4), 228–246.

CDC 1982 Perspectives in Disease Prevention and Health Promotion. Aquatic Deaths and Injuries — United States. *MMWR Weekly*, **31**(31), 427–419.

Hill, V. 1984 History of diving accidents. In: *Proceedings of the New South Wales Symposium on Water Safety*. Department of Sport and Recreation, Sydney, New South Wales, 28–33.

Keatinge, W.R. 1979 *Survival in cold water: the physiology and treatment of immersion hypothermia and of drowning*. Blackwell Scientific Publishers, Oxford. 135.

Poyner, B. 1979 How and when drownings happen. *The Practitioner,* **222,** 515–519.

Quan, L., Gore, E.J., Wentz, K., Allem, N.J. and Novak, A.H. 1989 Ten year study of paediatric drownings and near drownings in Kings County, Washington: lessons in injury prevention. *Pediatrics*, **83**(6), 1035–1040.

WHO 1998 *Guidelines for Safe Recreational Water Environments: Coastal and Fresh-waters*. Draft for consultation. World Health Organization, Geneva.

Chapter 8*

SANITARY INSPECTION AND MICROBIOLOGICAL WATER QUALITY

Sanitary inspection, water quality determination and data analysis and interpretation are essential elements in characterising the microbiological safety of water in recreational areas. Sanitary inspection is a necessary adjunct to water microbiological analysis. A well-conducted sanitary inspection can identify sources of microbiological hazards, microbiological water quality data confirm the presence of hazards, and the two together allow an estimation of the risk of illness to bathers and other users. In assessing the microbiological quality of recreational waters, it will normally be necessary to conduct:

- An intensive sanitary inspection (only once as part of an assessment or annually in monitoring programmes).
- Periodic appraisal visits in which water quality analysis and shortened inspections are undertaken.
- Follow-up appraisals to investigate abnormal events, new sources of pollution and extreme values of pollution indicators.

One of the most important aspects of aquatic microbiology is related to several human diseases transmitted via water. The design and development of epidemiological surveillance studies described in Chapter 13 have led to the awareness of the magnitude of human morbidity and mortality associated with waterborne infectious diseases. The most relevant micro-organisms and the associated waterborne infectious diseases are summarised in the *WHO Guidelines for Safe Recreational Water Environments* (WHO, 1998). The derivation of guideline values for microbiological quality are also discussed in the *WHO Guidelines for Safe Recreational Water Environments* (WHO, 1998).

This present chapter deals with sanitary inspection, microbiological analytical methods and data handling and reporting. Strategies to implement sanitary inspections and recommendations for selection of the site and frequency of water sampling are given in Chapter 9. Specific methods for sampling and analysis are detailed in the following sections together with the

* *This chapter was prepared by M.J. Figueras, J.J. Borrego, E.B. Pike, W. Robertson and N. Ashbolt*

different statistical procedures to express the overall microbiological water quality at a specific recreational water use area. It should be noted that a single beach or recreational area may vary widely in relation to microbiological measures of health risk within relatively short periods of time and thus the commonly used methods of defining a recreational water as passing or failing a defined microbiological standard has inherent limitations; these are discussed in this chapter and also in Chapter 9.

8.1 Sanitary inspection and sampling programmes

A sanitary inspection is a search for, and evaluation of, existing and potential microbiological hazards that could affect the safe use of a particular stretch of recreational water or bathing beach. It provides the foundation required to design and implement an effective water quality sampling programme and provides valuable information to assist in the interpretation of water quality data. In particular, it provides public health authorities with information to aid the selection of sampling locations, times and frequencies, in order to estimate more accurately water quality and therefore to allow for sound risk management decisions (see Chapter 9).

A comprehensive sanitary inspection of an existing recreational area should be conducted annually, just prior to the bathing season. The annual inspection should not only look for new sources of microbiological hazards but also review the adequacy of any sampling programme and corrective measures in place to deal with existing hazards. Further inspections should be conducted along with routine sampling during the bathing season, in order to identify recent events and their impact on water quality. During the peak bathing season additional inspections at different days and times of the day may provide a more complete picture of the bathing area.

Comprehensive inspections should be conducted prior to any major new or proposed activity which could significantly alter the microbiological quality of the water in an existing recreational water use area. A sanitary inspection should therefore be carried out as part of, or in response to, any proposal to expand or develop a new recreational bathing area. The findings of the inspection should receive prime consideration in any decision to proceed with development. A comprehensive annual sanitary inspection consists of four steps:

- Pre-inspection preparations.
- An on-site visit.
- The preparation of a preliminary report including recommendations on location of sampling sites and changes to the sampling frequency if necessary.
- The preparation of a final assessment report often in combination with water quality data.

While sampling, important field data can be obtained at each bathing area by inspecting specific sources of pollution. Microbial contamination may be suspected, for example, when inspection reveals abnormal colouration or odour of the water at the bathing site. In the Mediterranean coastal area where the influence of tides is minimal, changes in microbiological quality are mainly due to riverine and direct, especially urban, discharges at the bathing site. The microbiological contamination produced by long sea outfalls, if well designed, is normally diluted and should not influence the microbiological quality of the bathing area. Land-based sources of contamination are normally associated with smaller discharges or with the likelihood of heavy rain events, characteristic of the Mediterranean climate at the end of the summer period, where a great amount of water falls in a very short period of time. Heavy rain may wash out faeces from pastures or other agricultural land and directly influence microbiological water quality. Studies in other regions also document pulses of poor water quality associated with rainfall events (O'Shea and Field, 1992; Vonstille *et al.,* 1993; Armstrong *et al.,* 1997; Wyer *et al.,* 1994, 1995, 1997). In inland recreational waters the main sources of pollution are water inlets (PHLS, 1995). Therefore, influences from rivers, natural watercourses and, particularly around populated areas, combined sewer overflows, produce important changes in the microbiological quality of bathing waters. Sporadic malfunctioning of sewerage systems can produce similar problems (Davis *et al.,* 1995; Marsalek *et al.,* 1996). These events, if recent, can sometimes be recognised visually at the recreational site by changes in the appearance of the water. In marine recreational waters, a field analysis of the salinity can indicate the discharges of freshwater at the bathing site. Such measurements indicate indirectly, that land-borne discharges are occurring.

8.1.1 Pre-inspection preparations

The collection and review of any existing data or reports on the area, including reports of previous inspections, will allow a thorough and efficient on-site evaluation. Topographical maps and aerial photographs are useful tools for locating activities and features that could affect water quality and for establishing sampling sites. Historical data on tides, currents, prevailing winds, rainfall and discharges of sewage, storm overflows and combined sewer overflows, and urban and agricultural effluents should be collected and reviewed to determine the impact of these events, (either singly or collectively) on water quality. Depending on the availability of water quality data, experts conducting the annual inspection may need to collect samples for microbiological analyses. Therefore, adequate numbers of sterile sample bottles and sampling equipment should be readily available and prior arrangements should be made with the microbiology laboratory to process samples promptly after collection. Arrangements should be made to meet with user groups and with individuals in charge of any facility or activity that affects, or has the potential to affect, water quality in the recreational area. It will be essential to obtain the trust and co-operation of the groups or individuals if the survey is to provide an accurate assessment of water quality and to identify and remedy unacceptable water quality (see Chapter 6).

8.1.2 On-site visit

The purpose of the on-site visit is to identify and evaluate all existing and potential sources of microbiological contamination that could affect the safe use of the area. Attention should be paid to the presence of sewage disposal facilities, including long sea outfalls, industrial outfalls, seabird colonies, sanitary sewers, and rivers, tributaries, streams or ditches receiving sewage, storm water or agricultural runoff. All data recorded should be added to the catalogue of basic characteristics to form a catalogue of inspections that would enable the tracking of trends and influences (see Chapter 2).

Visual faecal pollution (including sanitary plastics), sewage odour and suspicious water colour should also be considered as an immediate indication of unacceptable water quality. Adjacent industries should also be identified and their impact assessed. The impact of local geography and meteorological conditions on water quality should also be evaluated. In most cases it will be necessary to collect representative water samples to confirm the presence of faecal pollution, to establish its variability and to identify the source. Non-toxic fluorescent tracer dyes, bacteriophages (such as PDR-1) or faecal sterol biomarkers (coprostanol and 24-ethylcoprostanol) may also be helpful to identify sources of contamination.

Epidemiological studies have shown that bathers can be a significant source of pathogenic micro-organisms (Seyfried *et al.*, 1985; Calderon *et al.*, 1991; Cheung *et al.*, 1991). In small bathing areas with a lot of bathers and a low rate of water turnover, the person to person disease transmission has to be considered, even if there is no source of faecal pollution from the outside. The assessment may therefore need to consider measures to control microbiological water pollution by bathers in the area. This is especially important in shallow, enclosed areas used by young children where water circulation and flushing rates are low. Intensive studies to locate sources of pollution and to propose remedial actions have been undertaken successfully (Wyer *et al.*, 1994; Tsanis *et al.*, 1995; Marsalek *et al.*, 1996). A specially designed form can assist in the process of comprehensive sanitary inspection (Box 8.1).

Box 8.1 Sanitary inspection form

Background information
Area name and code number: ______________________________
Location: ____________________
Type of water: Fresh ❑ Marine ❑ Estuarine ❑
Responsible authority: ______________________________
Address: ______________________________
Tel. __________ Fax. __________ E-mail __________

Laboratory of analysis:
Name: ______________________________
Distance (km) ________ Sample transport time (h) ________
Person responsible for samples during transport: __________

What land or human activity surrounds the bathing area? (check all that apply)
Forest ❑ Fields ❑ Desert ❑ Hills ❑ Swamp ❑ River/stream/ditch ❑
Agriculture (specify) ❑ __________ Urban ❑ Commercial ❑
Residential ❑ Industry (specify) ❑ __________ Hotel ❑
Harbour ❑ Airport ❑ Road/rail ❑ Military ❑ Waste tip ❑ Other ❑
Is the area surrounding the bathing area urban? _____

Additional details (historical information, reason for assessment, other contacts, etc.):
Size of bathing area: Area (m^2) ____ Length (m) ____ Mean width (m) ____
Is there a beach? ____ Average area (m^2) ____ Length (m) ____
Width (m) at high tide ____ Width (m) at low tide ____
Prevailing onshore winds: Direction _____ Typical speed (km/h) _____
Prevailing water currents: Direction _____ Typical speed ($m\ s^{-1}$) ___
Shoreline configuration ________ Presence of sandbars _____
Average wave heights: ________
Rainfall: Total annual _____ Seasonal patterns ____________
Temperature: *Water:* Average ___ Annual low ___ Annual high ___
Air: Average ___ Annual low ___ Annual high ___

Public facilities: No. of toilets ___ Showers ___ Drinking water fountains ___
Litter bins ___ (are they animal and/or bird-proof? ___)
Are methods in place to warn the public of danger? _____
Are the above facilities adequate?

Accessibility: Road ❑ Path ❑ No access ❑
Is there an adequate parking area? ______
Additional details ______________________________

Continued

Box 8.1 Continued

Microbiological hazards

a) Sewage and animal wastes. Is the water quality affected, or likely to be affected, by discharges from:
On-site or other private sewage disposal systems ❑ Communal sewage disposal or treatment facilities ❑ Long sea outfalls ❑ Agricultural activities ❑ Aquacultural activities ❑ Unconfined domestic or wild animals and birds ❑ Confined animals or birds (i.e. feedlots) ❑
Are discharges continuous or sporadic? __________
Is wastewater from toilets, showers, etc. likely to contaminate the bathing area? _____
Will typical bather densities impair water quality? ________

b) Storm water runoff. Is the water quality affected or likely to be affected by non-point discharges from:
Municipal storm drains or combined sewer overflows? ❑
Agricultural fields? ❑ Natural drainage? ❑
Are onshore winds likely to carry polluted water into the bathing area? _____
Are currents likely to carry polluted water into the bathing area? _____
Are tides likely to affect water quality in the bathing area? _____
Microbiological water quality data or additional information:_______________

Note: Any of the above with a "yes" answer require a detailed investigation and risk analysis. This investigation should include:

- Proximity of potential contamination source to bathing area.
- Background and contamination incident flow rates.
- Effective rainfall which triggers contamination events (and typical duration of contamination).
- For discharges from sewage systems or treatment facilities, include what type of treatment is used, the system capacity, flow rates and variability, and indicator standards.
- For animals/birds, stocking densities and types of animals, indicator data will be necessary to support and supplement this information.

Continued

8.1.3 Sampling location, time and frequency

The first step in planning a sampling activity is to define clearly the objective; in most cases the objective will be either exploratory (assessment) or monitoring (surveillance). While the former is designed to provide preliminary or "one-off" information about a site, the latter is undertaken for regulatory or non-regulatory purposes (Keith, 1990). For a recreational water use area,

Box 8.1 Continued

Chemical and other hazards
Water quality
Is the water likely to be affected by: Discharges from industrial sources? ❑
Agricultural drainage? ❑ Water craft mooring or use? ❑
Urban surface runoff? ❑
Are onshore winds likely to carry polluted water into the bathing area? _____
Are currents likely to carry polluted water into the bathing area? _____
Are tides likely to affect water quality in the bathing area? _____
Sand quality
Is the sand likely to be affected by: Discharges from industrial sources? ❑
Agricultural drainage? ❑ Water craft mooring or use? ❑ Urban surface runoff? ❑
Are plastic residues present? _____ Are tar residues present? _____
Are algae present? _____ Are other residues present? ____
Supporting chemical water quality data or additional comments: ___________
Note: Any of the above with a "yes" answer require a detailed investigation and risk analysis. This investigation should include:

- Proximity of potential contamination source to bathing area.
- For boats, densities and pumpouts.
- For urban surface runoff, the effective rainfall.
- For discharge from industrial sites, the type of discharge, treatment being used, flow rates and variability, system capacity and chemical/indicator standards.

Please attach a map of the beach area included in this sanitary inspection, with possible contamination sources (rivers, storm drains, outfalls, etc.) marked. If possible, maps of the entire catchment area indicating land-use, topography, and infrastructure networks (i.e. wastewater and storm drain systems, etc.) should also be attached.

Reporting systems
Are there formal mechanisms for reporting waste discharges, spills, treatment bypasses, etc. to the local health authorities? _____
Is there an illness or injury reporting mechanism in place that would be effective for epidemiological investigations? _____
Sampling or posting recommendations. This section should describe circumstances which indicate the need to post warning notices or close beaches and provide information such as sampling locations, times and frequencies.

both objectives may initially coincide. Exploratory sampling will be required to define subsequent sampling. Special requirements for epidemiological studies will be necessary as highlighted in Chapter 13.

The selection of sampling sites, time and frequency of sample collection should attempt to capture the overall microbiological quality of the water at the recreational water use area. These choices should be based upon the information gathered during the sanitary inspection. The selection of sampling stations and time of sampling should take into consideration, variables known to affect water quality, such as the length of the bathing area, presence and periodicity of point and non-point sources of faecal contamination, influences of local weather, the physical characteristics of the bathing area and the presence of bathers. For example, at bathing areas with no detectable sources of external faecal contamination, samples should be collected at the places with the greatest bather densities. Bathing areas known to be influenced by direct or indirect faecal contamination will require additional sampling sites to help define the degree and extent of pollution. The time of day can be an important source of variation (Brenniman *et al.,* 1981; Fleisher, 1985; Tillett, 1993; PHLS, 1995) especially at beaches with significant tides (Cheung *et al.,* 1991). Consideration should also be given to collecting samples at times when bather densities are greatest for example, afternoons at weekends (Cheung *et al.,* 1991; APHA, 1995). Chapter 9 gives an example of an approach to a sampling programme.

Sampling frequency can also influence the acquisition of reliable information on microbiological pollution in a bathing area (Fleisher, 1990; Tillett, 1993). For those laboratories with limited economic or human resources it is better to direct efforts towards increasing sampling frequency instead of confirming presumptive results for *Escherichia coli* and faecal streptococci. The sampling frequency adopted in many programmes and assessments is fortnightly during the bathing season. Some authors have advocated more frequent samplings such as weekly or more, especially in the peak season in temperate climates (Figueras *et al.,* 1997) and others maintain lower intensity monitoring (e.g. monthly) outside the bathing season. Evidence suggests that once an understanding of quality behaviour has been developed through relatively intensive monitoring and sanitary surveys, then reduced sampling frequencies may be justifiable and can contribute to reducing the burden of monitoring (Chapter 9). For colder climates where the bathing season is restricted by weather, water sampling should be concentrated in that period where historical data show a higher probability of favourable weather conditions for recreational activities. If abnormal favourable weather conditions appear, more frequent sampling should be carried out, especially in freshwater resources with poor water circulation that may be overcrowded under those circumstances.

Monitoring a bathing area or a site to reconfirm repeated failure to meet a guideline or poor water quality has little value. Equally, sampling frequency can be reduced when an area is known, through historical microbiological data, to have consistently good microbiological quality and when it is known from the catalogue of basic characteristics that it will not be subjected to pollution influences because potential sources of contamination are absent.

In these situations only occasional confirmatory sampling will be required. Such an approach will direct resources to those beaches known to have variable water quality (see Chapter 9).

Resampling and new sanitary inspection, following the detection of unexpected peak values, is essential to establish the cause of the observed peak. An exhaustive investigation, including an inspection of the site and possible collection of additional samples to locate the source or sources of pollution, is also essential where the cause is known not to be due to a sporadic event. The effect of episodic events, such as heavy rainfall, on the water quality of bathing beaches, and the management response to such events, is discussed in Chapter 9.

8.2 Sampling

8.2.1 Sampling procedures

Sampling in chest depth water, typically 1.2–1.5 m depth, represents areas of greatest bather density although sampling at ankle depth may be appropriate to determine risk to young children.

Microbiological counts from surface samples have been shown to have a tendency to be higher than those beneath the surface (PHLS, 1995), but the epidemiological significance of this has yet to be studied. Therefore samples should be collected from beneath the surface. Precise sampling recommendations vary, for example 30 cm below the surface is indicated by the American Public Health Association (APHA) (APHA/AWWA/WPCF, 1992) and the European Community (EC) Directive (EEC, 1976), while the World Health Organization (WHO) and the United Nations Environment Programme (UNEP) (WHO/UNEP, 1994a) have proposed 25 cm. Every sample within a monitoring programme should be taken as near as possible to the defined sampling location.

Care must be taken to avoid external contamination during sample collection. Sterilised sample bottles should be opened with the opening facing downward and should be held by the base and submerged in the water. At the appropriate depth, the bottle should be turned upwards with the mouth facing the current (if any). After retrieving the bottle, some water should be discarded to leave an air space of at least 2.5 cm to allow mixing by shaking before examination (APHA/AWWA/WPCF, 1989; Bartram and Ballance, 1996). The utmost care must be taken at all times not to touch the top of the bottle during removal or replacement of the cap.

The sample volume should be sufficient to carry out all the required tests. In practice, 300–500 ml are adequate. If *Salmonella, Vibrio cholerae* or enteroviruses are to be analysed, as required by some authorities or under certain circumstances, greater volumes of water will be necessary (1.5 litres, 10 litres and 10 litres respectively). Bottles of borosilicate glass or suitable autoclavable plastic (PHLS, 1994; Bartram and Ballance, 1996) are recommended. They should have screw caps that withstand repeated sterilisation at 121 °C or 180 °C. Quality assurance procedures, as described in Chapter 4, should be followed. All sampling bottles should be correctly labelled with the reference of the sampling point. Additional information of the time of collection, temperature of water and other observations should be recorded on sample record sheets designed for this purpose.

8.2.2 Sample storage

There is little published information available that gives a consensus on the time limit for storage of samples to avoid changes in the concentrations of indicator organisms (Gameson and Munro, 1980; Tillet and Benton, 1993). Storage times should be as short as possible and it is recommended here that samples should be analysed as soon as possible, preferably within 8 hours of collection. If samples cannot be analysed within 24 hours field analysis should be considered. Immediately after collection, the samples should be stored in insulated boxes with cooling packs (prefrozen packs) and/or ice. Samples should be kept in the dark and the temperature of the cooling box maintained below 10 °C where possible (APHA/AWWA/WPCF, 1995). This temperature may be difficult to reach and so in practice samples should be kept as cold as possible, but not frozen. In practical terms these storage conditions can have at best only a limited effect on reducing variations in bacterial populations. It is generally accepted that changes in microbial populations in water samples will begin to occur around 2 hours after collection; within 6 hours the samples are likely to have altered significantly particularly if no cooling mechanism was available and the samples were exposed to light. The key factor to consider in storage and transport of samples is time between collection and analysis rather than the time between collection and receipt at the laboratory. Ideally, the temperature of the insulated box should be controlled and recorded, as should the storage time. This information should be considered in the interpretation of results. Storage under these conditions should be as short as possible, and samples should be analysed promptly after collection.

8.3 Index and indicator organisms

Natural waters are subject to important changes in their microbial quality that arise from agricultural use, discharges of sewage or wastewater resulting from human activity or storm water runoff. Sewage effluents contain a wide variety of pathogenic micro-organisms that may pose a health hazard to the human population when the effects are discharged into recreational waters. The density and variety of these pathogens are related to the size of the human population, the seasonal incidence of the illness, and dissemination of pathogens within the community (Pipes, 1982). Appropriate indicators of faecal contamination under various conditions are discussed in Chapter 9 and in the *Guidelines for Safe Recreational Water Environments* (WHO, 1998).

Many waterborne pathogens are difficult to detect and/or quantify and the specific methodology to detect them in environmental water samples has still to be developed (Borrego, 1994). While faecal streptococci are suggested as the recommended indicator for salt water, either faecal streptococci or *Escherichia coli* can be used for monitoring freshwaters. Additional variables can be investigated if they are considered relevant, such as the spores of *Clostridium perfringens* in tropical waters where the traditional indicators may increase in number in soil and water (Hardina and Fujioka, 1991; Anon, 1996). Staphylococci are generally assumed to serve as indicators of water pollution deriving from bathers themselves (i.e. by shedding from the body surface). The epidemiological significance of the recovery of staphylococci remains unclear.

8.3.1 Thermotolerant coliforms and *E. coli*

Thermotolerant (faecal) coliforms constitute the subset of total coliforms that possess a more direct and closer relationship with homeothermic faecal pollution (Geldreich, 1967). These bacteria conform to all the criteria used to define total coliforms (all are aerobic and facultatively anaerobic, Gram-negative, non-spore forming rod-shaped bacteria that ferment lactose with gas and acid production in 24–48 hours at 36 ± 1 °C), but in addition they grow and ferment lactose with production of gas and acid at 44.5 ± 0.2 °C within the first 48 hours of incubation. For this reason, the term "thermotolerant coliforms" rather than "faecal coliforms" is a more accurate name for this group (WHO, 1993). The physiological basis of the elevated temperature phenotype in the thermotolerant coliforms has been described as a thermotolerant adaptation of proteins to, and their stability at, the temperatures found in the enteric tracts of animals (Clark, 1990). Thermotolerant coliforms include strains of the genera *Klebsiella* and *Escherichia* (Dufour, 1977). The thermotolerant coliform definition is not based on strictly taxonomic criteria, but on specific biochemical reactions or on the appearance of characteristic colonies on selective and/or differential culture media. Certain *Enterobacter* and *Citrobacter* strains are also able to grow under the conditions defined for thermotolerant coliforms (Figueras *et al.*, 1994; Gleeson and Gray, 1997). *E. coli* is, however, the only biotype of the family Enterobacteriaceae that is almost always faecal in origin (Bonde, 1977; Hardina and Fujioka, 1991). Therefore, the thermotolerant coliform group when used should ideally be replaced by *E. coli* as an indicator of faecal pollution. For the purpose of water testing, most *E. coli* can be confirmed by a positive indole test and by their inability to use citrate (as the only carbon source) in the culture medium. Alternatively, *E. coli* can be distinguished easily enzymatically by the lack of urease or presence of ß-glucuronidase enzymes. The enzymes can be recognised easily using culture media that contain specific substrata (Gauthier *et al.,* 1991; Brenner *et al*., 1993; Walter *et al*., 1994).

However, several studies have indicated the limitation of both the thermotolerant coliform group and *E. coli* as ideal faecal indicators or pathogen index organisms. Several thermotolerant *Klebsiella* strains have been isolated from environmental samples with high levels of carbohydrates in the apparent absence of faecal pollution (Dufour and Cabelli, 1976; Knittel *et al.,* 1977; Niemi *et al.,* 1997). Similarly, other members of the thermotolerant coliform group, including *E. coli*, have been detected in some pristine areas (Rivera *et al.,* 1988; Ashbolt *et al.,* 1997) and have been associated with regrowth in drinking water distribution systems (Lechevallier, 1990). The principal disadvantages of this organism as an indicator in water are: (i) its detection in other environments without faecal contamination (Hazen and Toranzos, 1990; Hardina and Fujioka, 1991), and (ii) its low survival capability in aquatic environments when compared with faecal pathogens (Borrego *et al.,* 1983; Cornax *et al.,* 1990).

8.3.2 Faecal streptococci and enterococci

Faecal streptococci have received widespread acceptance as useful indicators of faecal pollution in natural aquatic ecosystems. These organisms show a close relationship with health hazards (mainly for gastrointestinal symptoms) associated with bathing in marine and freshwater environments, (Cabelli *et al.,* 1982, 1983; Dufour, 1984; Kay *et al.,* 1994; WHO, 1998). They are not as ubiquitous as coliforms (Borrego *et al.,* 1982), they are always present in the faeces of warm-blooded animals (Volterra *et al*., 1986), and it is believed that they do not multiply in sewage-contaminated waters (Slanetz and Bartley, 1965). Enterococci, however, have been shown to grow in freshly stored urine (Höglund *et al.,* 1998). Nonetheless, their die-off rate is slower than the decline in coliforms in seawater (Evison and Tosti, 1980; Borrego *et al*., 1983) and persistence patterns are similar to those of potential waterborne pathogenic bacteria (Richardson *et al.,* 1991). Reviews of all these aspects have been carried out by Sinton *et al.,* (1993 a,b).

The group called faecal streptococci includes species of different sanitary significance and survival characteristics (Gauci, 1991; Sinton and Donnison, 1994). In addition, the proportion of the species of this group is not the same in animal and human faeces (Rutkowski and Sjogren, 1987; Poucher *et al.,* 1991). The taxonomy of this group, comprising species of two genera *Enterococcus* and *Streptococcus* (Holt *et al.,* 1993), has been subject to extensive revision in recent years (Ruoff, 1990; Devriese *et al.,* 1993; Janda, 1994; Leclerc *et al.,* 1996). Although several species of both genera are included under the term enterococci (Leclerc *et al.* 1996), the species most predominant in polluted aquatic environments are *Enterococcus faecalis, E. faecium* and *E. durans* (Volterra *et al.,* 1986; Sinton and Donnison, 1994; Audicana *et al.,* 1995).

Enterococci, a term commonly used in the USA, includes all the species described as members of the genus *Enterococcus* that fulfil the following criteria: growth at 10 °C and 45 °C, resistance to 60 °C for 30 minutes, growth at pH 9.6 and at 6.5 per cent NaCl, and the ability to reduce 0.1 per cent methylene blue. The most common environmental species fulfil these criteria and thus in practice the terms faecal streptococci, enterococci, intestinal enterococci and *Enterococcus* group can be considered synonymous.

8.3.3 Alternative faecal indicators

The lack of a strong relationship between faecal indicators and health outcomes in a number of epidemiological studies in warm tropical waters may, in part, relate to the inappropriate nature of *E. coli* or faecal streptococci as indices of waterborne pathogens in these recreational waters. In this context an alternative index group, sulphite-reducing clostridia or spores of *Clostridium perfringens*, have been proposed and are used in Hawaii (Anon, 1996).

Spores of *C. perfringens* are largely faecal in origin (Sorensen *et al.,* 1989), they are always present in sewage (about 10^4–10^5 colony forming units (cfu) per 100 ml), they are highly resistant in the environment and appear not to reproduce in aquatic sediments (which appears to be the case with thermotolerant coliforms) (Davies *et al.,* 1995). It is interesting to note, however, that dog faeces may have some 9×10^8 cfu *C. perfringens* per gram dry weight (dw), whereas pig faeces are similar to humans (4.8×10^5 cfu *C. perfringens* per gram dw). *C. perfringens* is generally less common or absent in other warm blooded animals. Hence, although dogs have a similar number of thermotolerant coliforms and faecal streptococci to that found in humans, the relatively higher ratio of *C. perfringens* spores found in dog faeces may be a useful indicator when fresh faecal contamination is being investigated (Leeming *et al.,* 1998).

It is important to note that spores of *C. perfringens* do not act as an indicator for non-sewage or animal faecal contamination in general, and therefore they are only suitable as indicator organisms for parasitic protozoa and viruses from sewage-impacted waters (Payment and Franco, 1993; Ferguson *et al.,* 1996). Their resistance to disinfectants may also be an advantage for indexing disinfectant-resistant pathogens. Simple anaerobic culture is possible for *C. perfringens* spores after a short heat treatment to remove vegetative cells. Confirmation of their presence may be assisted by the addition of a methylumbelliferyl phosphate substrate to the growth medium (Davies *et al.,* 1995).

Other indicator organisms for sewage, but also specific for human sewage are the bacteriophages to *Bacteroides fragilis* HSP40. These *B. fragilis* phages appear to survive in a manner that is similar to the hardier human enteric viruses under a range of conditions (Jofre *et al.*, 1995; Lucena *et al.,* 1996). Their numbers in sewage such as the F-specific RNA bacteriophages may be an order of magnitude lower than various coliphages. Furthermore, only 1–5 per cent of humans may excrete these phages (Leeming *et al.,* 1998), and thus they may be unsuitable pathogen indicator organisms for small communities. The International Office for Standardization (ISO) standard methods for these phages are under final review (ISO, 1999c).

The ratio between thermotolerant coliforms and faecal streptococci has been proposed by Geldreich (1976) as a means of distinguishing between human and animal-derived faecal matter. However, this method is no longer recommended (Howell *et al.,* 1995) and none of the currently-used bacterial indicators distinguish different sources of faecal matter confidently when used alone (Cabelli *et al.,* 1983), although genetic typing of *E. coli* shows some potential (Muhldorfer *et al.,* 1996). Identification of human enteric viruses can identify specifically the presence of human faecal material although the necessary procedures are difficult and expensive, and not readily quantifiable. Other microbiological options include specific identification of phenotypes of *Bifidobacterium* spp. (Gavini *et al.,* 1991), *Bacteroides* spp. (Kreader, 1995), serotypes of F-specific RNA bacteriophages (Osawa *et al.,* 1981) or, as previously discussed, the bacteriophages to *Bacteroides fragilis* (Puig *et al.,* 1997). However none of these organisms are suitable for quantifying the proportion of human faecal contamination. Moreover, no one indicator or single approach is likely to represent all the facets and issues associated with faecal contamination of waters.

Recently, Leeming *et al.* (1994, 1996) demonstrated the ability to distinguish human from herbivore-derived faecal matter using a range of faecal sterol biomarkers (Table 8.1). The distribution of sterols found in faeces, and hence their source-specificity, is caused by a combination of diet, the animal's ability to synthesise its own sterols and the intestinal microbiota in the digestive tract. The combination of these factors determines "the sterol fingerprint". The principal human faecal sterol is coprostanol (5β(H)-cholestan-3β-ol), which constitutes about 60 per cent of the total sterols found in human faeces. The C29 homologue of coprostanol is 24-ethylcoprostanol (24-ethyl-5β(H)-cholestan-3β-ol). In large quantities (relative to coprostanol), this faecal sterol is indicative of faecal contamination from herbivores. It is possible to determine the contribution of faecal matter from these two sources relative to each other by calculating the ratio of coprostanol to 24-ethylcoprostanol in human and herbivore (sheep and cow)

Table 8.1 Examples of faecal sterol biomarkers

Systematic name	Common name	Comments
C_{27} sterols		
5ß-cholestan-3α-ol	Coprostanol	Human faecal biomarker; high relative amounts indicate fresh human faecal contamination
5β-cholestan-3α-ol	Epi-coprostanol	Present in sewage sludges; high relative amounts suggest older faecal contamination
cholest-5-en-3ß-ol	Cholesterol	C_{27} precursor to 5α- and 5β-stanols
5α-cholestan-3ß-ol	Cholestanol	The thermodynamically most stable isomer is ubiquitous; if the ratio of coprostanol to cholestanol is < 0.5, origin of 5β-stanols may not be faecal
C_{29} sterols		
24-ethyl-5ß-cholestan-3ß-ol	24-ethylcoprostanol	Herbivore faecal biomarker; high relative amounts indicate herbivore faecal contamination
24-ethyl-5ß-cholestan-3α-ol	24-ethyl-*epi*-coprostanol	Present in some herbivore faeces
24-ethylcholest-5-en-3ß-ol	24-ethylcholesterol	C_{27} precursor to 5α- and 5β-stanols
24-ethyl-5α-cholestan-3ß-ol	24-ethylcholestanol	The thermodynamically most stable isomer is ubiquitous

faeces (Leeming *et al.,* 1996) and comparing these to ratios obtained for water samples (Leeming *et al.,* 1998). Other animals that are ubiquitous in urban areas such as dogs and birds, either do not have coprostanol in their faeces or have it in trace amounts only (Leeming *et al.,* 1994).

Faecal sterols generally associate with particulate matter, and can be concentrated from 1–10 litres of water by simply filtering the water through a glass fibre filter (such as type GFF, Whatman). The lipids are extracted by acetone, concentrated, derivatised and quantified by gas chromatography. Thus the method requires a suitable chemistry laboratory and may cost ten times more than that for the analysis of *E. coli* and enterococci. Nonetheless, it is an appropriate method for specific studies investigating the proportion of human and animal faecal contamination.

Table 8.2 Recommended serial dilutions for water samples in relation to the degree of microbiological contamination and type of indicator

Type of water	Serial dilutions for thermotolerant coliforms[1]					Serial dilutions for faecal streptococci				
Sewage	10^{-2}	10^{-3}	10^{-4}	10^{-5}	10^{-6}	10^{-2}	10^{-3}	10^{-4}	10^{-5}	10^{-6}
Secondary effluent	10^{-1}	10^{-2}	10^{-3}	10^{-4}	10^{-5}	1	10^{-1}	10^{-2}	10^{-3}	
Contaminated bathing water	10	1	10^{-1}	10^{-2}	10^{-3}	10	1	10^{-1}		
Clean water	10	1	10^{-1}			100	10	1		

[1] *E. Coli*

Source: Anon, 1983

8.4 Analytical methods

8.4.1 Most Probable Number

Most Probable Number (MPN) analysis is a statistical method based on the random dispersion (Poisson) of micro-organisms per volume in a given sample. Classically, this assay has been performed as a multiple-tube fermentation test. Although the technique is rather time consuming (taking between five and seven days), several laboratories prefer it to other methods of water analysis because it is applicable to all sample types.

The MPN technique is generally conducted in three sequential phases (presumptive, confirmatory, and complete), each phase requiring 1 to 2 days of incubation. In the initial or presumptive phase, three volumes of samples (usually 10, 1, and 0.1 ml) (Table 8.2 and Figures 8.1 and 8.2) are inoculated into 3, 5, or 10 tubes containing the appropriate medium to allow the target bacteria to grow (Figure 8.2; Tables 8.3 and 8.4). In this test, it is assumed that any single viable target organism in the sample will result in growth or a positive reaction in the medium.

After the incubation period, all the inoculated presumptive positive tubes must be inoculated into a more selective medium to confirm the presence of the target bacteria (confirmatory phase). The confirmation test is reliable evidence, but not proof, that the target bacteria have been detected. Therefore, subsamples of the confirmed positive reactions should be inoculated onto a selective agar medium and several verification tests (Gram stain, and biochemical, serological or enzymatic tests) should be carried out (Tables 8.3 and 8.4). This completed test is generally conducted on 10 per cent of the positive tubes as a quality control measure. For practical purposes, the number of positive and negative tubes in the confirmatory phase of the technique is generally used to determine the MPN of the target bacteria by using tables of positive and negative tube reactions (WHO/UNEP, 1994a; APHA/AWWA/WPCF, 1995).

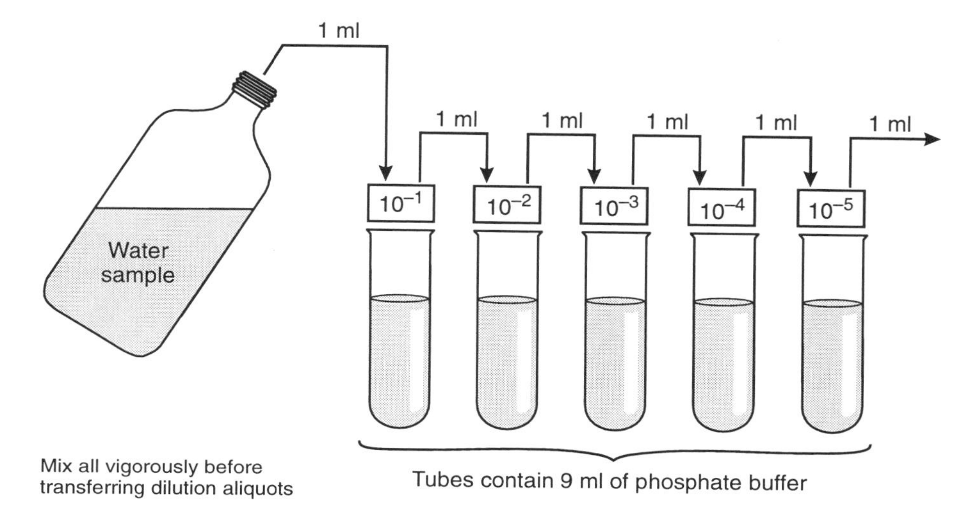

Figure 8.1 Preparation of a series of dilutions

The major advantages of the MPN technique are (Fujioka, 1997):

- It will accept both clear and turbid samples.
- It inherently allows the resuscitation and growth of injured bacteria.
- The results may be recorded by personnel with minimal skill.
- Minimal preparation time and effort are required to start the test, and therefore processing of samples can be initiated at any time of the day.

By contrast the MPN technique may also have several disadvantages, such as:

- The total time, labour, material and costs required to analyse one sample.
- The substantial increase in reagents, tubes, incubation space and cleanup requirements when multiple samples need to be analysed or when the sample volume must be increased to 100 ml.
- The multiphase nature of the technique, each phase requiring a 24 hour or 48 hour incubation period.
- The fact that MPN is a simple estimated number, while the true number (95 per cent confidence limit) may show extreme variation from the MPN.

The choice of precision level of the technique (using 3, 5 or 10 tubes of each dilution) depends on the required detection sensitivity, because the total volumes analysed by each are 33.3, 55.5, and 111 ml, respectively. Miniaturised MPN methods with 96 incubation wells (e.g. ISO 1996a,b) are more precise than traditional five-tube tests with three descending decimal dilutions and equivalent to membrane filtration (Hernandez *et al.,* 1991, 1995). The existing standardised procedures for the MPN technique are given in Tables 8.3 and 8.4.

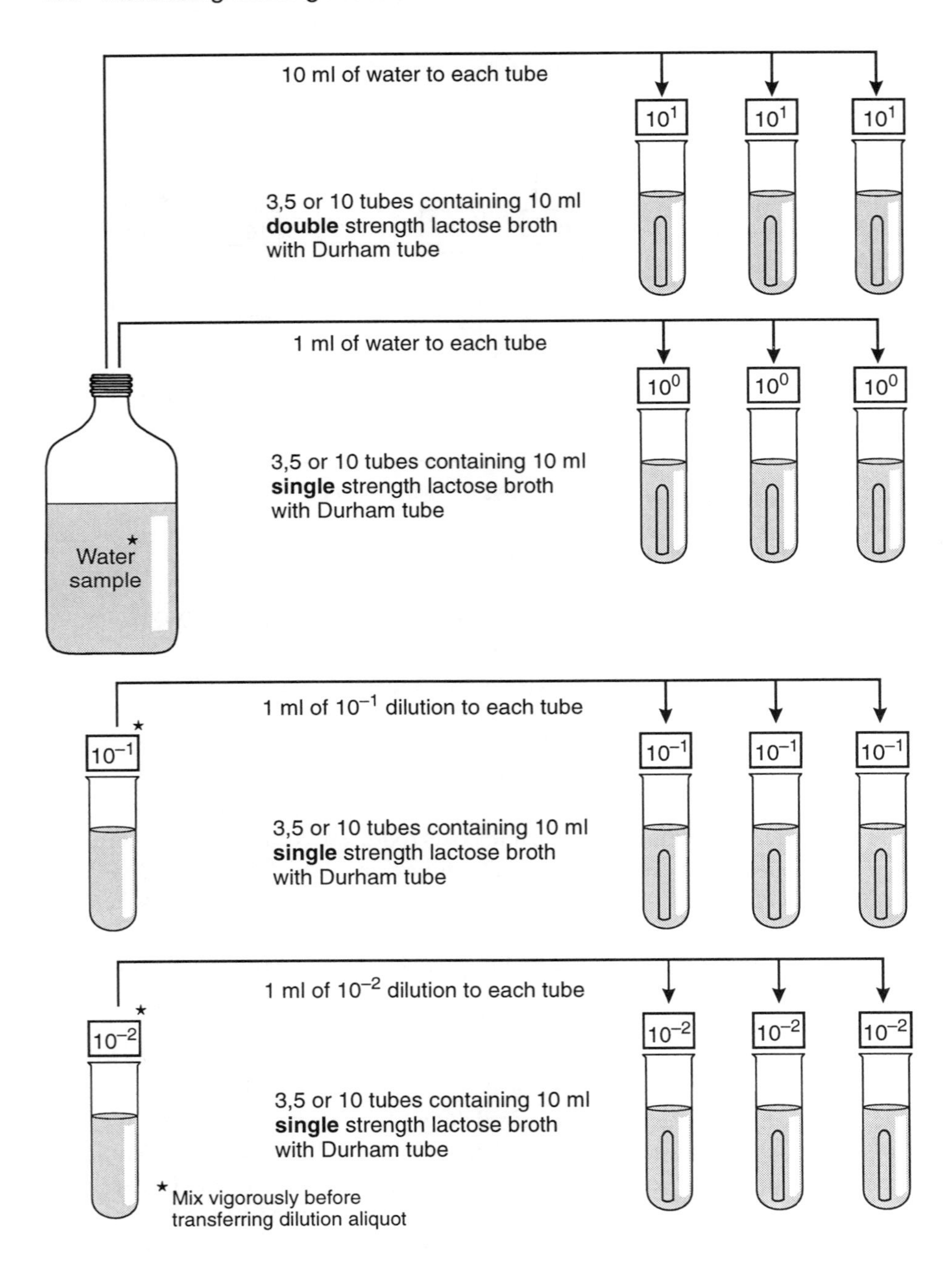

Figure 8.2 Inoculation scheme for the multiple test tube method

Table 8.3 Standard methods for the determination of thermotolerant coliforms (*E. coli* presumptive) — MPN methods

ISO 9308–2 (ISO, 1990c)[1]	ISO 9308–3 (96 wells)[2] (ISO, 1998)	APHA
Isolation media		
a) Lactose broth b) MacConkey broth c) Lauryl tryptose (lactose) broth d) Formate lactose glutamate medium	Tryptone salicine triton MUG broth (MU/EC)	a) EC medium, or b) A-1 medium
Incubation conditions	*Incubation conditions*	*Incubation conditions*
24–48 h at 35 ± 1 °C or 37 ± 1 °C	36–72 h at 44 ± 0.5 °C	a) 24 ± 2 h at 44.5 ± 0.2 °C b) 3 h at 35 ± 0.5 °C followed by 21 ± 2 h at 44.5 ± 0.2 °C
Reaction	*Reaction*	*Reaction*
Turbidity = (+)	Blue fluorescence = (+) for *E. coli*	a) and b) Gas production = (+) for thermotolerant coliforms
Confirmatory media tests		
Two confirmatory methods can be used:	Confirmatory tests are not required	If using EC medium, verify with the following test:
A. With two steps		
1. a) EC medium b) Brilliant green lactose (bile) broth		Brilliant green lactose (bile) broth
Incubation conditions		*Incubation conditions*
24 h at 44 ± 0.25 °C or 44.5 ± 0.25 °C		24 ± 2 h at 44.5 ± 0.2 °C
Reaction		*Reaction*
Gas production = (+) for thermotolerant coliforms		Gas production = (+) for thermotolerant coliforms
2. Tryptone water		If using A-1 medium, a confirmatory test is not required
Incubation conditions		
24 h at 44 ± 0.25 °C or 44.5 ± 0.25 °C		
Reaction		
Indol production with indol reagent Kovacs = (+) for *E. coli*		
B. With one step		
Lauryl tryptose mannitol broth with tryptophan		
Incubation conditions		
24 h at 44 ± 0.25 °C or 44.5 ± 0.25 °C		
Reactions		
Gas production = (+) and indol = (+/–) for thermotolerant coliforms; gas production = (+) and indol = (+) for *E. coli*		

MPN Most probable number
APHA American Public Health Association
MUG 4-methylumbelliferyl-ß-D-glucoside

[1] ISO 9308–2 is at an early stage of revision by an ISO working group
[2] Not suitable for drinking water — lower limit of detection is 15 counts per 100 ml

Table 8.4 Standard methods for the determination of faecal streptococci (enterococci) — MPN methods

ISO 7899–1(ISO, 1984a)[1]	APHA
Isolation media	
Azide dextrose broth	Azide dextrose broth
Incubation conditions	*Incubation conditions*
22 ± 2 h at 35 ± 1 °C or 37 ± 1 °C; negative tubes may be re-incubated for 22 ± 2 h	24 ± 2 h at 35 ± 0.5 °C; negative tubes may be re-incubated until 48 ± 3 h
Reaction	*Reaction*
Turbidity = (+)	Turbidity = (+)
Confirmatory media tests	
Two tests are recommended:	Two tests are recommended:
1. BEAA	1. PSE agar
Incubation conditions	*Incubation conditions*
44 ± 4 h at 44 ± 0.5 °C	24 ± 2 h at 35 ± 0.5 °C
	Reaction
	Brownish-black colonies with brown halos (+) for faecal streptococci
2. Catalase	2. BHIB containing 6.5% NaCl
	Incubation conditions
	24 h at 45°C
Reaction	*Reaction*
Dark brown to black colonies surrounded by black halos are (+) BEAA, with a (–) catalase test = faecal streptococci	Turbidity = (+) A (+) PSE with a (+) BHIB (6.5% NaCl) = enterococcus group

MPN Most probable number
APHA American Public Health Association
BEAA Bile esculin azide agar
PSE Pfizer selective enterococcus
BHIB Brain heart infusion broth

[1] ISO 7899–1 has been replaced recently with a new methodology proposed under the same ISO reference as in Table 8.7

8.4.2 Membrane filtration

The membrane filtration (MF) technique is based on the entrapment of the bacterial cells by a membrane filter (pore size of 0.45 μm) (Figure 8.3). After the water is filtered, the membrane is placed on an appropriate medium and incubated (Tables 8.5, 8.6 and 8.7). Discrete colonies with typical appearance are counted after 24–48 hours, and the population density of the target bacteria, usually described as cfu per 100 ml in the original sample, can be calculated from the filtered volumes and dilutions used. This technique is more precise than the MPN technique, but the MF test can only be used for low-turbidity waters with low concentrations of background micro-organisms.

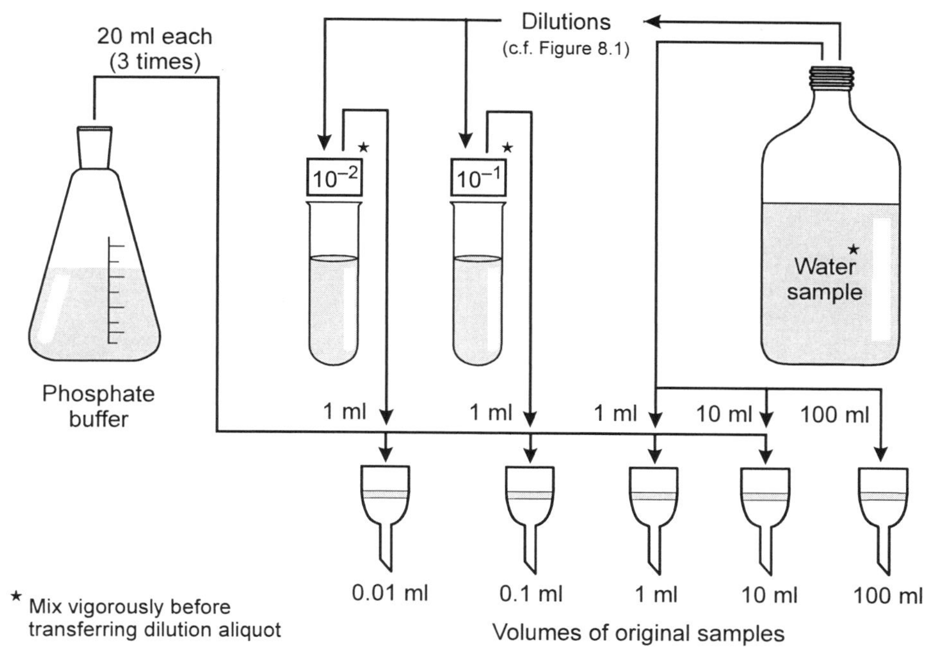

Figure 8.3 Preparation of dilution series and procedure for the membrane filtration method

The advantages of the MF technique include (Fujioka, 1997):

- Savings in terms of time, labour, and cost compared with the MPN technique.
- Direct determination of the concentrations of bacteria with high precision and accuracy.
- The formation of the target bacteria as colonies which can be purified for further identification and characterisation.
- The ability to process large volumes of water samples to increase greatly the sensitivity of this method.

Several disadvantages of the MF technique have also been reported:

- Inapplicability of the method to turbid samples which can clog the membrane or prevent the growth of the target bacteria on the filter.
- False negative results due to the inability of viable but non-culturable bacteria present in environmental waters to grow with standard MF methods.
- False positive results when non-target bacteria form colonies similar to the target colonies (Figueras *et al.,* 1994, 1996; Hernandez-Lopez and Vargas-Albores, 1994).

E. coli has been demonstrated to be a more specific indicator for the presence of faecal contamination than the thermotolerant coliform group (Dufour, 1977). Improvements in both MPN and MF techniques have been

Table 8.5 Standard methods for the determination of thermotolerant coliforms (*E. coli* presumptive) — MF methods

ISO 9308–1(ISO, 1990b)[1]	APHA
Isolation media	
a) TTC agar with Tergitol-7 or Teepol b) Lactose agar with Tergitol-7 or Teepol c) Membrane enrichment Teepol broth d) m-FC medium with 1% rosolic acid in 0.2N NaOH added e) Laurylsulphate broth	m-FC medium with 1% rosolic acid in 0.2N NaOH added (if there is interference with background growth)
Incubation conditions	*Incubation conditions*
18–24 h at 44 ± 0.25°C or 44.5 ± 0.25°C; a pre-incubation of 4 h at 30°C is recommended	24 ± 2 h at 44.5 ± 0.2°C
Reaction	*Reaction*
Depends on the media selected	Blue colonies = (+)
Confirmatory media tests	
Verify a representative number of colonies. Two confirmatory methods can be used:	Verify by picking at least 10 typical colonies; by two tests:
A. With two steps	
1. Lactose peptone water	1. Lauryl tryptose broth
Incubation conditions	*Incubation conditions*
24 h at 44 ± 0.25°C or 44.5 ± 0.25°C	24–48 h at 35 ± 0.5°C
Reaction	*Reaction*
Gas production = (+) for thermotolerant coliforms	Gas production = (+)
2. Tryptone water	2. EC broth
Incubation conditions	*Incubation conditions*
24 h at 44 ± 0.25°C or 44.5 ± 0.25°C	24 h at 44.5 ± 0.2°C
Reaction	*Reaction*
Indol production with indol reagent Kovacs = (+) for *E. coli*	Growth (+) and gas production = (+) for thermotolerant coliforms
B. With one step	
Lauryl tryptose mannitol broth with tryptophan	
Incubation conditions	
24 h at 44 ± 0.25°C or 44.5 ± 0.25°C	
Reaction	
Gas production = (+) and indol = (+/–) for thermotolerant coliforms; gas production = (+) and indol = (+) for *E. coli*	

MF Membrane filtration
APHA American Public Health Association
TTC Triphenyl-tetrazolium chloride

[1] ISO 9308–1 is under revision. Only one culture medium (a) has been chosen and is proposed under the same ISO reference in Table 8.7

Table 8.6 Standard methods for the determination of faecal streptococci (enterococci) — MF methods

ISO 7899–2 (ISO, 1984b)[1]	APHA
Isolation media	
a) KF streptococcus agar with 1% sterile solution of TTC added to cooled basal medium b) Slanetz-Bartley agar[2] with 1% sterile solution of TTC added to cooled basal medium	a) m-E agar for enterococci, or b) m-Enterococcus agar[2] for faecal streptococci
Incubation conditions	*Incubation conditions*
44 ± 4 h at 35 ± 1 °C or 37 ± 1 °C; however, if other types of micro-organisms are expected use 5 ± 1 h at 37 ± 1 °C followed by 44 ± 0.5 °C until 48 h	a) 48 h at 41 ± 0.5 °C; transfer membrane fil- to esculin iron agar for 20 min ± 2 h at 41 ± 0.5 °C b) 48 h at 35 ± 0.5 °C
Reaction	*Reaction*
a) and b) Colonies: red, brown or pink (+)	a) and b) Colonies: pink to red (+)
Confirmatory media tests	
Verify a representative number of colonies by two tests:	Verify at least 10 well-isolated typical colonies by sub-culturing on:
1. BEAA	BHIA
Incubation conditions 48 h at 44 ± 0.5 °C	*Incubation conditions* 24–48 h at 35 ± 0.5 °C
2. Catalase	Transfer a loop-full of growth to:
Reaction	BHIB
Dark brown to black colonies surrounded by black halos are (+) BEAA, with a (–) catalase test = faecal streptococci	*Incubation conditions* 24 h at 35 ± 0.5 °C
	A series of five tests are recommended for confirmation:
	1. Catalase
	2. Gram
	3. BEA
	Incubation conditions 48h at 35 ± 0.5 °C
	Reaction Growth = (+)
	4. BHIB
	Incubation conditions 48 h at 45 ± 0.5 °C
	Reaction Growth = (+)

Continued

Table 8.6 Continued

ISO 7899–2 (ISO, 1984b)	APHA
	5. BHIB containing 6.5% NaCl *Incubation conditions* 48 h at 35 ± 0.5 °C *Reaction* Turbidity =(+) *Final reaction* A (+) BEA with a (+) BHIB 45 °C (test number 4) = faecal streptococci; a (+) BEA with (+) BHIB 45 °C (test number 4) and (+) BHIB 6.5% NaCl (test number 5) = enterococci

MF Membrane filtration
KF KF streptococcus agar
BEAA Bile esculin azide agar
BHIB Brain heart infusion broth
APHA American Public Health Association
TTC Triphenyl-tetrazolium chloride
BHIA Brain heart infusion agar
BEA Bile esculin agar

[1] ISO 7899–2 is currently under revision; the new proposed version under the same ISO reference is given in Table 8.7

[2] Slanetz-Bartley has the same formulation as m-Enterococcus but the latter already includes TTC

carried out for the rapid and selective enumeration of *E. coli*. Barnes *et al.* (1989) designed a rapid seven hour membrane filter test for quantification of thermotolerant coliforms from drinking water samples and other freshwaters and salt waters, although it is not suitable for salt water due to the high proportion of false positives obtained. Fluorogenic and chromogenic tests using 4-methylumbelliferyl-ß-D-glucuronide (MUG) have been applied in MPN and MF techniques, for the detection of ß-glucuronidase that is specific to *E. coli* (Manafi and Kneifel, 1989; Balebona *et al.*, 1990; Gauthier *et al.*, 1991; Rice *et al.*, 1991). A miniaturised MPN method with a 96-well micro-plate has been developed for *E. coli* (Hernandez *et al.*, 1991; ISO, 1996b) (Table 8.3). Based on this principle a number of different media have been developed for the use in MF and MPN techniques (Frampton *et al.*, 1988; McCarty *et al.*, 1992). Commercially available media include Colisure (formerly Millipore, now IDEXX) (McFeters *et al.*, 1995), Colilert (IDEXX) (Edberg *et al.*, 1988; Palmer *et al.*, 1993), m-ColiBlue (Hach), ColiComplete (BioControl), Chromocult (Merck) and MicroSure (Gelman). Similar media for the detection of *E. coli* in water have also been described (Sartory and Howard, 1992; Brenner *et al.*, 1993; Walter *et al.*, 1994). Molecular methods have also been designed to detect specifically *E. coli* from water samples, such as PCR-gene probes for the *uid* gene (Bej *et al.*, 1991a,b; Tsai *et al.*, 1993; McDaniels *et al.*, 1996). In addition, other alternative techniques, i.e.

Table 8.7 Recently proposed modifications to ISO standard methods

MF methods for coliforms and ***E. Coli*** ISO DIS 9308–1 (ISO, 1997)[1]	**MF methods for intestinal enterococci** ISO DIS 7899–2 (ISO, 1999b)[2]	**MPN methods for intestinal enterococci** ISO DIS 7899–1 (96 wells) (ISO, 1999a)[3]
Isolation media	***Isolation media***	***Isolation media***
For a standard test use a lactose TTC agar with Tergitol-7, incubate for 21 ± 3 h at 36 ± 2 °C [4]; typical colonies will turn the medium yellow For a rapid test use tryptone soya agar and incubate for 4–5 h at 36 ± 2 °C [4]	Use a m-Enterococcus agar (Slanetz-Bartley) with a 1% sterile solution of TTC added to the cooled basal medium, incubate for 44 ± 4 h at 36 ± 2 °C [4]; typical colonies are light and dark red	Use a medium with tryptose, nalidixic acid, TTC thallium acetate and MUD (MUD/SF medium); incubate for 36–72 h at 44 ± 0.5 °C ; fluorescence indicates intestinal enterococci
Confirmatory media tests	***Confirmatory media tests***	***Confirmatory media tests***
In the case of the standard test, verify all or a representative number of typical colonies (at least 10), using the following series of tests: 1. Non selective agar (i.e. tryptone soya agar); incubate for 21 ± 3 h at 36 ± 2 °C [4] 2. Oxidase test; the non-appearance of a dark purple colour within 5–10 s indicates a negative result; a (–) oxidase = coliform bacteria 3. Tryptophane broth, incubate for 21 ± 3 h at 44 ± 0.5 °C [4] Add indol reagent; indol production (i.e. a red ring) indicates a positive result A (–) oxidase and (+) indol = *E. coli* For a rapid test, transfer the membrane filter to tryptone bile agar, incubate for 19–20 h at 44 ± 0.5 °C [4]; place the membrane filter on a filter paper saturated with indol reagent; the appearance of red colonies = *E. coli*	Transfer the membrane filter to a bile esculin azide agar, preheated at 44 °C; incubate at 44 ± 0.5 °C for 1 h The appearance of dark brown to black colonies surrounded by black halos = intestinal enterococci	Tests are not required

MF Membrane filtration
MPN Most probable number
TTC Triphyl-tetrazolium chloride
MUD 4-methylumbelliferyl-β-D-glucoside

1 Suitable for drinking water with low background growth
2 Suitable for drinking water, swimming pools and other water with low intestinal enterococci
3 Not suitable for drinking water; lower limit of detection is 15 counts per 100 ml
4 These conditions substitute and standardise those from the previous ISO 9308–1 and 7899–2

enzyme capture (Kaspar *et al.,* 1987) and radioisotopes (Reasoner and Geldreich, 1989) have been proposed.

8.5 Laboratory procedures

8.5.1 Faecal streptococci and enterococci

Early attempts to quantify faecal streptococci relied on enrichment tube procedures and the MPN technique; Rothe Azide Dextrose broth followed by a confirmation in Ethyl Violet Azide (Litsky) broth being the procedure most widely accepted by researchers. A rapid system for enumeration of faecal streptococci or enterococci in water samples using a miniaturised fluorogenic assay based on a 96-well microplate MPN system has been described by several workers (Hernandez *et al.,* 1991; Poucher *et al.,* 1991; Budnicki *et al.,* 1996) and the technique has recently been proposed as an ISO method (Table 8.7). In addition, Enterolert (IDEXX) is available for the MPN technique with up to 100 ml of sample, and has been shown to be reliable (Fricker and Fricker, 1996).

The enumeration of faecal streptococci by a MF procedure using a selective medium was first reported by Slanetz and Bartley (1957). Since then, several media have been proposed, including Thallous Acetate agar (Barnes, 1959), KF agar (Kenner *et al.,* 1961), PSE agar (Isenberg *et al.,* 1970), Kanamycin Aesculin Azide (KEA) agar (Mossel *et al.,* 1973), mSD agar (Levin *et al.,* 1975), and mE agar (APHA/AWWA/WPCF, 1989). The accepted standardised procedures for the MF method are given in Table 8.6. Other media formulations and incubation procedures for faecal streptococci have been proposed for specific situations (Lin, 1974), such as increasing the membrane incubation period from 48 hours to 72 hours to recover stressed faecal streptococci. Rutkowski and Sjogren (1987) developed a medium, designated M2, to distinguish between human and animal pollution sources.

The methods for enumeration of faecal streptococci from natural waters have been compared by different authors (Volterra *et al.,* 1986; Yoshpe-Purer, 1989). Dionisio and Borrego (1995) compared eight methods for the specific recovery of faecal streptococci from natural freshwater and marine waters on the basis of the following characteristics: accuracy, specificity, selectivity, precision and relative recovery efficiency. The results obtained indicated that none of the tested methods showed perfect selectivity. The methods that showed the best performance characteristics were the MPN technique (with Rothe and Litsky media) and the m-Enterococcus agar in conjunction with the MF technique. The latter is the only technique recommended in the "Standard Methods" for faecal streptococci in conjunction with membrane filtration (APHA/AWWA/WPCF, 1995). A rapid confirmation technique, based on the transplantation of the membrane from m-Enterococcus agar after incubation for 48 hours at 36 ± 1 °C to Bilis-Esculin-Agar (BEA) for 4 hours additional incubation, improves the low specificity of the m-Enterococcus agar, enabling the confirmation of 100 per cent of the colonies (Figueras *et al.,* 1996). A similar procedure has been proposed by ISO (Table 8.7).

Recently, Audicana *et al.* (1995) designed and tested a modification of the KEA agar, named Oxolinic acid-Aesculin-Azide (OAA) agar, to improve the selectivity in the enumeration of faecal streptococci from water samples by the MF technique. The OAA agar showed higher specificity, selectivity and relative recovery efficiencies than those obtained when using m-Enterococcus and KF agars. In addition, no confirmation of typical colonies was needed when OAA agar was used, which shortens the time taken significantly and increases the accuracy of the method. The excellent performance of this culture medium was recently reconfirmed in a routine monitoring programme for bathing waters (Figueras *et al*., 1998). A Europe-wide standardisation trial demonstrated that the m-Enterococcus agar with total confirmation of the colonies (Figueras *et al.,* 1996), the OAA medium (Audicana *et al.,* 1995), and the miniaturised MPN method (Hernandez *et al.,* 1991) produced the best results (Hernandez *et al.,* 1995).

8.5.2 Thermotolerant coliforms and *E. coli*

The presumptive detection of thermotolerant coliforms can be considered sufficient to give an estimation for the presence of *E. coli*. The EC and A-1 media are the most widely recommended for the presumptive detection of thermotolerant coliforms with the MPN technique (APHA/AWWA/WPCF, 1992). The differences between the two approaches are based on the incubation periods: 44.5 ± 0.2 °C for 24 hours for the EC medium, and 36 ± 1 °C for 3 hours and transfer to 44.5 ± 0.2 °C for 21 hours for the A-1 medium. The tubes containing gas and acid in EC medium are confirmed in the same medium by subsequent incubation at 44.5 ± 0.2 °C for 24 hours. The A-1 medium does not require a confirmation test. Table 8.3 details accepted media. Thermotolerant coliform density and the 95 per cent confidence limits can be estimated with the use of MPN tables (APHA/AWWA/WPCF 1995; Bartram and Ballance, 1996).

The mFC agar is the most frequent medium used to quantify thermotolerant coliforms in water samples when the MF technique is used. Petri dishes containing filters are incubated at 44.5 ± 0.2 °C for 24 hours. Typical thermotolerant coliform colonies appear various shades of blue, atypical *E. coli* may be pale yellow, and non-thermotolerant coliform colonies are grey to cream in colour. Table 8.5 details accepted standardised media.

The existing ISO methods are now under revision. Table 8.7 shows the proposed modifications. The unification of temperature precision has been introduced by ISO as 36 ± 2 °C and 44 ± 0.5 °C.

8.6 Field analyses

Field analysis embraces all the tests that can be performed completely or partially at the site of sampling. Several field analysis techniques have been developed for drinking waters, where the principal requirement is the absence of indicator organisms (Manja *et al.*, 1982; Bernard *et al.,* 1987; Dange *et al.,* 1988; Dutka and El-Shaarawi, 1990; Smoker, 1991; Ramteke, 1995; Grant and Ziel, 1996). Quantitative on-site analysis using the MF technique is also possible (Bartram and Ballance, 1996).

The microbiological quality of bathing waters is presently assessed by the techniques described previously for indicator organisms. Field analyses may be preferred when the time between the collection of the sample and its examination will be long. Field laboratory equipment with filtration and incubation devices are being marketed. The time taken to obtain presumptive results will be the same as at a standard laboratory.

On-site filtration with a delayed incubation is another possibility when conventional procedures are impractical, i.e. when it is not possible to maintain the desired temperature during transport, and when the time between sample collection and analysis will exceed the optimum time limit. With this procedure, filters are placed in water tight plastic Petri dishes with a transport medium, and in conditions that maintain viability but will not allow visible growth. The test is completed at the laboratory by transferring the membranes to appropriate selective media and incubating them for the period of time required. It has to be recognised that growth will start if high temperatures are encountered during transport. Delayed incubation has been found to produce results consistent with those from immediate standard tests (Chen and Hickey, 1983, 1986; APHA/AWWA/WPCF, 1995; Brodsky *et al.,* 1995).

The continuous demands for more rapid techniques that can be performed on site and provide direct results have yet to be satisfied, despite the advances in analytical methods, particularly those based on DNA chips or arrays (Eggers *et al.*, 1997). By contrast, a one-hour assay for thermotolerant coliforms has been demonstrated for marine bathing beaches, based on MUG detection of ß-glucuronidase activity with a portable fluorometer (Davies and Apte, 1996, 1999).

8.7 Data recording, interpretation and reporting

Analysis may be performed as part of a regulatory monitoring programme, as part of a survey of an area used for water recreation, or as part of an epidemiological study in which water quality is related to risks to health from infectious diseases. Each approach has its own requirements which are specified at the outset by the regulatory authority or the study director. One of the most important functions of the analyst is to provide reliable and accurate results in a form that can be recorded for statistical interpretation and reporting, as described in Chapter 3. The following guidance will help the microbiological analyst achieve this aim.

8.7.1 Forms and records

An individual record must be produced for each site inspection or sample and should include the location and reference of the sampling site (that should ideally be equal to the code number of the sample), the date, the time, weather conditions, tide, water temperature, results of visual inspection for abnormal conditions, sources of contamination and the name of the inspector or sampler. Sampling records should also list the laboratory procedures and results (method of analysis, dilutions or volumes analysed, time of analysis, results for each step, any anomalies in the analysis of results and the name of the analyst). Ideally, record forms should occupy a single page. Forms should be conveniently archived, because they will be used later by the data handlers for transcription to the database and they will be analysed for the purposes of preparing the report of the monitoring or survey programme. Great care should be taken in preparing the report and its contents and format should be agreed by those responsible for analysis, data handling and for reporting results and, if necessary, by those responsible for co-ordinating results of regional, national and international programmes. For quality assurance, it should be possible to conduct an "audit trail" through the whole process of visiting the site, analysing the sample and filing the results on the database (see Chapter 4). An example of a record form for site inspections is given in Box 8.1.

8.7.2 Recording results of microbiological analyses

The results of microbiological analyses of water quality must always be regarded as an estimate of the water quality at the time and site of sampling, rather than as an absolute determination (PHLS 1994, APHA/AWWA/WPCF, 1995). All enumeration methods depend on the assumption that bacteria and other micro-organisms are randomly distributed in water samples and that the samples conform to the Poisson distribution (see Chapter 3). In reality, the clumping of bacteria and their aggregation on particles cause samples to depart from the Poisson distribution, thereby introducing additional error. There is little that can be done to reduce this error, apart from taking representative samples (free of sediment and other solid matter) and mixing the contents of the sampling bottles vigorously before taking sub-samples for analysis. It has been shown that two halves of the same sample can vary widely in the counts observed (PHLS, 1994).

A historical record of water quality in a bathing area, in normal and extreme situations, enables the selection of the most appropriate dilutions and facilitates the correct enumeration of the final density of indicator organisms that has to be reported as the total number of cfu per 100 ml.

Great care should go into the counting and recording of analytical results in order to avoid recording results wrongly, leading to errors that can result in statistical misinterpretations of water quality at the recreational area. Typical sources of error in the laboratory are caused by operator fatigue and mistakes, such as mislabelling of bottles, Petri dishes and tubes, and errors in preparing and transferring volumes and dilutions of samples. Because labels attached to the lids of Petri dishes, tubes and bottles can be transposed, the labels should be placed on the dish, tube or bottle itself, because these contain the culture medium. Although it may not be obvious, operators vary in the accuracy with which they count colonies and, in addition, unsuspected partial colour blindness can interfere with the interpretation of biochemical reactions of target colonies or in tubes of diagnostic media. Mistakes and errors can be minimised by proper training and supervision of samplers and analysts in their duties. In addition, laboratory quality controls and the careful application of standard procedures for analysis are essential. Standard procedures should be written correctly and copies should always be available for reference in the laboratory.

The multiple tube method is very sensitive for the detection of a small number of indicator organisms, but the MPN is not a precise value. Confidence intervals, i.e. most probable range (MPR), are often published with the MPN and are meant to indicate the imprecision of the method (Tillett, 1995). However, it should be stated clearly that the range applies to the sample and not to the water source (PHLS, 1994).

For thermotolerant coliforms and faecal streptococci membrane filters with 20–60 typical colonies are recommended for counting, with the provision that filters with no more than 200 colonies of all types should be considered; if the counts of colonies on the membranes are all below the minimum recommended, they should be totalled (APHA/AWWA/WPCF, 1995). Counting colonies on all filters has been shown to improve precision (Gameson, 1983) and is the only method specified in other standard procedures (ISO, 1988, 1990b). The count, in cfu per 100 ml, becomes the sum of all colonies counted multiplied by 100 ml and divided by the total volume (in ml) of water filtered.

If confirmation tests have been applied to a number of typical and atypical colonies, the initial count should be adjusted by multiplying it by the percentage of verified colonies. This percentage will be calculated by dividing the number of verified colonies by the total number of colonies subject to verification and then by multiplying the result by 100 (PHLS, 1994; WHO/UNEP, 1994a; APHA/AWWA/WPCF, 1995). This procedure has a considerable effect in reducing the precision of the count, depending on the total number of presumptive colonies selected and the fraction confirming (PHLS, 1994). Nevertheless, all colonies should be selected if there are ten or fewer colonies on a membrane. New proposed ISO methods try to overcome the imprecision by proposing verification methods for all the colonies grown on the filter (Table 8.7).

If the total number of bacteria colonies, including the specific target colonies, exceed 200 per membrane or are not distinctive enough to enable counting, results should be reported as “too numerous to count” (APHA/AWWA/WPCF, 1995) or “count too high to be estimated at the dilution employed” (PHLS, 1994). A new sample should be requested immediately if possible and more appropriate volumes should be selected for filtration (Table 8.2). The resultant data however, represent the results from a different sample; nevertheless the data may help to investigate the event. If this approach is not possible, it is preferable to try to count a sector of the original filter, before the information is lost thereby estimating the total number of colonies (even though these counts may lack precision). This technique is facilitated by the grid printed on the membranes. The details of the estimate should be recorded, for example “Count in n squares = x; diameter of filtration circle = d squares; estimated count = $\pi d^2 x/4n$ colonies”. If 135 colonies were counted in 10 squares and the diameter of the filtration area was 11.5 squares, the count would be estimated to two significant figures as $3.142 \times 11.5^2 \times 135/40 = 1{,}400$ colonies.

Although the statistical reliability of membrane filter results is higher than that of the MPN procedure, membrane counts are not absolute numbers; 95 per cent confidence limits can be calculated using a normal distribution equation (Fleisher and McFadden, 1980; PHLS, 1994; APHA/AWWA/WPCF, 1995).

8.7.3 Statistical procedures

A single water sample from a recreational area gives very little useful information. However, when individual results are accumulated and analysed statistically, then the trend in water quality will become apparent. Statistical analysis also enables evaluation of the improvement of water quality after remedial actions have been applied (e.g. to sewage contamination sources) and enables achievement of comparability between different regions within the same country or across countries. To establish comparability for the concentrations of thermotolerant coliforms from different regions, it is essential to agree what will be analysed (i.e. all the thermotolerant coliforms or only *E. coli*) otherwise comparison is impossible (Figueras *et al.*, 1994,

Table 8.8 Comparability of methods for assessing compliance with microbiological quality criteria in two bathing areas

Method of assessing compliance	Interim criteria of quality[1] = 100 cfu per 100 ml 50%	
	Bathing area A	Bathing area B
Percentage compliance	Non-compliance (only 1 result complies)	Compliance (10 results comply)
Ranking method	590 non-compliance	8 compliance
Geometric mean[2] (95% confidence intervals)	597 non-compliance (216, 1,646)	16 compliance (4, 66)
Log normal distribution method[3]	680 non-compliance	13 compliance
50 percentile point 90 percentile point	597 non-compliance	15 compliance
50th percentile 90th percentile	595 non-compliance	20 compliance

The sets of data for faecal coliforms obtained consecutively from the two bathing areas are as follows (in units of cfu per 100 ml):
A (16; 170; 3,390; 450; 450; 590; 740; 190; 1,180; 6,700; 2,800; 600) and
B (92; 1,600; 36; 0; 140; 4; 0; 36; 4; 8; 0; 32)

1 According to the UNEP/WHO (1985) Interim Criteria for Recreational Waters, the concentrations of thermotolerant coliforms in at least 10 water samples should not exceed 100 cfu per 100 ml in 50% of the samples and 1,000 cfu per 100 ml in 90% of the samples

1997). The type of data analysis needed depends on the nature of the study (see Chapter 3).

For regulatory monitoring programmes, the objective of data analysis is to demonstrate compliance with a standard. The definition of the standard specifies the type of statistical analysis required. Most recreational water quality standards derive from those of the US EPA (Dufour and Ballentine, 1986), UNEP/WHO (1985) or the European Bathing Water Directive (EEC, 1976). More recently, WHO has published the *Guidelines for Safe Recreational Water Environments* (WHO, 1998). Microbiological standards typically specify the frequency of analysis and the number or proportion of samples that must not exceed given limiting values of the target organism. The rules for interpreting compliance differ with each standard. For microbiological surveys or epidemiological studies, the type of data analysis is decided at the planning stage. The procedures and statistics that are most often used to assess compliance during or after a bathing season are described below, together with worked examples from two sets of data from the two different bathing areas shown in Table 8.8, with the calculation of their basic statistics in Table 8.9.

Table 8.8 Continued

Methods of assessing compliance	Interim criteria of quality[1] = 1,000 cfu per 100 ml 90%	
	Bathing area A	Bathing area B
Percentage compliance	Non-compliance (only 8 results comply)	Compliance (11 results comply)
Ranking method	3,390 non-compliance	140 compliance
Geometric mean[2] (95% confidence intervals)	597 non-compliance (216, 1,646)	16 compliance (4, 66)
Log normal distribution method[3]	4,700 non-compliance	530 compliance
50 percentile point 90 percentile point	4,618 non-compliance	288 compliance
50th percentile 90th percentile	5,707 non-compliance	1,162 non-compliance

[2] The regulation that applies the geometric mean has only one standard in the interim criteria and not two as in the example and the geometric mean is calculated from at least 5 samples equally spaced over a 30 day period (running geometric mean) (US EPA, 1986)

[3] Data extracted from WHO/UNEP (1994b) and Anon (1983)

Percentage compliance

To assess percentage compliance (EEC, 1976) the regulatory percentage of the total number of data "n" obtained from a sampling station has to be calculated. The individual results of the set of data that comply with the established standards have to be counted in order to see if they are higher or lower than the compliance level. This is the approach used in the European Union. For thermotolerant coliforms the guideline standard is 100 cfu per 100 ml in 80 per cent of the samples and the mandatory standard is 2,000 cfu per 100 ml in 95 per cent of the samples (EEC, 1976). This approach is very easy to calculate but does not take into account the absolute values of all the microbiological counts and does not produce any average numerical value of the concentration of micro-organisms in the bathing area. For the worked example of Table 8.8, 50 per cent and 90 per cent of 12 samples is 6 and 10.8 (= 11) respectively, which signifies the number of samples that must have counts under or equal to the standards associated to those percentages. In the case of bathing area A, only one result complies with the 50 per cent standard instead of the six needed, and only eight results comply with the 90 per cent standard when there should be 11. This bathing area is therefore failing the

Table 8.9 Basic statistics of worked examples from two sets of data obtained for thermotolerant coliforms from two bathing areas

	Bathing area A			Bathing area B		
Sample	Count[1]	Rank[2]	Log count	Count	Rank	Log (count+1)[3]
1	16	1	1.20	92	10	1.97
2	170	2	2.23	1,600	12	3.20
3	3,390	11	3.53	36	8	1.56
4	450	4	2.65	< 1	1	0.00
5	450	5	2.65	140	11	2.15
6	590	6	2.77	4	4	0.70
7	740	8	2.87	< 1	2	0.00
8	190	3	2.28	36	9	1.57
9	1,180	9	3.07	4	5	0.70
10	6,700	12	3.83	8	6	0.95
11	2,800	10	3.45	< 1	3	0.00
12	600	7	2.78	32	7	1.52
Total	17,276	78	33.31	1,952	78	14.32
Average	1,440	–	2.7758	162.7	–	1.1933
SD[4]	1,965	–	0.6934	455	–	0.9883

SD Standard deviation
[1] cfu per 100 ml
[2] Ranks are given in ascending order
[3] This transformation has been used to enable the geometric mean and the log standard deviation to be calculated, given that three of the values are below the limit of detection, i.e. < 1 cfu per 100 ml
[4] Calculated as $s = \sqrt{\{[\Sigma x^2 - (\Sigma x)^2/n]/(n-1)\}}$

compliance assessment. Bathing area B complies with both the 50 and 90 per cent standards (Table 8.8).

Ranking method

The ranking method (WHO/UNEP, 1994b) is a very simple method because it involves ordering and multiplication operations, making the use of any complex formulae or laborious graphical analysis unnecessary. The interim UNEP/WHO Mediterranean criteria for recreational waters specify that thermotolerant coliform counts in at least ten samples taken during the bathing season must not exceed 100 cfu per 100 ml in 50 per cent of samples and 1,000 cfu per 100 ml in 90 per cent of samples (UNEP/WHO, 1985). The "n" values obtained are first ranked in ascending order of concentration (by definition, the order number, "i", takes values of 1 to n) (Table 8.9). Then the appropriate order numbers for a given percentage, P (i.e. 50 and 90 per cent) are calculated as $i = n \times \mathrm{P}/100$. If ten samples have been taken, then the 50 per cent is measured directly against the fifth value of cfu per 100 ml in the ranking and the 90 per cent against the ninth value. If the number of samples taken does not give a whole number value, the result should be rounded to the nearest whole number to obtain the order. This concentration has to be lower than or equal to the specified standards to comply with the interim criteria. The order point in the rank for the 50 per cent criterion of 12 samples (examples of Table 8.8 and 8.9) is 50 × 12/100 = 600/100 = 6. The order point for the 90 per cent criterion is 90 × 12/100 = 1,080/100 = 10.8 = 11th position. Thus for bathing area A, the sixth point in rank order corresponds to a concentration of 590 cfu per 100 ml while the eleventh corresponds to a concentration of 3,390 cfu per 100 ml. For bathing area B, the corresponding values are 8 and 140 respectively. Whereas A fails both standards, B complies with both (Table 8.8).

Geometric mean

The other systems of interpreting water quality, i.e. the geometric mean with confidence intervals (US EPA, 1986) and the lognormal distribution method (WHO/UNEP, 1994b), are based on the fact that sets of microbiological data from sampling a recreational area are found to conform to a skewed positive distribution, because normally there are many low values and only a few high values.

The transformation of the microbiological counts obtained into decimal logarithms often produces a more symmetrical distribution. The proper descriptive statistic for central tendency is the geometric mean (equal to the median in the case of a normal distribution) with two associated measures of dispersion: the standard deviation of the logarithms of the values (the log standard deviation) and the 95 per cent confidence interval of the geometric mean.

The geometric mean is equal to the antilogarithm of the arithmetic mean of the logarithms of individual concentrations. In the USA it is considered to be the best estimate of the central tendency and the preferred statistic for summarising microbiological results (APHA/AWWA/WPCF, 1995). If there are values less than 1cfu per 100 ml (i.e. 0 cfu per 100 ml) in the data set it will be impossible to calculate the geometric mean, because the logarithm of zero does not exist. In this instance one has to be added to all the results (+1) before their logarithmic transformation and then after the average of the logarithms is calculated and the antilog has been taken, the added value has to be subtracted. The calculation of the measures of dispersion are given in Box 8.2.

Log-normal distribution method

The log-normal distribution method involves the ranking of results and the transformation of data into logarithms to determine the lognormal distribution that fits most closely the experimental results. This can be done by hand fitting the data directly onto log-normal probability paper together with their corresponding cumulative frequencies. The concentration of

Box 8.2 Calculation of the measures of dispersion

The logarithmic standard deviation, s_l is calculated by entering the log values, x, into a scientific calculator, programmed to calculate the sample standard deviation. This can also calculated manually as $s_l = \sqrt{\{[\Sigma x^2 - (\Sigma x)^2/n]/(n - 1)\}}$. The 95 per cent confidence intervals of the geometric mean are calculated in two stages by the method below:

- The standard error (se) of the logarithmic mean "*m*", is calculated as $s_l/\sqrt{n}$.
- The 95 per cent confidence intervals are defined as $m \pm t_{(0.025)} \times$ se, where $t_{(0.025)}$ is the value of Student's t for $\alpha = 0.025$ and for $n - 1$ degrees of freedom (from statistical tables).
- For bathing area A the antilog of the log mean is antilog 2.7758 = 597 cfu per 100 ml (Tables 8.8 and 8.9), the standard error (se) of the logarithmic mean 2.7758 is $0.6934/\sqrt{12} = 0.2001$ and the 95 per cent confidence intervals of the log mean where $t_{(0.025)}$ is the value of Student's t for $\alpha = 0.025$ and for $n - 1$ degrees of freedom (from statistical tables; for 11 degrees of freedom = 2.201) is $2.7758 \pm 2.201 \times 0.2001$, giving $2.7758 - 0.4405 = 2.3353$ and $2.7758 + 0.4405 = 3.2163$ respectively. The antilogs of these values are 216 and 1,646 respectively.
- For bathing area B three counts are below the limit of detection, so the transformation count +1 had been applied to all counts before taking logarithms. The antilog of the log average 1.1933 is 15.6; and the estimated geometric mean is obtained by subtracting 1 from this, giving 14.6, rounded-off to 15. The standard error is $0.9883/\sqrt{12} = 0.2853$ and the 95 per cent confidence intervals of the log mean are thus $1.1933 \pm 2.201 \times 0.2853$, giving results of $1.1933 - 0.6279 = 0.5654$ and $1.1933 + 0.6279 = 1.8212$. The antilogs of these are 4 and 66 respectively.

micro-organisms corresponding to certain specified percentile points on the frequency distribution cumulative frequencies (50 per cent or 90 per cent) can be deduced by the graphic representation (WHO/UNEP, 1994b). This approach is quite similar to the calculation of percentile points on a continuous distribution of an infinite number of samples.

Percentile points, "*p*", on the distribution of the *n* data values from the mean "*m*", (or log mean) and the standard deviation "*s*" (or log standard deviation), as $p = m + zs$, where z is the standard normal variable for the desired percentile, obtained from tables of the quantiles (percentage points) of the standard normal distribution. The values of z for the 80-, 90- and 95-percentage points are 0.8416, 1.2816 and 1.6449 respectively. The value for the 50-percentage point will be equal to the mean "*m*". For completeness, it can be noted that the standard normal distribution is symmetrical, so that the values of z for the 20, 10 and 5 percentile points are respectively −0.8416, −1.2816 and −1.6449. This approach can only be applied when the data follow a normal distribution, whereas calculation of classical percentiles does not require normality. For bathing area A, in the example, the 50 percentile point corresponds to the value of "*m*" (597), while the 90 percentile point is estimated by the log-normal distribution method from the log mean 2.7758 and the log standard deviation 0.6934 (Table 8.9). Hence, the log 90 percentile point is $2.7758 + 1.2816 \times 0.6934 = 3.6645$ and the antilog is 4,618 cfu per 100 ml. This value is quite similar to that obtained by plotting the data on the log-normal–normal distribution probability paper (4,700 cfu per 100 ml). For bathing area B, the 50 percentile point corresponds to a value of $m = 15$ cfu per 100 ml while the 90 percentile point is $1.1933 + 1.2816 \times 0.9883 = 1.1933 + 1.2666 = 2.4599$ and its antilog is 288 cfu per 100 ml.

The classical statistical calculation of percentiles estimates the variability of the distribution of a set of results (after ordering them in ascending order) independently of whether they are normally distributed and indicates the concentration of micro-organisms that embraces a specific percentile. For example, 80 per cent and 95 per cent of the data will be below the value (cfu per 100 ml) of the 80th and 95th percentiles, respectively, and 20 per cent and 5 per cent of the data will be above those concentrations, respectively. Some computer programs will not calculate the 95th percentile unless 20 records are available, although they will do so if the individual records are specifically considered the midpoint of an interval. An example of the calculation by hand is given in Box 8.3.

8.7.4 Interpretation and reporting

Interpretation of results and reporting do not normally involve the analyst. Nevertheless, strict adherence to analytical control procedures make it possible for queries about unusual or anomalous results to be referred back to the analyst and sampler, through an audit trail.

Although the absolute values may differ with different approaches described, there is a high level of agreement on the water microbiological quality qualification of the beach in relation to compliance. Notice from Table 8.8 how, in bathing area A, only the geometric mean allows compliance in relation to a 1,000 cfu per 100 ml standard. However, in the regulations that govern the use of the geometric mean there is only one standard and the mean is calculated from five consecutive results taken over a period of one month (i.e. the running geometric mean). In this approach there is a strong influence from the most frequent values (8 of the 12 samples rank from 16 to 740 cfu per 100 ml). In the second example, bathing area B, the 90th percentile is the most restrictive, because this approach is highly influenced by the single high value obtained.

One of the features of microbiological studies of water quality at recreational areas is the wide variations in results, temporally and spatially

Box 8.3 Calculation of percentiles

$Pr = x_i + (j - i)(x_{i+1} - x_i)$
where:
Pr is the percentile required (i.e. P_{50}, P_{80}, P_{90} or P_{95})
x_i is the concentration that corresponds with an i position in the ranking corresponding to that Pr
j is the next position in the ranking (calculated as $j = r(n + 1)/100$)
x_{i+1} is the next concentration in the ranking

- For bathing areas A and B of Table 8.8 and 8.9, the calculation of j is:
  50 (12 + 1)/ 100 = 650/100 = 6.5 = j for P_{50}, and
  90 (12 + 1) = 1,170/100 = 11.7 = j for P_{90}
- For bathing area A:
  x_i = 590 is the concentration at position 6 while 600 is the concentration at position x_{i+1} (= 7)
  Then P_{50} = 590 + (6.5–6) (600–590) = 590 + 5 = 595 cfu per 100 ml, and
  P_{90} = 3,390 + (11.7 – 11) × (6,700 – 3,390) = 3,390 + 2,317 = 5,707 cfu per 100 ml
- For bathing area B, applying the same criteria:
  P_{50} = 8 + (6.5 – 6) × (32 – 8) = 8 + 12 = 20 cfu per 100 ml, and
  P_{90} = 140 + (11.7 – 11) × (1,600 – 140) = 140 + 1,022 = 1,162 cfu per 100 ml

(Fleisher and McFadden, 1980; Gameson, 1982; Tillett, 1993; PHLS, 1994) that are much greater than those caused by laboratory procedures. With effective quality control in the laboratory, variability caused by procedures, such as sub-sampling from the same bottle or by the enumeration procedures themselves, are little greater than expected from the assumption of random distribution of bacteria in the sample and of the Poisson theory, particularly when the mean number is low.

A single limit standard leads to water of borderline quality and with low variability, consistently passing, whereas water which is usually of high quality, but is occasionally affected by intermittent pollution, would fail, even though the former arguably poses a greater risk to health. More detailed study of the results in the latter case might identify the causes of failure and enable advice to be given to the public not to use the water when poor conditions are expected, or enable remedial action to be taken. Bathing area A shows consistently bad quality, with a higher geometric mean (597) and a lower log standard deviation (0.69) than bathing area B (15 and 0.98 respectively). The failure of bathing area B is caused by a single, unusually high count (1,600 cfu per 100 ml) that may be due to a rain effect on the second sampling date.

8.7.5 Control charts

One way of identifying systematic changes in water quality, or of pinpointing sudden deterioration of water quality at a recreational area, is to create control charts (see Chapter 4). Water quality data points are plotted sequentially, as they are obtained, against time on a chart. Any existing historical data from specific bathing areas can be used to create control charts and helps to identify patterns of behaviour. The occurrence of high values is used to initiate investigation. Such values may be set to coincide to a guideline standard. More conventionally, two upper limit values are set on a control chart, at the mean plus twice and three-times the sample standard deviation, i.e. at $m + 2s$ and $m + 3s$ respectively. These represent values that would be expected to be exceeded only once in 20 or 100 samples respectively, and that would indicate a warning (for example, the need for checking the results and/or for resampling) and then the need for taking remedial action (such as closing the beach until conditions improve and identifying and removing the source of pollution). The bacteriological data in Table 8.8 are presented as two control charts in Figure 8.4, showing the generally poor quality at bathing area A and the greater variability at B, which fails the 90 percentile criterion solely through the high count at the second sampling, even though it is otherwise of good quality.

8.7.6 Technical assessment report

Once the investigation (generally of sanitary conditions and water quality) or monitoring programme is completed, the information is assembled into a comprehensive report. The main body of the report should state the objectives; the manner in which the programme was conducted, with a full description of the recreational area; a historical account of problems and developments; the strategy and reports of inspection, sampling and analysis; significant results obtained; a discussion of the results; and conclusions and recommendations for action. Many of the readers will not have a technical background and therefore an easily readable and accurate "executive summary" should be provided at the front of the report. This gives such readers the main points of the report and invites them to follow-up areas of interest. The report itself should enable the technical reader to understand fully the way in which the study was carried out. Larger bodies of data should be placed in an appendix, so as not to interrupt the flow of the report. A typical report of a sanitary inspection and microbiological analysis includes the following: a description of the survey area(s) and sampling stations, and of any identified hazard(s) and source(s) of pollution (photographs and maps

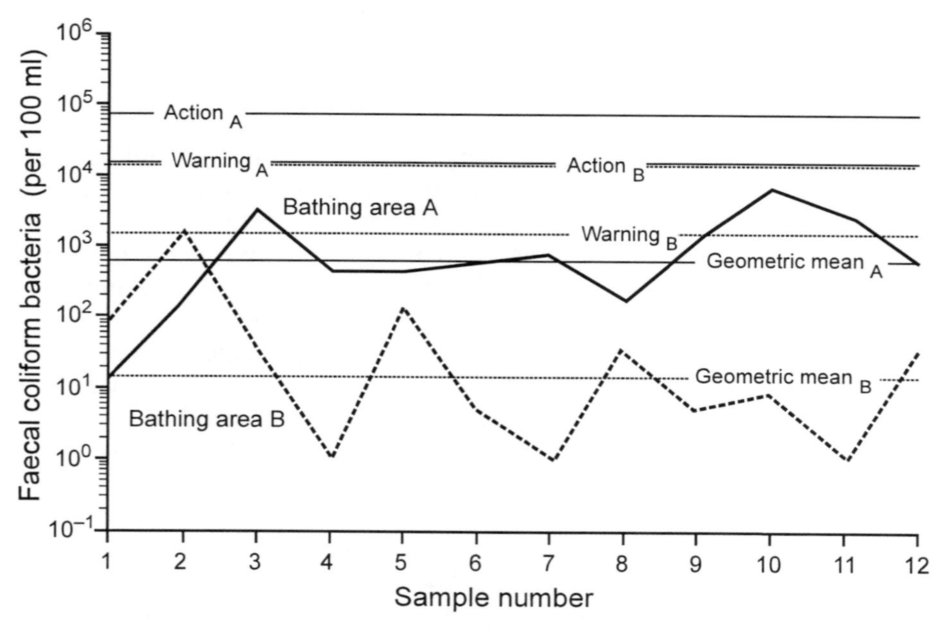

Figure 8.4 Counts of faecal coliform bacteria per 100 ml at bathing areas A and B (see Table 8.8) displayed as control charts, with warning and action limits at two and three logarithmic standard deviations respectively above geometric mean counts

would be useful); the results of the study, including those of the sanitary inspections and microbiological analysis; an in-depth assessment of the risks associated with identified hazards and/or poor water quality; recommendations about the suitability of the area for recreational water use; description and evaluation of various options for improving conditions and thereby for reducing aesthetic and health hazards to users; and recommendations for action, including modifications if necessary, to the monitoring programmes.

The results and recommendations should be discussed with any interested parties before the report is released formally. A contingency plan should also be developed, with the assistance of any interested parties, to investigate and respond to cases of illness or to any unforeseen event or condition that could lead to a deterioration in water quality and possibly increase the risk of illness or danger to bathers. Consideration should also be given to preparing a non-technical report for the general public.

8.8 Quality control

All laboratories should guarantee that the results of the microbiological analysis of a water sample actually originated from the sample and were not introduced accidentally during sampling or analysis. To support this guarantee, internal and external quality controls should be implemented (Tillett and Lightfoot, 1995). Quality control is described in detail in Chapter 4. Internal quality control includes constant monitoring of equipment (pH meters, balances, pipettes, sterilising equipment, incubators, etc.) and reagents (membrane filters, culture media, buffer solutions, etc.) using controls and reference materials (PHLS, 1994; APHA/AWWA/WPCF, 1995; Janning *et al.*, 1995). Working practice should also be included in this quality control, as well as the precision of the techniques (MPN and MF techniques). In addition, controls have to be made at regular intervals (PHLS, 1994; WHO/UNEP, 1994c). External quality controls are meant to establish good performance by comparing laboratory results with those of other laboratories testing the same artificially prepared sample (Tillet *et al,* 1993; PHLS, 1994).

8.9 Presenting information to the public

Aspects of public information are considered in Chapter 6. It is sufficient to note here that the quality of recreational water is of great public concern and is often used in publicity to attract visitors to recreational areas. Several countries have developed regular information services, using television, teletext, newspapers and radio (EEC, 1996) to supplement bulletins in municipal buildings and on public notice boards at the recreational areas. The implementation of a monitoring programme with these characteristics is described by Figueras *et al.* (1997). Generally, the public simply want to know if it is safe to use the water and most people have little understanding of the meaning of bacterial counts, let alone their variability. The information presented to the public, therefore, should be direct and unequivocal, up-to-date and not open to misinterpretation.

8.10 Elements of good practice

- Sanitary inspection should be undertaken as a necessary adjunct to microbiological analysis of waters to identify all real and potential sources of microbiological contamination. It should assess the impact of any microbiological contamination present on the quality of the recreational water and on the health of bathers. During the inspection, the temporal and spatial influences of pollution on water quality should receive full consideration.
- An exhaustive sanitary inspection should be carried out immediately prior to the main bathing season. Inspections of specific conditions should be conducted in conjunction with routine sampling during the bathing season. Pertinent information should be recorded on standardised checklists and used to update the catalogue of basic characteristics. If a problem is identified, it may be necessary to collect supplementary samples or information to characterise the problem.

- Visual faecal pollution or sewage odour should be considered a definite sign of elevated microbiological pollution and the necessary steps should be taken to prevent health risks to bathers.
- Standard operating procedures for sanitary inspections, water sampling (including depth) and analyses should be well described to ensure uniform assessments.
- Sample point location and the distance between each location should reflect local conditions (overall water quality, bather usage, predicted sources of faecal pollution, temporal and spatial variations due to tidal cycles, rainfall, currents, onshore winds and point or non-point discharges) and may vary widely between sites.
- Sterile sample containers should be used for microbiological samples. Scrupulous care should be taken to avoid accidental contamination during handling and during sample collection. Every sample should be identified clearly with the time of collection, date and location.
- The most appropriate depth for sampling should be selected and adhered to consistently.
- The sample should be kept in the dark and maintained as cool as possible within a chilled insulated container and returned to the laboratory promptly after collection. Samples should be analysed as soon as possible and preferably within 8 hours of collection. It is recommended that samples should not be stored for more than 24 hours at 5 °C.
- Additional information should be collected at the time of sampling, including: water temperature, weather conditions, water transparency, presence of faecal material, abnormal colouration of the water, floating debris, cyanobacterial or algal blooms, flocks of sea birds and any other unusual factors. All information should be recorded on standardised checklists.
- The minimum microbiological variables that should be investigated are faecal streptococci or enterococci and thermotolerant coliforms or *E. coli.* While the former is a recommended indicator for salt water both can be used for freshwater. Additional variables should be investigated if considered relevant and if resources allow.
- The influence of specific events, such as the influence of rain on the recreational water use areas, should be established particularly in relation to the duration of the peak contamination period.
- Extreme events, such as epidemics and natural disasters, may require additional measures to ensure there is no additional risk associated with recreational water use areas.
- The procedures to be used for transformation of raw data, to meet the statistical requirements, should be agreed with the statistical expert prior to analysis. It is usually necessary to transform bacterial counts to

logarithms and to convert their approximately log-normal frequency distribution to normality.

- When unexpectedly high microbiological results are obtained, resampling should be carried out to determine whether the unexpected results were due to sporadic events or persistent contamination. In the latter case, the source of pollution should be established and appropriate action taken.

8.11 References

Anon 1983 *La Qualitat de les Aigües Litorals. Informe.* Generalitat de Catalunya, Departament de Sanitat i Seguritat Social, Serie Sanejament Ambiental, Generalitat de Catalunya, Barcelona, 66 pp.

Anon 1996 Proposed amendments to the Hawaii Administrative Rules Chapter 11-54-08, Recreational Waters. In: *Water Quality Standards.* Department of Health, State of Hawaii, Honolulu, 54–86.

APHA/AWWA/WPCF 1989 *Standard Methods for the Examination of Water and Wastewater* 17th Edition, American Public Health Association, Washington, D.C.

APHA/AWWA/WPCF 1992 *Standard Methods for the Examination of Water and Wastewater* 18th Edition, American Public Health Association, Washington, D.C.

APHA/AWWA/WPCF 1995 *Standard Methods for the Examination of Water and Wastewater* 19th Edition, American Public Health Association, Washington, D.C.

Armstrong, I., Higham, S., Hudson, G. and Colley, T. 1997 The Beachwatch pollution monitoring programme: Changing priorities to recognize changed circumstances. *Marine Pollution Bulletin,* **33**(7–12), 249–259.

Ashbolt, N.J., Dorsch, M.R., Cox, P.T. and Banens, B. 1997 Blooming *E. coli,* what do they mean? In: D. Kay and C. Fricker [Eds] *Coliforms and E. coli, Problem or Solution?* The Royal Society of Chemistry, Cambridge, 78–85.

Audicana, A., Perales, I. and Borrego, J.J. 1995 Modification of kanamycin-esculin-azide agar to improve selectivity in the enumeration of fecal streptococci from water samples. *Applied and Environmental Microbiology,* **61,** 4178–4183.

Balebona, M.C., Moriñigo, M.A., Cornax, R., Borrego, J.J., Torregrossa, V.M. and Gauthier, M.J. 1990 Modified Most-Probable-Number technique for the specific determination of *Escherichia coli* from environmental samples using a fluorogenic method. *Journal of Microbiological Methods,* **12,** 235–245.

Barnes, E.M. 1959 Differential and selective media for the faecal streptococci. *Journal of the Science of Food and Agriculture*, **10**, 656–659.

Barnes, R., Curry, J.I., Elliott, L.M., Peter, C.R., Tamplin, B.R. and Wilcke, B.W. Jr. 1989 Evaluation of the 7-h membrane filter test for quantitation of fecal coliform in water. *Applied and Environmental Microbiology,* **55,** 1504–1506.

Bartram, J. and Ballance, R. [Eds] 1996 *Water Quality Monitoring. A Practical Guide to the Design and Implementation of Freshwater Quality Studies and Monitoring Programmes.* E & FN Spon, London.

Bej, A.K., DiCesare, J.L., Haff, L. and Atlas, R.M. 1991 Detection of *Escherichia coli* and *Shigella* spp. in water by using the polymerase chain reaction and gene probes for *uid. Applied and Environmental Microbiology,* **57,** 1013–1017.

Bej, A.K., McCarthy, S.C. and Atlas, R.M. 1991b Detection of coliform bacteria and *Escherichia coli* by multiplex polymerase chain reaction: Comparison with defined substrate and plating methods for water quality monitoring. *Applied and Environmental Microbiology,* **57,** 2429–2432.

Bernard, C., Boissonnade, G., Regnat, Y., Colin Jourdain, M.J. and Bouchinet, J. 1987 Automatic detection of coliform bacteria for industrial control of drinking water quality. *Water Research,* **21,** 1089–1099.

Bonde, G.J. 1977 Bacterial indication of water pollution. In: M.R. Droop and H.W. Jannasch [Eds] *Advances in Aquatic Microbiology*, Vol. 1, Academic Press, Inc., New York, 273–364.

Borrego, J.J. 1994 Diseño de medios de enumeración de bacterias patógenas alóctonas de aguas naturales. *Microbiología SEM,* **10,** 169–180.

Borrego, J.J., Arrabal, F., Vicente, A. de, Gomez, L.F. and Romero, P. 1983 Study of microbial inactivation in the marine environment. *Journal Water Pollution Control Federation,* **55**, 297–302.

Borrego, J.J., Vicente, A. de and Romero, P. 1982 Study of the microbiological pollution of a Malaga littoral area. I. Relationship between faecal coliforms and coliphages. *Journe Estudes Pollutions,* **4,** 551–560.

Brenner, K.P., Rankin, C.C., Roybal, Y.R., Stelma, G.N., Scarpino, P.V. and Dufour, A.P. 1993 New medium for the simultaneous detection of total coliforms and *E. coli* in water. *Applied and Environmental Microbiology,* **59,** 3534–3544.

Brenniman, G.R., Rosenberg, S.H. and Northrop, R.L. 1981 Microbial sampling variables and recreational water quality standards. *American Journal of Public Health,* **71,** 283–289.

Brodsky, M.H., Davidson, C.A., Dickson, J.S., Pettis, M.J. and Tieso, T.L. 1995 Delayed incubation method for microbiological analysis of environmental specimens and samples. *American Journal of Public Health,* **58,** 884–889.

Budnicki, G.E., Howard, R.T. and Mayo, D.R. 1996 Evaluation of enterolert for enumeration of enterococci in recreational waters. *Applied and Environmental Microbiology,* **62,** 3883–3884.

Cabelli, V.J., Dufour, A.P., McCabe, L.J. and Levin, M.A. 1982 Swimming-associated gastroenteritis and water quality. *American Journal of Epidemiology,* **115,** 606–616.

Cabelli, V.J., Dufour, A.P., McCabe, L.J. and Levin, M.A. 1983 A marine recreational water quality criterion consistent with indicator concepts and risk analysis. *Journal Water Pollution Control Federation,* **55**, 1306–1324.

Calderon, R.L., Mood, E.W. and Dufour, A.P. 1991 Health effects of swimmers and nonpoint sources of contaminated water. *International Journal of Environmental Health Research,* **1,** 21–31.

Chen, M. and Hickey, P.J. 1983 Modification of delayed-incubation procedure for detection of fecal coliforms in water. *Applied and Environmental Microbiology,* **46**, 889–893.

Chen, M. and Hickey, P.J. 1986 Elimination of overgrowth in delayed-incubation membrane filter test for total coliforms by m-ST holding medium. *Applied and Environmental Microbiology,* **52,** 778–781.

Cheung, W.H.S., Chang, K.C.K and Hung, R.P.S. 1991 Variations in microbial indicator densities in beach waters and health-related assessment of bathing water quality. *Epidemiology and Infection*, **106,** 329–344.

Clark, J.A. 1990 The presence-absence test for monitoring water quality. In: G.A. McFeters [Ed.] *Drinking Water Microbiology.* Springer-Verlag, New York, 399–411.

Cornax, R., Moriñigo, M.A., Romero, P. and Borrego, J.J. 1990 Survival of pathogenic microorganisms in seawater. *Current Microbiology,* **20,** 293–298.

Dange, V., Jothikumar, N. and Khanna, P. 1988 One-hour portable test for drinking water. *Water Research*, **22,** 133–137.

Davies, C.M. and Apte, S.C. 1996 Rapid enzymatic detection of faecal pollution. *Water Science and Technology,* **34**(7), 169–171.

Davies, C.M. and Apte, S.C. 1999 Field evaluation of a rapid portable test for monitoring faecal coliforms in coastal waters. *Environmental Toxicology and Water Quality,* **14,** (In Press).

Davies, C.M., Long, J.A., Donald, M. and Ashbolt, N.J. 1995 Survival of fecal microorganisms in aquatic sediments of Sydney, Australia. *Applied Environmental Microbiology,* **61,** 1888–1896.

Davis, E.M., Garrett, M.T. and Skinner, T.D. 1995 Significance of indicator bacteria changes in an urban stream. *Water Science and Technology*, **31**, 243–246.

Devriese, L.A., Pot, B. and Collins, M.D. 1993 Phenotypic identification of the genus Enterococcus and differentiation of phylogenetically distinct enterococcal species and species groups. *Journal of Applied Bacteriology*, **75**, 399–408.

Dionisio, L.P.C. and Borrego, J.J. 1995 Evaluation of media for the enumeration of faecal streptococci from natural water samples. *Journal of Microbiological Methods*, **23,** 183–203.

Dufour, A.P. 1977 *Escherichia coli* : the fecal coliform. In: A.W. Hoadley and B.J. Dutka [Eds] *Bacterial Indicators/Health Hazards Associated with Water.* American Society for Testing and Materials, Philadelphia, 48–58.

Dufour, A.P. 1984 *Health effects criteria for fresh recreational waters.* EPA-600/1-84-004, US Environmental Protection Agency, Washington, D.C.

Dufour, A.P. and Ballentine, R. 1986 *Bacteriological Ambient Water Quality Criteria for Marine and Freshwaters.* Report no. EPA-440/5-84-002, United States Environmental Protection Agency, Washington D.C. (available from National Technical Information Service, Springfield VA, Document no. PB86-158045).

Dufour, A.P. and Cabelli, V.J. 1976 Characteristics of *Klebsiella* from textile finishing plant effluents. *Journal Water Pollution Control Federation,* **48**, 872–876.

Dutka, B.J. and El-Shaarawi, A.H. 1990 *Use a Simple Unexpensive Microbial Water Quality Tests: Results of a Three Continent, Eight Country Research Project.* IDRC Report IDRC-MR247e, Rivers Research Branch, National Water Research Institute, Canada Centre for Inland Waters, Burlington, Ontario, Canada.

Edberg, S.C., Allen, M.J. and Smith, D.B. 1988 National field evaluation of a defined substrate method for the simultaneous detected of total coliforms and *Escherichia coli* from drinking water: comparison with the standard multiple-tube fermentation method. *Applied and Environmental Microbiology*, **54,** 1559–1601.

EEC 1976 *Council Directive concerning the Quality of Bathing Water* (76/160/ECC). European Economic Community, Brussels.

EEC 1996 *Quality of Bathing Water* (EUR 16755). European Economic Community, Brussels.

Eggers, M.D., Balch, W.J., Mendoza, L.G., Gangadharan, R., Mallik, A.K., McMahon, M.G., Hogan, M.E., Xaio, D., Powdrill, T.R., Iverson, B., Fox, G.E., Willson, R.C., Maillard, K.I., Siefert, J.L. and Singh, N. 1997 Advanced approach to simultaneous monitoring of multiple bacteria in space. In: 27th International Conference on Environmental Systems, Lake Tahoe, Nevada, July 14–17, 1997. SAE Technical Series 972422, The Engineering Society for Advancing Mobility Land Sea Air and Space, SAE International, Warrendale, PA, 1–8.

Evison, L.M. and Tosti, E. 1980 An appraisal of bacterial indicators of pollution in seawater. *Progress in Water Technology*, **12,** 591–599.

Ferguson, C.M., Coote, B.G., Ashbolt N.J. and Stevenson I.M. 1996 Relationships between indicators, pathogens and water quality in an estuarine system. *Water Research.,* **30**(9), 2045–2054.

Figueras, M.J., Inza, I., Polo, F., Feliu, M.T. and Guarro, J. 1996 A fast method for the confirmation of fecal streptococci from m-Enterococcus medium. *Applied and Environmental Microbiology*, **62**, 2177–2178.
Figueras, M.J., Inza, I., Polo, F. and Guarro, J. 1998 Evaluation of the oxolinic acid-esculin-azide medium for the isolation and enumeration of faecal streptococci in a routine monitoring programme for bathing waters. *Canadian Journal of Microbiology*, **44**, 998–1002.
Figueras, M.J., Polo, F., Inza, I. and Guarro, J. 1994 Poor specificity on m-Endo and m-FC culture media for the enumeration of coliform bacteria in sea water. *Letters in Applied Microbiology*, **19,** 446–450.
Figueras, M.J., Polo, F., Inza, I. and Guarro, J. 1997 Past, present and future perspectives of the EU bathing water directive. *Marine Pollution Bulletin*, **34,** 148–156.
Fleisher, J.M. 1985 Implications of coliform variability in the assessment of the sanitary quality of recreational waters. *Journal of Hygiene*, **94,** 193–200.
Fleisher, J.M. 1990 Conducting recreational water quality surveys: some problems and suggested remedies. *Marine Pollution Bulletin*, **21,** 562–567.
Fleisher, J.M. and McFadden, R.T. 1980 Obtaining precise estimates in coliform enumeration. *Water Research*, **14,** 477–483.
Frampton, E.W., Restaino, L. and Blaszko, N. 1988 Evaluation of the ß-glucuronidase substrate 5-bromo-4-chloro-3-indolyl-ß-D-glucuronide (X-GLUC) in a 24-h direct plating method for *Escherichia coli*. *American Journal of Public Health*, **51,** 402–404.
Fricker, E.J. and Fricker, C.R. 1996 Use of defined substrate technology and a novel procedure for estimating the numbers of enterococci in water. *Journal of Microbiological Methods,* **27,** 200–210.
Fujioka, R.S. 1997 Indicators of marine recreational water quality. In: C.J. Hurst, G.R. Knudsen, M.J. McInerney, L.D. Stetzenbach and M.V. Walter [Eds] *Manual of Environmental Microbiology*. ASM Press, Washington, D.C., 176–183.
Gameson, A.L.H. 1982 *Investigations of Sewage Discharges to Some British Coastal Waters. Chapter 5. Bacterial distributions, Part 1.* Technical report TR79, Water Research Centre, Medmenham, 40 pp.
Gameson, A.L.H. 1983 *Investigations of Sewage Discharges to Some British Coastal Waters. Chapter 3. Bacteriological Enumeration Procedures, Part 2.* Technical report TR193, Water Research Centre, Medmenham, 60 pp.
Gameson, A.L.H. and Munro, D. 1980 *Investigations of Sewage Discharges to Some British Coastal Waters. Chapter 5. Bacterial distributions, Part 2.* Technical Report TR147, Water Research Centre, Medmenham, 115 pp.
Gauci, V. 1991 Enumeration of faecal streptococci in seawater. In: *Development and Testing of Sampling and Analytical Techniques for Monitoring of*

Marine Pollutants (Activity A): Final Reports on Selected Microbiological Projects, MAP Technical Reports Series No. 54, United Nations Environment Programme, Athens, 47–59.

Gauthier, M.J., Torregrossa, V.M., Balebona, M.C., Cornax, R. and Borrego, J.J. 1991 An intercalibration study of the use of 4-methylumbelliferyl-ß-D-glucuronide for the specific enumeration of *Escherichia coli* in seawater and marine sediments. *Systematic and Applied Microbiology*, **14,** 183–189.

Gavini, F., Poucher, A.M., Neut, C., Monget, D., Romond, C., Oger C. and Izard, D. 1991 Phenotypic differentiation of bifidobacteria of human and animal origin. *International Journal of Systematic Bacteriology,* **41,** 548–557.

Geldreich, E.E. 1967 Fecal coliform concepts in stream pollution. *Water and Sewage Works,* **114**, 98–110.

Geldreich, E.E. 1976 Faecal coliform and faecal streptococcus density relationships in waste discharges and receiving waters. *Critical Reviews Environ. Contr.*, **6,** 349–369.

Gleeson, C. and Gray, N. 1997 *The Coliform Index and Waterborne Disease. Problems of Microbial Drinking Water Assessment.* Chapman and Hall, London, 191 pp.

Grant, M.A. and Ziel, C.A. 1996 Evaluation of a simple screening test for faecal pollution in water. *Journal of Water Supply Research and Technology AQUA,* **45**, 13–18.

Hardina, C.M. and Fujioka R.S. 1991 Soil: The environmental source of *Escherichia coli* and enterococci in Hawaii's streams. *Environmental Toxicology Water Quality,* **6,** 185–195.

Hazen, T.C. and Toranzos, G.A. 1990 Tropical source water. In: G.A. McFeters [Ed.] *Drinking Water Microbiology*, Springer-Verlag, New York, 32–54.

Hernandez, J.F., Delattre, J.M. and Maier, E.A. 1995 *BCR Information. Sea Water Analysis. Sea Water Microbiology. Performance of Methods for the Microbiological Examination of Bathing Water, Part 1.* EUR 16601 EN. Directorate-General, Science, Research and Development, Commission of the European Communities, Brussels.

Hernandez, J.F., Guibert, J.M., Delattre, J.M., Oger, C., Charriere, C., Hughes, B., Serceau, R. and Sinegre, F. 1991 Miniaturized fluorogenic assays for enumeration of *E. coli* and enterococci in marine water. *Water Science and Technology,* **24,** 137–141.

Hernandez-Lopez, J. and Vargas-Albores, F. 1994 False-positive coliform readings using membrane filter techniques for seawater. *Letters in Applied Microbiology*, **19,** 483–485.

Holt, J.G., Krieg, N.R., Sneath, P.H.A., Staley, J.T. and Williams, S.T. 1993 *Bergey's Manual of Determinative Bacteriology*, 9th edition, Williams & Wilkins, Co., Baltimore.

Höglund, C., Stenström, T.A., Jönsson, H. and Sundin, A. 1998 Evaluation of faecal contamination and microbial die-off in urine separating sewage systems. *Water Science Technology,* **38**(6), 17–25.

Howell, J.M., Coyne, M.S. and Cornelius, P.L. 1995 Faecal bacteria in agricultural waters of the bluegrass region of Kentucky. *Journal Environmental Quality,* **24,** 411–419.

Isenberg, H.D., Goldberg, D. and Sampson, J. 1970 Laboratory studies with a selective enterococcus medium. *Applied Microbiology*, **20,** 433–436.

ISO 1984a *Water Quality—Detection and Enumeration of Faecal Streptococci - Part 1: Most Probable Number Method.* ISO 7899-1, International Organization for Standardization, Geneva.

ISO 1984b *Water Quality—Detection and Enumeration of Faecal Streptococci - Part 2: Membrane Filtration Method.* ISO 7899-1, International Organization for Standardization, Geneva.

ISO 1988 *Water Quality—General Guide to the Enumeration of Microorganisms by Culture.* ISO 8199, International Organization for Standardization, Geneva.

ISO 1990a *Water Quality—Sampling - Part 6: Guidance on Sampling of Rivers and Streams.* ISO 5667-8, International Organization for Standardization, Geneva.

ISO 1990b *Water Quality—Detection and Enumeration of Coliform Organisms, Thermotolerant Coliform Organisms and Presumptive* Escherichia coli - *Part 1: Membrane Filtration Method.* ISO 9308-1, International Organization for Standardization, Geneva.

ISO 1990c *Detection and Enumeration of Coliform Organisms, Thermotolerant Coliform Organisms and Presumptive* Escherichia coli - *Part 2: Most Probable Number Method.* ISO 9308-2, International Organization for Standardization, Geneva.

ISO 1992 *Water Quality—Sampling - Part 9: Guidance on Sampling from Marine Waters.* ISO 5667-9, International Organization for Standardization, Geneva.

ISO 1996a *Water Quality - Enumeration of Intestinal Enterococci in Surface and Waste Water - Part 1: Miniaturized Method (Most Probable Number) by Inoculation in Liquid Medium.* ISO/DIS 7899-1, International Organization for Standardization, Geneva.

ISO 1996b *Water Quality - Enumeration of* Escherichia coli *in Surface and Waste Water - Part 3: Miniaturized Method (Most Probable Number) by Inoculation in Liquid Medium.* ISO/DIS 9308-3, International Organization for Standardization, Geneva.

ISO 1997 *Water Quality - Detection and Enumeration of* Escherichia coli *and Coliform Bacteria - Part 1: Membrane Filtration Method.* ISO/DIS 9308-1, International Organization for Standardization, Geneva.

ISO 1999a *Water Quality—Detection and Enumeration of Intestinal Enterococci - Part 1: Miniaturized Method (Most Probable Number) by Inoculation in Liquid Medium.* ISO 7899-1, International Organization for Standardization, Geneva.

ISO 1999b *Water Quality—Detection and Enumeration of Intestinal Enterococci - Part 2: Membrane Filtration Method.* ISO 7899-2, International Organization for Standardization, Geneva.

ISO 1999c *Water Quality. Detection and enumeration of bacteriophages infecting* Bacteriodes fargilis. ISO 10705-4, International Organization for Standardization, Geneva.

Janda W.M. 1994 Streptococci and "*Streptococcus*-like" bacteria: old friends and new species. *Clinical Microbiology Newsletter,* **16,** 61–71.

Janning, B., Tveld, P.H., Mooijman, K.A. and Havelaar, A.H. 1995 Development, production and certification of microbiological reference materials. *Fresenius Journal of Analytical Chemistry*, **352,** 240–245.

Jofre, J., Olle, E., Ribas, F., Vidal, A. and Lucena, F. 1995 Potential usefulness of bacteriophages that infect *Bacteriodes fragilis* as model organisms for monitoring virus removal in drinking water treatment plants. *Applied Environmental Microbiology*, **61**(9), 3227–3231.

Kaspar, C.W., Hartman, P.A. and Benson, A.K. 1987 Coagglutination and enzyme capture tests for detection of *Escherichia coli* ß-galactosidase, ß-glucuronidase, and glutamate decarboxylase. *Applied and Environmental Microbiology*, **53,** 1073–1077.

Kay, D., Fleischer, J.M., Salmon, R.L., Jones, F., Wyer, M.D., Godfree, A.F., Zelenauch-Jacquotte, Z. and Shore, R. 1994 Predicting likelihood of gastroenteritis from sea bathing: results from randomized exposure. *Lancet,* **344,** 905–909.

Keith, L.H. 1990 Environmental sampling: A Summary. *Environmental Science and Technology*, **24,** 610–617.

Kenner, B.A., Clark, H.F. and Kabler, P.W. 1961 Faecal streptococci. I. Cultivation and enumeration of streptococci in surface waters. *Applied Microbiology*, **9,** 15–20.

Knittel, M.D., Seidler, R.J., Elby, C. and Cabe, L.M. 1977 Colonization of the botanical environment by *Klebsiella* isolates of pathogenic origin. *Applied and Environmental Microbiology*, **34,** 557–563.

Kreader, C.A. 1995 Design and evaluation of Bacteroides DNA probes for the specific detection of human fecal pollution. *Applied Environmental Microbiology,* **61,** 1171–1179.

Lechevallier, M.W. 1990 Coliform regrowth in drinking water: a review. *Journal American Water Works Association*, **82,** 74–86.

Leclerc, H., Devriese, L.A. and Mossel, D.A.A. 1996 Taxonomical changes in intestinal (faecal) enterococci and streptococci: consequences on their use as indicators of faecal contamination in drinking water. *Journal of Applied Bacteriology,* **81,** 459–66.

Leeming, R., Ball, A., Ashbolt, N. and Nichols, P. 1994 Distinguishing between human and animal sources of faecal pollution. *Chem. Australia,* **61**(8), 434–435.

Leeming, R., Ball, A., Ashbolt, N. and Nichols, P. 1996 Using faecal sterols from humans and animals to distinguish faecal pollution in receiving waters. *Water Research,* **30**(12), 2893–2900.

Leeming, R., Nichols, P.D. and Ashbolt, N.J. 1998 *Distinguishing Sources of Faecal Pollution in Australian Inland and Coastal Waters using Sterol Biomarkers and Microbial Faecal Indicators.* Research Report No. 204, Water Services Association of Australia, Melbourne, 46 pp.

Levin, M.A., Fischer, J.R. and Cabelli, V.P. 1975 Membrane filter technique for enumeration of enterococci in marine waters. *Applied Microbiology,* **30**, 66–71.

Lin, S.D. 1974 Evaluation of fecal streptococci tests for chlorinated secondary sewage effluents. *Journal of Environmental Engineering, ASCE,* **100,** 253–267.

Lucena, F., Araujo, R. and Jofre, J. 1996 Usefulness of bacteriophages infecting *Bacteriodes fragilis* as index microorganisms of remote faecal pollution. *Water Research,* **30**(11), 2812–2816.

Manafi, M. and Kneifel, W. 1989 A combined chromogenic-fluorogenic medium for the simultaneous detection of total coliforms and *Escherichia coli* in water. *Zentralblatt fur Hygiene,* **189,** 225–234.

Manja, K.S., Maurya, M.S. and Rao, K.M. 1982 A simple field test for the detection of faecal pollution in drinking water. *Bulletin WHO,* **10,** 797–801.

Marsalek, J., Dutka, B.J., McCorquodale, A.J. and Tsanis, I.K. 1996 Microbiological pollution in the Canadian upper great lakes connecting channels. *Water Science and Technology,* **33,** 349–356.

McCarty, S.C., Standridge, J.H. and Stasiak, M.C. 1992 Evaluating commercially available define-substrate test for recovery of *E. coli*. *Journal American Water Works Association,* **84,** 91–97.

McDaniels, A.E., Rice, E.W., Reyes, A. L., Johnson, C.H., Haugland, R.A. and Stelma, G.N. 1996 Confirmational identification of *Escherichia coli,* a comparison of genotypic and phenotypic assays for glutamate decarboxylase and ß-D-glucuronidase. *Applied and Environmental Microbiology,* **62,** 3350–3354.

McFeters, G.A., Broadaway, S.C., Pyle, B.H., Pickett, M. and Egozy, Y. 1995 Comparative performance of Colisure and accepted methods in the detection of chlorine-injured total coliforms and *E. coli. Water Science and Technology,* **31,** 259–261.

Mossel, D.A.A., Harrewjn, G.A. and Berdien, J.M. 1973 *Recommended Routine Monitoring Procedures for the Microbiological Examination of Foods and Drinking Water*. United Nations Children's Fund (UNICEF), Geneva.

Muhldorfer, I., Blum, G., Donohue-Rolfe, A., Heier, H., Olschlager, T., Tschape, H., Wallner, U. and Hacker, J. 1996 Characterization of *Escherichia coli* strains isolated from environmental water habitats and from stool samples of healthy volunteers. *Res. Microbiol.,* **147**(8), 625–635.

Niemi, R.M., Niemelä, S.I., Lahti, K. and Niemi, J.S. 1997 Coliforms and *E. coli* in Finnish surface waters. In: D. Kay and C. Fricker [Eds] *Coliforms and E. coli. Problems or Solution?* The Royal Society of Chemistry, Cambridge, 112–119.

Osawa, S., Furuse, K. and Watanabe, I. 1981 Distribution of ribonucleic acid coliphages in animals. *Applied Environmental Microbiology,* **41**, 164–168.

O'Shea, M.L. and Field, R. 1992 Detection and disinfection of pathogens in storm-generated flows. *Canadian Journal of Microbiology*, **38,** 267–276.

Palmer, C.J., Tsai, Y., Lang, A.L. and Sangermano, L. 1993 Evaluation of Colilert-Marine Water detection of total coliforms and *Escherichia coli* in the marine environment. *Applied and Environmental Microbiology*, **59,** 786–790.

Payment, P. and Franco, E. 1993 *Clostridium perfringens* and somatic coliphages as indicators of the efficiency of drinking water treatment for viruses and protozoan cysts. *Applied Environmental Microbiology,* **59**, 2418–2424.

Pipes, W.O. 1982 Indicators and water quality. In: W.O. Pipes [Ed.] *Bacterial Indicators of Pollution*. CRC Press Inc., Boca Raton, 83–95.

Poucher, A.M., Devriese, L.A., Hernandez, J.F. and Delattre, J.M. 1991 Enumeration by a miniaturized method of *Escherichia coli, Streptococcus bovis* and enterococci as indicators of the origin of faecal pollution of waters. *Journal of Applied Bacteriology*, **70,** 525–530.

PHLS 1994 *The Microbiology of Water. Methods for the Examination of Waters and Associated Materials*. Report No. 71, Her Majesty's Stationery Office, London.

PHLS 1995 Water Surveillance Group Preliminary study of microbiological parameters in eight inland recreational waters. *Letters in Applied Microbiology*, **21,** 267–271.

Puig, A., Jofre, J. and Araujo, R. 1997 Bacteriophages infecting various *Bacteroides fragilis* strains differ in their capacity to distinguish human from animal faecal pollution. *Water Science Technology,* **35**(11–12), 359–362.

Ramteke, P.W. 1995 Comparison of standard most probable number method with three alternate tests for detection of bacteriological water quality indicators. *Environmental Toxicology and Water Quality*, **10,** 173–178.

Reasoner, D.J. and Geldreich, E.E. 1989 Detection of fecal coliforms in water using [^{14}C]-mannitol. *Applied and Environmental Microbiology*, **55,** 907–911.

Rice, E.W., Allen, M.J., Brenner, D.J. and Edberg, S.C. 1991 Assay for ß-glucuronidase in species of the genus *Escherichia* and its applications for drinking-water analysis. *Applied and Environmental Microbiology,* **57,** 592–593.

Richardson, K.J., Stewart, M.H. and Wolfe, R.L. 1991 Application of gene probe technology to the water industry. *Journal American Water Works Association,* **83,** 71–81.

Rivera, S.C., Hazen, T.C. and Toranzos, G.A. 1988 Isolation of fecal coliforms from pristine sites in a tropical rain forest. *Applied and Environmental Microbiology*, **54**, 513–517.

Ruoff, K.L. 1990 Recent taxonomic changes in the genus *Enterococcus*. *European Journal of Clinical Microbiology and Infectious Diseases*, **9**, 75–79.

Rutkowski, A.A. and Sjogren, R.E. 1987 Streptococcal population profiles as indicators of water quality. *Water Air and Soil Pollution*, **34,** 273–284.

Sartory, D.P. and Howard, L. 1992 A medium detecting ß-glucuronidase for the simultaneous membrane filtration enumeration of *Escherichia coli* and coliforms from drinking water. *Letters in Applied Microbiology*, **15,** 273–276.

Seyfried, P.L., Tobin, R., Brown, N.E. and Ness, P.E. 1985 A prospective study of swimming related illness. I. Swimming associated health risk. *American Journal of Public Health*, **75,** 1068–1070.

Sinton, L.W. and Donnison, A.M. 1994 Characterization of faecal streptococci from some New Zealand effluents and receiving waters. *New Zealand Journal of Marine and Freshwater Research*, **28,** 145–158.

Sinton, L.W., Donnison, A.M. and Hastie, C.M. 1993a Faecal streptococci as faecal pollution indicators: a review. Part I: Taxonomy and enumeration. *New Zealand Journal of Marine and Freshwater Research*, **27,** 101–115.

Sinton, L.W., Donnison, A.M. and Hastie, C.M. 1993b Faecal streptococci as faecal pollution indicators: a review. Part II: Sanitary significance, survival, and use. *New Zealand Journal of Marine and Freshwater Research*, **27,** 117–137.

Slanetz, L.W. and Bartley, C.H. 1957 Numbers of enterococci in water, sewage and faeces determined by the membrane filter technique with an improved medium. *Journal of Bacteriology*, **74,** 591–595.

Slanetz, L.W. and Bartley, C.H. 1965 Survival of fecal streptococci in seawater. *Health Laboratory Sciences*, **2,** 142–148.

Smoker, D. 1991 Capturing the coliform. *Water Services*, **95**, 20–22.

Sorensen, D.L., Eberl, S.G. and Diksa, R.A 1989 *Clostridium perfringens* as a point source indicator in non-point polluted streams. *Water Research,* **23**, 191–197.

Tillet, H.E. 1993 Potential inaccuracy of microbial counts from routine water samples. *Water Science and Technology,* **27,** 15–18.

Tillett, H.E. 1995 The most probable number estimates and the usefulness of confidence intervals: comment. *Water Research*, **199,** 1213–1214.

Tillet, H.E. and Benton, C. 1993 Effects of transit time on indicator organism counts from water samples. *Public Health Laboratories Services*, **10**(2), 116–117.

Tillett, H.E. and Lightfoot, N.F. 1995 Quality control in environmental microbiology compared with chemistry: what is homogeneous and what is random. *Water Science and Technology* **31,** 471–477.

Tillet, H.E., Lightfoot, N.F. and Eaton, S. 1993 External quality assessment in water microbiology: statistical analysis of performance. *Journal of Applied Bacteriology,* **74,** 497–502.

Tsai, Y.L., Palmer, C.J. and Sangermano, L.R. 1993 Detection of *Escherichia coli* in sewage and sludge by PCR. *Applied and Environmental Microbiology*, **59,** 353–357.

Tsanis, I.K., Wu, J. and Marsalek, J. 1995 Feasibility of modelling remedial measure for microbiological pollution of the St. Clair River at Sarnia Bay. *Journal of Great Lakes Research*, **21,** 138–154.

UNEP/WHO 1985 *Assessment of the Present State of Pollution of the Mediterranean Sea and Proposed Control Measures.* Document UNEP/WG.118/6, United Nations Environment Programme, Athens.

US EPA 1986 *Ambient Water Quality Criteria for Bacteria.* EPA 440/5-84-002, United States Environmental Protection Agency, Washington, D.C.

Volterra, L., Bonadona, L. and Aulicino, F.A. 1986 Fecal streptococci recoveries in different marine areas. *Water Air and Soil Pollution*, **29,** 403–413.

Vonstille, W.T., Stille, III, W.T. and Sharer R.C. 1993 Hepatitis A epidemics from utility sewage in Ocoee, Florida. *Archives in Environmental Health,* **48,** 120–124.

Walter, K.S., Fricker, E.J. and Fricker, C.R. 1994 Observation on the use of a medium detecting ß-glucuronidase activity and lactose fermentation for the for the simultaneous detection of *E. coli* and coliforms. *Letters in Applied Microbiology*, **19,** 47–49.

WHO 1993 *Guidelines for Drinking Water Quality.* 2nd Edition, World Health Organization, Geneva.

WHO 1998 *Guidelines for Safe Recreational Water Environments.* Draft for consultation. World Health Organization, Geneva.

WHO/UNEP 1994a *Guidelines for Health-Related Monitoring of Coastal Recreational and Shellfish Areas. Part II. Bacterial Indicator Organisms.* Document EUR/ICP/CEH 041(3). World Health Organization, Regional Office for Europe, Copenhagen.

WHO/UNEP 1994b *Guidelines for Health-Related Monitoring of Coastal Recreational and Shellfish Areas. Part IV. Statistical Methods.* Document EUR/ICP/CEH 041(5). World Health Organization, Regional Office for Europe, Copenhagen.

WHO/UNEP 1994c *Guidelines for Health-Related Monitoring of Coastal Recreational and Shellfish Areas. Part V. Quality Control*. EUR/ICP/CEH 041(6), Copenhagen.

Wyer, M.D., Jackson, G., Kay, D., Yeo, J. and Dawson, H. 1994 An assessment of the impact of inland surface water input to the bacteriological quality of coastal waters. *Journal of the Institution of Water and Environmental Management*, **8**, 459–467.

Wyer, M.D., Kay, D., Jackson, G.F., Dawson, H.M., Yeo, J. and Tanguy, L. 1995 Indicator organism sources and coastal water quality: a catchment study on the Island of Jersey. *Journal of Applied Bacteriology*, **78,** 290–296.
Wyer, M.D., O'Neill, G., Kay, D., Crowther, J. Jackson, G. and Fewtrell, L. 1997 Non outfall sources of faecal indicator organisms affecting the compliance of coastal waters with directive 76/160/EEC. *Water Science and Technology*, **35,** 151–156.
Yoshpe-Purer, Y. 1989 Evaluation of media for monitoring fecal streptococci in seawater. *Applied and Environmental Microbiology*, **55,** 2041–2045.

Chapter 9*

APPROACHES TO MICROBIOLOGICAL MONITORING

Despite evident successes in the protection of public health, present approaches to the regulation of recreational water quality suffer several limitations. A modified approach to regulation of recreational water quality could provide for improved protection of public health with the minimum necessary monitoring effort and could provide greater scope for interventions, especially for those within the resources of local authorities. This chapter describes the principal issues discussed at a meeting of experts, who concluded that such an alternative was possible and should be tested and promoted.

9.1 Issues

9.1.1 Current regulatory schemes

Recreational water standards have had some success in driving cleanups, increasing public awareness, contributing to informed personal choice and contributing to a public health benefit. These successes are difficult to quantify, but the need to control and minimise adverse health effects has been the principal concern of regulation. Present regulatory schemes for the microbiological quality of recreational water are primarily or exclusively based on percentage compliance with faecal indicator counts. Examples of compliance criteria currently in use are given in Table 9.1. A number of constraints are evident in the current standards and guidelines:

- Management actions are retrospective and can only be deployed after human exposure to the hazard.
- The risk to health is primarily from human excreta, the traditional indicators of which may also derive from other sources.
- There is poor inter-laboratory and international comparability of microbiological analytical data.

* *This chapter represents the conclusions of a meeting of experts organised by WHO in co-operation with the US EPA and held in Annapolis, MD, USA, in November 1998 (for further details, see the Acknowledgements section). The US EPA has not conducted a policy and legal evaluation of this chapter.*

Table 9.1 Examples of guidelines and standards for microbiological quality of water (number of organisms per 100 ml)

	Primary contact recreation		
Country	TC	FC	Other
Brazil	80% < 5,000[1]	80% < 1,000[1]	
Colombia	1,000	200	
Cuba	1,000[2]	200[2] 90% < 400	
Ecuador	1,000	200	
Europe, EEC[3]	80% < 500[4] 95% < 10,000[5]	80% < 100[4] 95% < 2,000[5]	Faecal streptococci 100[4] *Salmonella* 0 per litre[5] Enteroviruses 0 PFU per litre[5] Enterococci 90% < 100
France	< 2,000	< 500	Faecal streptococci < 100
Israel	80% < 1,000[6]		
Japan	1,000		
Mexico	80% < 1,000[7] 100% < 10,000[9]		
Peru	80% < 5,000[7]	80% < 1,000[7]	
Poland			*E. coli* < 1,000
Puerto Rico		200[10] 80% < 400	
USA			
California	80% < 1,000[11,12] 100% < 10,000[9]	200[2,12] 90% < 400[13]	
US EPA			Enterococci 35[2] (marine) 33[2] (fresh) *E. coli* 126[2] (fresh)
Former USSR			*E. coli* < 100
UNEP/WHO		50% < 100[14] 90% < 1,000[14]	
Uruguay		< 500[14] < 1,000[15]	
Venezuela	90% < 1,000 100% < 5,000	90% < 200 100% < 400	
Yugoslavia	2,000		

TC Total coliforms
FC Faecal or thermotolerant coliforms

1 "Satisfactory" waters, samples obtained in each of the preceding 5 weeks
2 Logarithmic average for a period of 30 days of at least 5 samples
3 Minimum sampling frequency – fortnightly
4 Guide
5 Mandatory
6 Minimum 10 samples per month
7 At least 5 samples per month
8 Monthly average

Table 9.1 Continued

	Shellfish harvesting		Protection of indigenous organisms		
Country	TC	FC	TC	FC	Reference(s)
Brazil		100% < 100			Ministerio del Interior, 1976
Colombia					Ministerio de Salud, 1979
Cuba					Ministerio de Salud, 1986
Ecuador					Ministerio de Salud Publica, 1987
Europe, EEC[3]					EEC, 1976
					CEPPOL/UNEP, 1991
France					WHO/UNEP, 1977
Israel					INCYTH, 1984
Japan	70		1,000		Environmental Agency, 1981
Mexico	70[8] 90% < 230		10,000[8] 80% < 10,000 100% < 20,000		SEDUE, 1983
Peru	80% < 1,000	80%< 200 100% < 1,000	80% < 20,000	80% < 40,000	Ministerio de Salud, 1983
Poland					WHO, 1975
Puerto Rico	70[10] 80% < 230				JCA, 1983
USA					
California	70[8]				California State Water Resources Board, undated
US EPA		14[2] 90% < 43			US EPA, 1986; Dufour and Ballentine, 1986
Former USSR					WHO/UNEP, 1977
UNEP/WHO		80% < 10 100% < 100			WHO/UNEP, 1978
Uruguay					DINAMA, 1998
Venezuela	70[2] 90% < 230	14[2] 90% < 43			Venezuela, 1978
Yugoslavia					INCYTH, 1984

9 No sample taken during the verification period of 48 hours should exceed 10,000 per 100 ml
10 At least 5 samples taken sequentially from the waters in a given instance
11 Period of 30 days
12 Within a zone bounded by the shoreline and a distance of 1,000 feet from the shoreline or the 30 foot depth contour, whichever is further from the shoreline
13 Period of 60 days
14 Geometric mean of at least 5 samples
15 Not to be exceeded in at least 5 samples

Source: Adapted from Salas, 1998

Table 9.2 Outbreaks of disease associated with recreational waters in the USA, 1985–1994

Etiological agent	No. of cases	No. of outbreaks
Shigella	935	13
E. coli	166	1
Leptospira	14	2
Giardia	65	4
Cryptosporidium	418	1
Norwalk virus	41	1
Adenovirus 3	595	1
Acute gastro-intestinal infections	965	11

Sources: Morbidity and Mortality Weekly Report, 1988, 1990, 1991, 1993; Kramer, *et al.*, 1996

- While beaches are classified as safe or unsafe, there is a gradient of increasing severity, variety and frequency of health effects with increasing sewage pollution and it is desirable to promote incremental improvements prioritising "worst failures".

The present form of regulation tends to focus upon sewage treatment and outfall management as the principal or only effective interventions. Because of the high costs of these measures, local authorities may be disenfranchised and few options for effective local intervention in securing bather safety from sewage pollution may be available. The limited evidence available from cost-benefit studies of pollution control alone rarely justifies the proposed investments. The costs may be prohibitive or may detract resourcing from greater public health priorities (such as securing access to a safe drinking water supply), especially in developing countries. If pollution abatement on a large scale is the only option available to local management, then many will be unable to undertake the required action.

Considerable concern has been expressed regarding the burden (cost) of monitoring, primarily but not exclusively to developing countries, especially in light of the precision with which the monitoring effort assesses the risk to the health of water users and the effectiveness with which it supports decision-making to protect public health.

9.1.2 Pathogens

There is a broad spectrum of illnesses that have been associated with swimming in marine and fresh recreational waters. Table 9.2 is a list of microbes that have been linked to swimming-associated disease outbreaks in the USA between 1985 and 1994. Two bacterial pathogens, *Escherichia coli* and *Shigella*, and two pathogenic protozoans, *Giardia* and *Cryptosporidium,* are

Table 9.3 Serological response to Norwalk virus and rotavirus in individuals with recent swimming-associated gastro-enteritis

Antigen	No. of subjects	Age range	No. with 4-fold titer increase
Norwalk virus	12	3 months – 12 years	4
Rotavirus	12	3 months – 12 years	0

of special interest because of the circumstances under which the associated outbreaks occurred. These outbreaks usually occurred in very small, shallow bodies of water that were frequented by children. Epidemiological investigations of the outbreaks found that the source of the etiological agent was usually the bathers themselves, most likely children. Each outbreak affected a large number of bathers, as might be expected in unmixed, small bodies of water containing large numbers of pathogens.

Outbreaks caused by *Leptospira,* Norwalk virus and Adenovirus 3 were more typical because the sources of pathogens were external to the beaches and, except for *Leptospira*, associated with faecal contamination. *Leptospira* are usually associated with animals that urinate into surface waters. Swimming-associated outbreaks attributed to *Leptospira* are very rare. Conversely, outbreaks of acute gastro-intestinal infections with an unknown aetiology are more common. Although the cause is unknown, the symptoms associated with the illness are frequently similar to those observed in viral infections.

Very few studies, other than those associated with outbreaks, have been conducted to determine the etiological agents related to swimming-associated illness. Some previously unpublished data shown in Table 9.3 do confirm that viruses are candidate organisms for the gastro-enteritis observed in epidemiological studies conducted at bathing beaches. The data in the table are from acute and convalescent sera obtained from swimmers who suffered from acute gastro-enteritis after swimming at a very contaminated beach in Alexandria, Egypt. The sera were obtained from 12 subjects, all of whom were less than 12 years old, on the day after the swimming event and again about 15 days later. The sera were tested with Norwalk virus and rotavirus antigens. None of the subjects showed a fourfold increase in titre to rotavirus antigen. However, 33 per cent did show a fourfold increase in titre to the Norwalk virus antigen. This reactivity indicated that Norwalk virus is a pathogen that has the potential to cause swimming-associated gastro-enteritis. These data also show a possible approach for linking specific pathogens to swimming-associated illness.

The types and numbers of various pathogens in sewage vary depending on the incidence of disease in the contributing population and known seasonality in human infections. Hence, numbers vary greatly across different parts

Table 9.4 Examples of pathogens and indicator organisms commonly found in raw sewage

Pathogen or indicator[1]	Disease or role	No. per litre
Bacteria		
Campylobacter spp.	Gastro-enteritis	37,000
Clostridium perfringens[2]	Indicator organism	6×10^5–8×10^5
E. coli	Indicator organism	10^7–10^8
Salmonella spp.	Gastro-enteritis	20–80,000
Shigella	Bacillary dysentery	10–10,000
Viruses		
Polioviruses	Indicator	1,800–5,000,000
Rotaviruses	Diarrhoea, vomiting	4,000–850,000
Parasitic protozoa		
Cryptosporidium parvum oocysts	Diarrhoea	1–390
Entamoeba histolytica	Amoebic dysentery	4
Giardia lamblia cysts	Diarrhoea	125–200,000
Helminths		
Ascaris spp.	Ascariasis	5–110
Ancylostoma spp.	Anaemia	6–190
Trichuris spp.	Diarrhoea	10–40

1 Many important pathogens in sewage have yet to be adequately enumerated, such as adenoviruses, Norwalk/SRS viruses and Hepatitis A

2 From Long and Ashbolt, 1994

Source: Adapted from Yates and Gerba, 1998

of the world and times of year, but a general indication of incidence is given in Table 9.4.

9.1.3 Indicators

The risk of exposure to pathogens in recreational waters has been well described in the literature (WHO, 1998) and this information has been noted and used by risk managers. However, it is very difficult to detect pathogens, especially viral and protozoan pathogens, in water samples obtained from bathing beaches. Methods for detecting and identifying infectious viruses or parasites are either very difficult to perform or do not exist at all. Bacterial pathogens can be detected, but their fastidious nutritional requirements and susceptibility to different types of environmental stress also can make the task very difficult. One hundred years ago, Escherich and other creative bacteriologists who were concerned with cholera and typhoid fever, proposed the use of a harmless organism that was always found in faeces and which was easy to detect on simple bacteriological media, as an indicator of the presence of faecal material in water. By implication, these indicator organisms would signal the potential presence of organisms that cause gastro-intestinal disease.

The indicator concept has been used successfully for a long time. The faecal bacteria most commonly used today are thermotolerant coliforms, *E. coli* and enterococci or faecal streptococci, which are described in detail in Chapter 8. However, there are still many questions concerning the effectiveness of the way in which water quality is measured and monitored; a number of environmental and physical factors may influence the usefulness of faecal bacteria as indicators. No single indicator or approach is likely to represent all the facets and issues associated with contamination of waterways with faecal matter. Table 9.5 provides an overview of possible indicators, together with the strengths and drawbacks of each.

Die-off in marine and freshwater environments

The differential die-off of indicators in marine and freshwater environments is illustrated for coliforms in Figure 9.1. The figure, which was adapted from Chamberlain and Mitchell (1978), shows that in marine waters the mean T_{90} (the time taken for 90 per cent of organisms to die) for total coliforms is about 2.2 hours, whereas in freshwaters the mean T_{90} is about 58 hours. These results were obtained from *in situ* studies at wastewater outfalls where die-off was determined after accounting for dispersion and dilution. Similar behaviour is exhibited by thermotolerant coliforms and *E. coli*. Although similar studies have not been conducted with enterococci, laboratory studies by various investigators, particularly Hanes and Fragala (1967) suggest that enterococci also die-off more rapidly in sea water than in freshwater environments (Table 9.6). The differential die-off for enterococci is not as great as that for *E. coli,* which may account for their superior effectiveness as indicators of health risk. Very few similar studies have been conducted for viral indicators. One study, conducted in marine and freshwaters in Italy (Table 9.7), showed that Polio, ECHO and Coxsackie viruses decayed at approximately the same rate in these two environments (Cioglia and Loddo, 1962). If, as appears likely, indicators have different die-off characteristics in marine and freshwater, whereas viral indicators die-off at similar rates in these two environments, then viral pathogens may be present at higher levels in these waters relative to the bacterial indicator numbers. The conclusion that can be drawn is that higher levels of exposure to viral pathogens may occur in marine waters at similar bacterial indicator levels and may require reconsideration of guideline levels in the two environments.

Solar radiation

The effect of sunlight on *E. coli* and enterococci is shown in Figure 9.2. The rate of *E. coli* die-off increases rapidly as solar radiation increases.

Table 9.5 The relative merits of selected indicators of sewage contamination

Indicator	Advantages	Disadvantages
Faecal streptococci/ enterococci	Marine and potentially freshwater human health indicator More persistent in water and sediments than coliforms Faecal streptococci may be cheaper than enterococci to assay	May not be valid in tropical waters due to potential growth in soils
Thermotolerant coliforms	Indicator of recent faecal contamination	Possibly not suited to tropical waters due to growth in soils and waters Confounded by non-sewage sources (e.g. *Klebsiella* spp. in pulp and paper wastewaters)
E. coli	Potential freshwater human health indicator Indicator of recent faecal contamination Potential for typing *E. coli* to aid in sourcing faecal contamination Rapid identification possible if defined as β-glucuronidase-producing bacteria	Possibly not suited to tropical waters due to growth in soils and waters
Sanitary plastics	Little training of staff required and immediate assessment can be made for each bathing day Can be categorised	May reflect old sewage contamination and thus be of little health significance Subjective and prone to variable description
Preceding rainfall (12, 24, 48 or 72 h)	Simple regressions may account for 30–60% of the variation in microbial indicators for a particular beach	Each beach catchment may need to have its rainfall response assessed Response may depend on the period before the event
Sulphite-reducing clostridia[1]	Inexpensive assay with H_2S production Always in sewage impacted waters Possibly correlated with enteric viruses and parasitic protozoa	Enumeration requires anaerobic culture May also come from dog faeces May be too conservative an indicator
Somatic coliphages	Standard method well established Similar physical behaviour to human enteric viruses	Not specific to sewage May not be as persistent as human enteric viruses May grow in the environment
F-specific RNA phages	More persistent than some coliphages Standard ISO method available Host does not grow in environmental waters below 30°C	WG49 host may lose plasmid (although F-amp is more stable) Not specific to sewage Not as persistent in marine waters

Continued

Table 9.5 Continued

Indicator	Advantages	Disadvantages
Bacteroides fragilis phages	More resistant than other phages in the environment and similar to hardy human enteric viruses Appears to be specific to sewage ISO method recently published	Because numbers in sewage are lower than for other phages and most do not excrete this phage, it is of limited value for small populations Requires anaerobic culture
Faecal sterols	Coprostanol largely specific to sewage Coprostanol degradation in water similar to die-off of thermotolerant coliforms A ratio of 5β: 5α stanols > 0.5 is indicative of faecal contamination; i.e. a ratio coprostanol:5α-cholestanol of > 0.5 indicates human faecal contamination, while C_{29} 5β(24-ethyl-coprostanol): 5α stanol ratio of > 0.5 indicates herbivore faeces Ratio of coprostanol:24-ethylco-prostanol can be used to indicate the proportion of human faecal contamination, which can be further supported by ratios with faecal indicator bacteria (Leeming *et al.*, 1996)	Requires expensive gas chromato-graphy (about US$ 100 per sample) Requires up to 10 litres of sample to be filtered through a glass fibre filter to concentrate particulate stanols
Caffeine	May be specific to sewage, but unproven to date Could be developed into a dipstick assay	Yet to be proven as a reliable method
Detergents	Relatively routine methods available	May not be related to sewage (e.g. industrial pollution)
Turbidity	Simple, direct and inexpensive assay available in the field	May not be related to sewage; correlation must be shown for each site type
Cryptosporidium[2]	Required for potential zoonoses, such as *Cryptosporidium* spp., where faecal indicator bacteria may have died out, or not present	Expensive and specialised assay (e.g. Method 1622, US EPA); human/animal speciation of sero-types is not currently defined

[1] *Clostridium perfringens* [2] Animal-sourced pathogens

Conversely, the rate of die-off of enterococci does not increase as the intensity of sunlight increases. Other investigators have observed similar effects of sunlight on indicators. Although human viruses have not been examined under similar experimental conditions, viruses of *E. coli* (coliphages) have been tested and they react very much in the same manner as enterococci. If

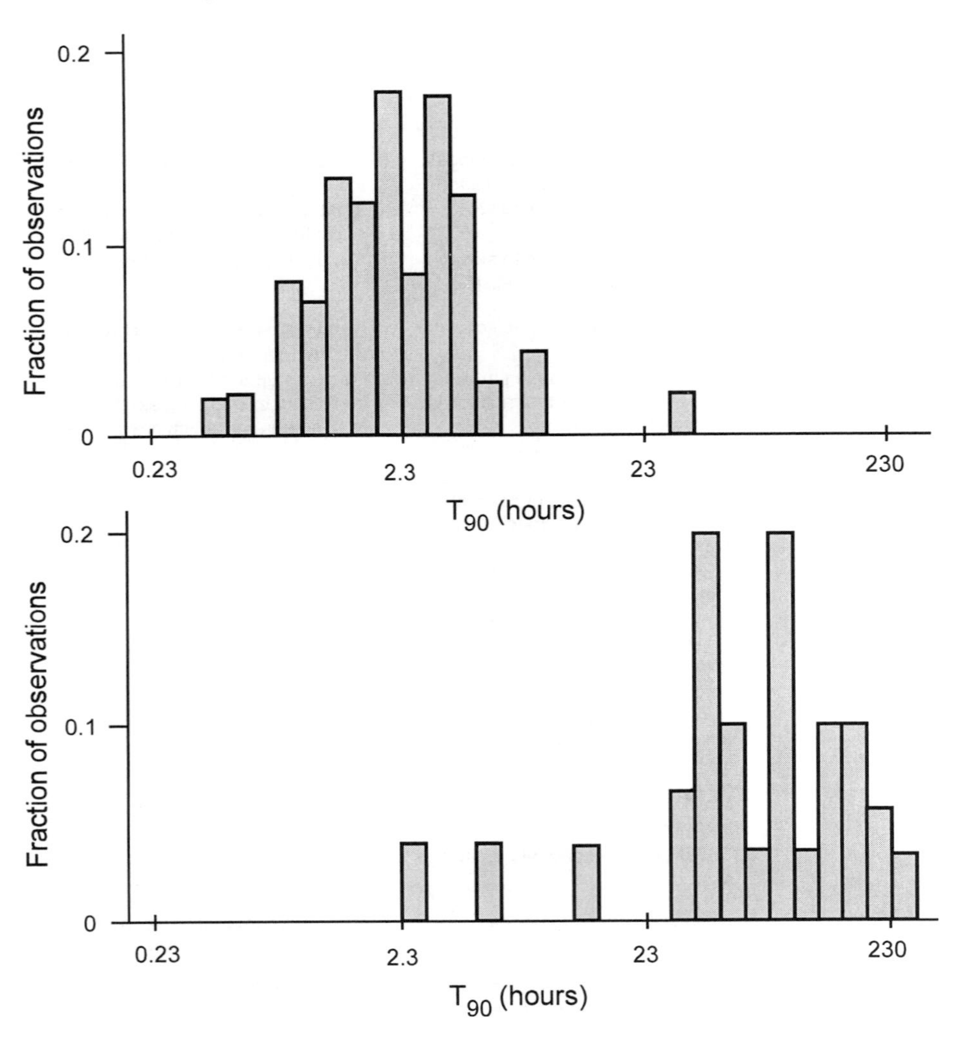

Figure 9.1 Survivial of coliforms in marine and fresh waters (Adapted from Chamberlain and Mitchell, 1978)

human viruses react to sunlight in a manner similar to bacterial viruses (phages) this would provide yet another explanation why enterococci are superior to *E. coli* as a predictor of human health risk at bathing beaches.

Effects of chlorine

Enterococci and *E. coli* are both sensitive to chlorine, although enterococci are somewhat more resistant to this disinfectant than *E. coli*. For example, to achieve a two-log removal (i.e. 99 per cent removal), reported calculated CT

Table 9.6 Decay rate estimates for *E. coli* and enterococci in sea water and fresh water

	Decay rate[1] (days)		
Indicator	Fresh water	Sea water	Reference
E. coli	3.9[2]	0.8[2]	
	6.3		Bitton *et al.*, 1983
	2.7		McFeters and Stuart, 1974
	3.1		Keswick *et al.*, 1982
	4.6	0.8	Hanes and Fragala, 1967
		0.7	Omura *et al.*, 1982
Enterococci	4.4[2]	2.5[2]	
	34.7		Bitton *et al.*, 1983
	4.2		McFeters and Stuart, 1974
	4.5		Keswick *et al.*, 1982
	3.0	2.4	Hanes and Fragala, 1967
		2.6	Omura *et al.*, 1982

[1] Time required for 90% of the population to die off, in days

[2] Median values

Table 9.7 Survival of enteroviruses in sea water and river water

	Die-off rate (days)[1]	
Virus strain	Sea water	River water
Polio I	8	15
Polio II	8	8
Polio III	8	8
ECHO 6	15	8
Coxsackie	2	2

[1] Maximum number of days required to reduce the virus population by 3 log values

Source: Adapted from Cioglia and Loddo, 1962

values for *E. coli* (Conc. of disinfectant (mg l^{-1}) × Contact time (mins)) are in the range of 5 mg min l^{-1} compared with 120 mg min l^{-1} for *S. faecalis*. Enterococci survival may therefore be more similar to that exhibited by faeces-carried pathogens than that of *E. coli*. This differential resistance to disinfection is another factor that influences the effectiveness of indicator bacteria in surface waters where disinfection of wastewaters by chlorine is practised.

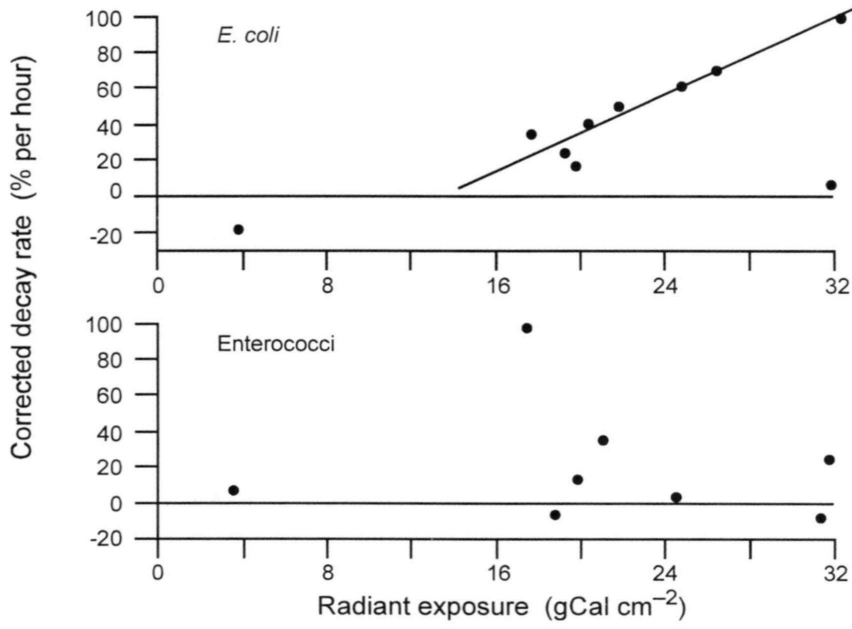

Figure 9.2 The effect of solar radiation on the die-off of *E. coli* and enterococci (Adapted from Sieracki, 1980)

Rainfall

Rainfall can have a significant effect on indicator densities in recreational waters increasing the densities to high levels because animal wastes are washed from forest land, pasture land and urban settings, or because treatment plants are overwhelmed causing sewage to by-pass the treatment process. In either case, the effect of rainfall on beach water quality can be quite dramatic (Figure 9.3) (Calderon, 1990 Pers. Comm.). The effect, illustrated in Figure 9.3, on a beach surrounded by forests, was very rapid and usually persisted for 1–2 days. The highly variable effect of rainfall on water quality can result in the frequent closing of beaches. The important question is whether high indicator levels that result from animal wastes carried to surface waters by rain water run-off, indicate the same level of risk to swimmers as would exist if the source of the indicators was a sewage treatment plant. There are conflicting reports in the literature with regard to risk associated with exposure to recreational water contaminated by animals.

Sources of indicators

Coliforms and thermotolerant coliforms are known to have sources other than mammalian enteric systems. These two indicator groups can grow to very high densities in industrial wastewaters, such as those discharged by pulp and

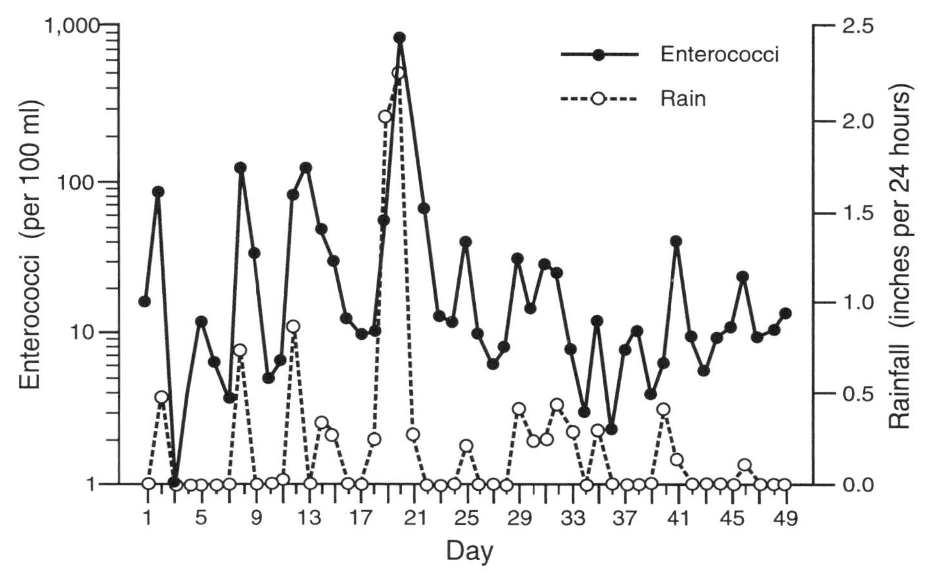

Figure 9.3 The effect of rainfall on enterococci densities in bathing beach waters (After Calderon, Pers. Comm.)

paper mills. *E. coli* and enterococci are not usually associated with industrial wastewaters, but some investigators believe that these indicators can grow in soil in tropical climates. Under any of these conditions, where the source of the indicator is other than the faeces of warm-blooded animals, it is questionable whether the indicator would have any value as a measure of faecal contamination of recreational waters.

The most commonly used indicators for surface water quality, *E. coli* or faecal coliforms and enterococci or faecal streptococci, can readily be detected in the faeces of humans, other warm-blooded animals and birds. The broad spectrum of animals in which enterococci can be found is shown in Table 9.8. This list is not exhaustive, but helps to illustrate that there are many non-human sources of enterococci. This issue is closely related to rainfall because, if it can be shown that the risk of exposure to water contaminated by animals is significantly less than that contaminated by humans, the way in which water quality is currently measured may have to be changed considerably. Methods for distinguishing human from animal-derived faecal matter are described in Chapter 8.

9.1.4 Pollution abatement and water quality

Beaches, especially near urban areas, are often subject to pollution from sewage and industrial discharges, combined sewer overflows (CSO) and urban run-off. Pollution abatement is, therefore, a key part of coastal zone

Table 9.8 Occurrence of enterococci in human faeces and in faeces of other warm-blooded animals

Species	Total number of subjects	% of subjects with samples containing	
		E. faecalis	*E. faecium*
Humans	32	41	88
Dogs	21	29	76
Puppies	2	100	100
Cats	1	–	–
Kittens	2	100	100
Pigs	22	77	100
Piglets	3	33	100
Horses	6	50	33
Sheep	4	100	100
Cows	15	–	73
Chickens	13	92	100
Goats	2	100	100
Beavers	3	–	–

management aimed at minimising health risks to bathers and ecological impacts. Pollution abatement measures for sewage may be grouped into three wastewater disposal alternatives: treatment, dispersion through sea outfalls and discharge to non-surface waters (i.e. reuse, in which wastewater is stored and then used for agricultural or other purposes, or groundwater injection). In practice, there are numerous anomalies to these general categories. In addition, CSOs and sanitary sewer overflows (SSO) usually occur as a result of excessive rainfall events and can result in high human health risks for certain beach zones. Pollution abatement alternatives for these overflows, such as holding tanks, separate storm overflow submarine outfalls, over-design of sewer systems for extreme storm events, etc., are often prohibitively expensive and difficult to justify. In view of the costs of control, it may be preferable for integrated beach zone management to focus on restricting beach use or, at the very minimum, warning the public of the potential health risks during and after high risk events.

Treatment

For large urban communities, at least secondary or tertiary sewage treatment plants with disinfection are necessary for onshore or near-shore discharges to protect nearby recreational areas. Public health risks can vary depending on the operation of the plant and the effectiveness of disinfection. Smaller communities with lesser population densities usually apply treatment by means of septic tank systems, latrines, etc. The ground acts as a filter for pathogenic organisms and, therefore, such disposal systems result in a very low health risk to recreational areas except where Karst topography occurs leading to the possibility of direct contamination.

The general removal levels of the major pathogen groups by conventional primary, secondary and tertiary sewage treatment are summarised in Table 9.9. The advent of new detection methods for a range of hardier enteric viruses may change views on the persistence of viruses that cannot be enumerated by culture-based methods. For example, identification of hepatitis A virus by antigen capture polymerase chain reaction (AC-PCR), followed by hybridisation on membranes, indicated their presence in raw sewage and secondary treatment effluent in 80 per cent and 20–30 per cent of samples respectively (Divizia *et al.*, 1998). Advanced sewage treatment based on ultra- and nanofiltration methods can also be an effective barrier to viruses, with over 10^6 removal (Otaki *et al.*, 1998), and other pathogens (Jacangelo *et al.*, 1995; Madireddi *et al.*, 1997). Additionally, revaluation of ultra violet (UV) (Oppenheimer *et al.*, 1997), ozone (Perezrey *et al.*, 1995) and disinfection kinetics (Haas *et al.*, 1996; Gyurek and Finch, 1998) are also changing the way in which engineers are evaluating disinfection and treatment processes.

Oxidation pond treatment may remove significant numbers of pathogens, particularly the larger protozoan cysts and helminth ova. However, short circuiting due to poor design, thermal gradients or hydraulic overloading may all reduce considerably the residence time from the typical 30–90 days. In addition to removal by sedimentation during long resident times, inactivation by sunlight and temperature, and predation by other micro-organisms may reduce faecal bacterial numbers by 90–99 per cent (Yates and Gerba, 1998). Inactivation of viral and parasitic protozoa is also influenced heavily by temperature. For example, poliovirus type 1 may be inactivated by 99 per cent in 5 days in summer but may take 25 days in winter (Funderburg *et al.*, 1978). The cysts and oocysts of *Giardia* and *Cryptosporidium* may take at least 37 days to achieve a 99.9 per cent reduction (Grimason *et al.*, 1992, 1996b), whereas the larger ova of helminths may be totally removed in 12–26 days (Grimason *et al.*, 1996a).

Long sea submarine outfalls

Long sea outfalls are assumed to be properly designed and of sufficient length, diffuser discharge depth and design to ensure a low probability of the sewage plume reaching designated beach zones. As such, the long sea outfall is a very low human health risk alternative because the bather is unlikely to come into physical contact with the sewage, whether treated or untreated. Modern diffusers are usually designed to achieve minimum, near-field,

Table 9.9 Pathogen removal during sewage treatment

Level of treatment	Enteric viruses	*Salmonella*	*C. perfringens*	*Giardia*
No treatment (raw sewage)				
No. remaining (per litre)	10^5–10^6	5,000–80,000	10^5	9,000–200,000
Primary treatment[1]				
% removal	50–98.3	95.5–99.8	30	27–64
No. remaining (per litre)	1,700–500,000	160–3,360	70,000	72,000–146,000
Secondary treatment[2]				
% removal	53–99.92	98.65–99.996	98	
No. remaining (per litre)	80–470,000	3–1,075	2,000	
Tertiary treatment[3]				
% removal	99.983–99.9999998	99.99–99.9999995	99.9	98.5 to 99.99995
No. remaining (per litre)	0.007–170	0.000004–7	100	0.099–2,951

1 Physical sedimentation
2 Primary sedimentation, trickling filter/activated sludge and disinfection
3 Primary sedimentation, trickling filter/activated sludge, disinfection, coagulation-sand filtration and disinfection; note that tertiary treatment does not involve coagulation–sand filtration and the second disinfection step in the case of *C. perfringens*

Sources: Long and Ashbolt, 1994; Yates and Gerba, 1998

immediate dilutions of 100 to 1 that would reduce the concentration of organics and nutrients in the sewage to levels that would have no adverse ecological effects in an open ocean situation. Higher dilutions are achieved most of the time, depending on the current structure. Under stratified conditions, complete sewage plume submergence can occur and can reduce further the possibility of sewage reaching designated beach zones. The diffuser length, depth and orientation, as well as the area and spacing of the discharge ports, are the key design considerations (Roberts, 1996). For pathogenic and indicator organisms, additional order-of-magnitude reductions may be required to meet established bathing beach water quality criteria, depending on the degree of treatment and disinfection. This far-field "dilution" is achieved through additional physical dilution and mortality in the ocean environment subsequent to discharge. The design distance required, i.e. length of the outfall, to achieve the additional far-field reduction is determined by the dominant current structure and mortality rates (T_{90}).

Pre-treatment with milliscreens with apertures of 1–1.5 mm, is considered to be the minimum treatment required to remove floating matter and thus avoid aesthetic impacts on the designated beach zones. For the same aesthetic considerations, removal of grease and oil should be implemented at source, especially if effluent concentrations are high and not reduced sufficiently after initial dilution. To avoid possible ecological impacts in the vicinity of the discharge, more advanced treatment may be justified.

Discharge to non-surface waters

Reuse of wastewater and groundwater recharge are two methods of sewage disposal that have minimal impact on recreational waters. In arid regions, sewage (after appropriate treatment) can be an important resource for agricultural purposes such as crop irrigation. Reuse has the dual benefit of the productive use of sewage while avoiding wasteful discharges to the marine environment with the inherent pollution potential. Direct injection of sewage below ground for groundwater recharge is practised in some regions of the world, usually in combination with advanced treatment. Groundwater injection is a no (or very low) human health risk option for designated beach zones except in areas with Karst topography.

9.1.5 Hydrological considerations

Rivers contribute a significant proportion of the bacterial load to coastal beach areas. In some regions, significant numbers of freshwater beaches are directly affected by river water quality. The bacterial concentration in river water is determined by faecal pollution from point and non-point or diffuse sources. Major point sources include sewage effluents, CSOs, industrial effluents and confined animal sources such as feedlots. Non-point sources

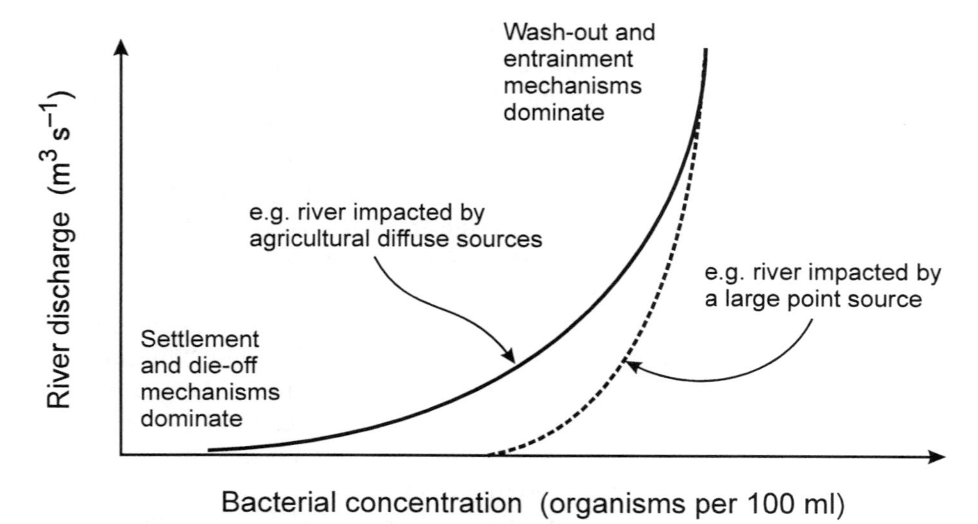

Figure 9.4 The relationship between river discharge and bacterial concentration

relate directly to agricultural activity within the watershed, and are influenced primarily by the type of livestock and its density. A significant contribution is also derived from urban surfaces.

The transport of microbial contaminants through the watershed to the river and subsequently through the river system to the marine environment is controlled by the flow of water and therefore, rainfall is a key influence on concentrations (see section 9.1.3). Faecal material is transported from the watershed surface to the river and changes in flow are determined by rainfall and by the hydrological characteristics of the basin (soils, bedrock, etc.) which therefore have a significant impact on the total flux of microbes transported. In river water, the decrease in bacterial concentrations downstream of a source, conventionally termed "die-off", largely reflects the settlement or sedimentation of organisms to the river bed. In riverbed sediments, survival times are increased significantly and the bacteria are readily resuspended when the river flow increases. All rivers demonstrate a close correlation between flow and bacterial concentration due to the increased supply of bacteria from watershed surfaces and some point sources (e.g. CSOs) during rainfall events (Figure 9.4). The two curves represent hypothetical examples. In reality, all rivers will exhibit individual relationships depending on their hydrological characteristics and bacterial sources. The shape of the flow relationship will be variable between different catchments and may also break down during prolonged high flows if the store of organisms in the bed-sediment (or the catchment surface) is exhausted. This phenomenon, however, has only been documented for small streams dominated by diffuse inputs and is less likely to occur on major rivers with multiple point and non-point sources. The processes controlling transport and fate of bacteria in watersheds are now well understood and river water bacteria concentrations can be modelled and predicted (see section 9.4.1).

9.2 Alternative approaches to monitoring and assessment programmes

The experts who met in Annapolis in November 1998 (see Acknowledgements) agreed that an improved approach to the regulation of recreational water, reflecting health risk more reliably and providing enhanced scope for effective management intervention, was necessary and feasible. The major output of the meeting was the development of such an approach, which is described in this section. Because this approach is so different to established practice it includes elements that require substantial testing. The description provides sufficient detail to enable field testing but should be amended to take account of specific local circumstances. The approach will be refined further as experience with implementation accumulates. This chapter also sets out principles for the design of an intensive assessment for evaluating the modified approach and studying relationships between factors that affect beach water quality and the ability of monitoring schemes to detect these changes. The Annapolis group would like to encourage pilot testing of this approach, and is interested in receiving the results of any such studies. The proposed approach leads to a classification scheme through which a beach would be assigned to a class (i.e. very poor, poor, fair, good or excellent) based upon health risk. By enabling local management to respond to sporadic or limited areas of pollution and thereby upgrade a beach's classification, it provides a significant incentive for local management actions as well as for pollution abatement. The classification scheme provides a generic statement of the level of risk and indicates the principal management and monitoring actions likely to be appropriate. The advantage of a classification scheme, as opposed to a pass or fail approach, lies in its flexibility. A large number of factors can influence the condition of a given beach. A classification system reflects this, and allows regulators to invoke mitigating approaches for beach management.

The most robust, accurate and feasible index of health risk is provided by a combination of a measure of a microbiological indicator of faecal contamination with an inspection-based assessment of the susceptibility of an area to direct influence from human faecal contamination. This reflects two principal factors. Firstly, high counts of faecal indicator bacteria may be caused by either human faecal contamination or contamination from other sources. In general, sources other than human faecal contamination present a significantly lesser risk to human health, and by adopting a combined classification it is possible to reflect this modified risk. Secondly, any microbiological

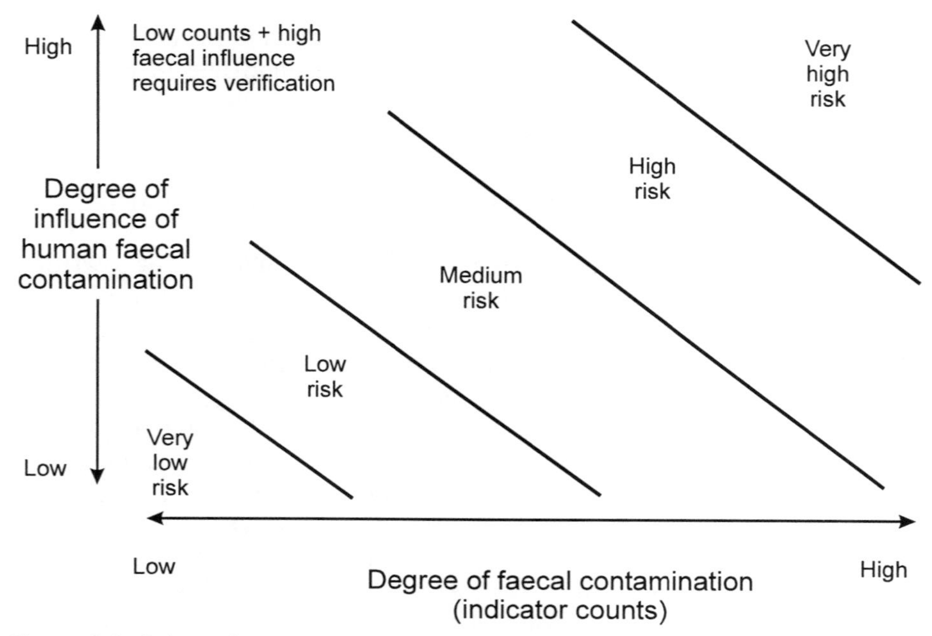

Figure 9.5 Schematic representation of classes of health risk

analytical result provides information on only a moment in time, whereas microbiological quality may vary widely and rapidly even within a small area (see section 9.1). It is possible to perform a large number of analyses to obtain an improved evaluation of the situation, but with concomitant cost. However, information concerning the existence of sources of contamination and their likely influence upon the recreational water use area provides a robust and rapid means of increasing the reliability of the overall assessment. This would lead to a series of classes of relative risk as presented schematically in Figure 9.5. The strengths of such an approach are demonstrated by the case study presented in Box 9.1.

Variation in water quality may occur in response to events (such as rainfall) with predictable outcomes, or the deterioration may be constrained to certain areas or sub-areas of a single beach. It is possible to discourage use of areas that are of poor quality, or to discourage use at times of increased risk. In addition, if success in discouraging bathing at times of risk can be demonstrated, it might be reasonable to up-grade the classification of a beach. Because measures to predict and discourage use at certain times or in areas of elevated risk may be inexpensive, greater cost-benefit and greater possibilities for effective local management intervention are possible.

Figure 9.6 illustrates the process for assigning a classification to a given beach. The two principal components of the scheme are:

Box 9.1 A risk management approach to beach closure in Southern California

During February 1992, a severe winter storm battered the southern California coastline. Winds, high surf and the deluge of rain led to much damage. One casualty of the storm was a pipe that carried treated wastewater from 200,000 homes and businesses to the ocean for disposal. Following the storm, divers confirmed that the 48-inch diameter pipe was broken and about 250,000 gallons per day of non-disinfected secondary treated wastewater were leaking into 10 feet depth of water approximately 90 feet from shore. Water samples were collected directly above the broken pipe and at the adjacent swimming and surfing beach which was used all year round. Coliform concentrations in the samples directly above the pipe break exceeded State standards for recreational water contact, whereas the samples at the beach did not.

In spite of the relatively low coliform densities at the beach, the local Health Officer closed the beach because of the discharge of non-disinfected wastewater. The Health Officer was of the opinion that even though State coliform standards were not exceeded at the beach it did not mean viruses that cause gastro-intestinal illness, hepatitis or polio were not present. The Health Officer's concern stemmed from the fact that activated sludge treatment alone is only between 90 and 95 per cent efficient in removal of human enteric viruses. Sampling at two local treatment facilities had demonstrated human enteric virus levels in secondary treated wastewater to be between 5 and 50 infectious units per gallon. Even with dilution and dispersion of the indicator bacteria to below State standards, a discharge known to contain human enteric viruses constituted an unacceptable risk to this particular Health Officer. In closing the beach, the Health Officer took a risk management approach to swimmer health protection, namely to prevent contact with waters known to contain faecal contamination, regardless of the density of wastewater "indicator bacteria" measured during water testing.

- A primary classification based upon the combination of evidence for the degree of influence of human faecal material (a sanitary inspection) alongside counts of suitable faecal indicator bacteria (a microbiological quality assessment).
- The possibility of "reclassifying" a beach to a higher (better) class if effective management interventions are deployed to reduce human exposure at certain times or in places of increased risk.

9.3 Primary classification

The primary classification is based upon the combination of an inspection-based assessment of the area's susceptibility to influence from human faecal contamination and a microbiological indicator measure of faecal contamination.

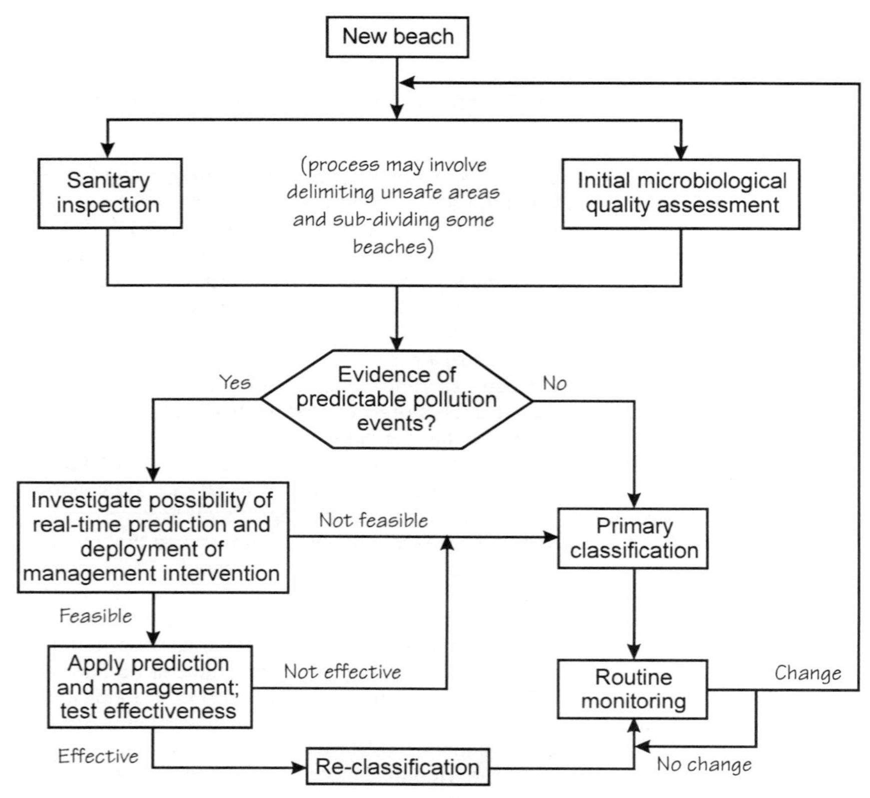

Figure 9.6 The steps to be taken for assigning a classification to a new beach or location

9.3.1 Sanitary inspection

Sanitary inspection is the evaluation of the principal sources of faecal pollution. The three most important sources of human faecal contamination of bathing beaches for public health purposes are:

- Sewage, including CSO and storm water discharges.
- Riverine discharges, where the river is a receiving water for sewage discharges and is used directly for recreation or discharges near a coastal or lake area used for recreation.
- Bather contamination, including excreta.

All of these sources lead to the presence of faecal indicators that may be recovered and that can provide a semi-quantitative estimate of health risk, as has been demonstrated by many epidemiological studies (WHO, 1998).

Sources of faecal indicators other than human sewage also exist, such as drainage from areas of animal pasture and intensive livestock rearing. In general, due to the "species barrier", the density of pathogens of public health importance is assumed to be less in aggregate in animal excreta than in human excreta and may therefore represent a significantly lower risk to human health. As a result, the use of faecal indicator bacteria alone as an index of risk to human health may overestimate risks significantly where the indicators derive from sources other than human excreta. Nevertheless, the human health risk associated with pollution of recreational waters from animal excreta is not zero and some pathogens, such as *Cryptosporidium,* can be transmitted through this route.

The experts at the Annapolis meeting ranked qualitatively the relative risk to human health through direct sewage discharge, riverine discharge contaminated with sewage and bather contamination. In doing so they took account of the likelihood of human exposure and the degree of treatment of sewage. In taking account of sewage discharges to recreational areas and of rivers, account was also taken of the pollutant load, using population as an index. While in many circumstances several contamination sources would be significant at a single location, the approach adopted was to categorise a beach according to the single most significant source of pollution. Even two sources of similar magnitude would, on aggregate, increase exposure by a factor of two which, in microbiological terms, is of very limited significance. The methodology for designing and conducting a sanitary inspection is described in Chapters 2 and 8.

Sewage discharges

Sewage discharges or outfalls may be classified into three principal types:

- Discharges directly to the beach (above low water level in tidal areas).
- Discharges through "short outfalls", where the discharge is into the water but sewage-polluted water is likely to contaminate the beach area.
- Discharges through long sea outfalls, where the sewage is diluted and dispersed and is unlikely to pollute bathing areas.

Although the terms "short" and "long" are often used in relation to outfalls, length is generally less important than proper location and effective diffusion that will ensure the pollution is unlikely to reach bathing areas. A short outfall is assumed to be a discharge to the inter-tidal zone, with a significant probability of the sewage plume reaching the designated beach zone. For short outfalls, the relative risk is increased, based upon the size of the contributing population. An effective outfall is assumed to be properly designed, with sufficient length and diffuser discharge depth to ensure low probability of the sewage plume reaching the designated beach zone. Urban storm water run-off and outputs from CSOs are included within the scheme under the category of direct beach outfalls.

The classification is based upon a qualitative assessment of risk of contact or exposure under "normal" conditions with respect to operation of sewage

Table 9.10 Potential human health risks arising from exposure to sewage

	Discharge type		
Level of treatment	Directly on beach	Short outfall[1]	Effective outfall[2]
None[3]	Very high	High	NA
Preliminary	Very high	High	Low
Primary (including septic tanks)	Very high	High	Low
Secondary	High	High	Low
Secondary plus disinfection	Medium	Medium	Very low
Tertiary	Medium	Medium	Very low
Tertiary plus disinfection	Very low	Very low	Very low
Lagoons	High	High	Low

1 The relative risk is modified by population size; relative risk is increased for discharges from large populations and decreased for discharges from small populations

2 Assumes that the design capacity has not been exceeded and that climatic and oceanic extreme conditions are considered in the design objective (i.e. no sewage on the beach zone)

3 Includes combined sewer overflows

treatment works, hydrometeorological and oceanographic conditions. The potential risk to human health through exposure to sewage can be categorised as shown in Table 9.10. The sewage effluent treatments listed in this table are classified as no treatment (raw sewage); preliminary (filtration with milli- or micro-screens); primary (physical sedimentation); secondary (primary sedimentation and high rate biological processes such as trickling filter or activated sludge); secondary with disinfection; tertiary (advanced wastewater treatment, including primary sedimentation, trickling filter or activated sludge, and coagulation or sand filtration); tertiary with disinfection; and lagoons (low rate biological treatment). Septic tank systems are assumed to be equivalent to primary treatment.

Riverine discharges

Riverine discharges are categorised with respect to the sewage effluent load and the degree of dilution, as illustrated in Table 9.11. Effluent load is characterised by the total human population in the watershed or catchment above the beach or estuary. The population of relevance is the peak population which, in many recreational water use areas, will be significantly greater than the resident population and is likely to occur during weekends and local holidays during the summer season. Dilution is defined by the "dry weather" river flow or discharge during the bathing season. Use of dry weather flow is a "worst case" approach and coincides with reality where the bathing season is

Table 9.11 Potential human health risks arising from exposure to sewage through riverine flow and discharge

	Treatment level				
Dilution effect[1,2]	None	Primary	Secondary	Secondary plus disinfection	Lagoon
High population with low river flow	Very high	Very high	High	Low	Medium
Low population with low river flow	Very high	High	Medium	Very low	Medium
Medium population with medium river flow	High	Medium	Low	Very low	Low
High population with high river flow	High	Medium	Low	Very low	Low
Low population with high river flow	High	Medium	Very low	Very low	Very low

1 The population factor takes account of all the population upstream from the beach and assumes no instream reduction in the hazard factor used to classify the beach

2 Stream flow is the 10 per cent flow during the period of active beach use; stream flow assumes no dispersion plug flow conditions to the beach

also the season of reduced flow. In many circumstances, the most significant sewage discharges are near to the coast and die-off during riverine travel is likely to be of limited significance for the travel times encountered in many rivers. Removal of pathogens through sedimentation may be of some significance but could not be accounted for reliably in a simple way. Resuspension of sediments and CSO discharges can be important during pollution episodes and in this context may be predictable (see section 9.4.1). Episodic input can dominate in areas subject to frequent summer rainstorms such as Northwest Europe.

In practice, several discharges into a single river course are likely to occur and where larger discharges are treated to a higher level, the smaller sources (including septic tank discharges) and CSOs may represent the principal source of concern. It is assumed that the discharge travels in a consolidated manner, with little mixing or dilution by the river water or little dispersion. The overall riverine discharge risk category is that accorded by the most significant single pollution source.

The classification can be used directly for freshwater beaches on the river and for beaches in estuarine areas or which are dominated by riverine pollution. For marine beaches the same classification may be used but it should be varied depending on the proximity of the river to the beach.

Table 9.12 Potential human health risks arising from exposure to sewage from bathers

Bather density/ dilution factor	Risk category	Bather density/ dilution factor	Risk category
High density		Low density	
High dilution[1]	Low	High dilution	Very low
Low dilution[1,2]	Medium	Low dilution[2]	Low

1 Move to the next highest risk category if no sanitary facilities are available at beach site

2 If no water movement

Bather shedding
While bather shedding is generally of lesser importance than sewage or riverine discharge, the resulting pollution is direct and fresh, and therefore potentially of great public health significance. Several studies (see section 9.1.2) have demonstrated accumulation of faecal material (as indicated by recovery of faecal indicator bacteria) during the course of a day, despite potentially enhanced die-off due to sunlight. Small volume areas of limited turnover are especially affected, such as bays and coastal and estuarine areas constrained by sandbars. The two principal factors of importance are therefore bather density and degree of dilution (Table 9.12). Low dilution is assumed to represent no water movement (such as occurs in lakes and lagoons and coastal embayments). The likelihood of bathers defaecating into the water is substantially increased if toilet facilities are not readily available. Where bather densities are high, the classification should therefore be increased to the next higher class if no sanitary facilities are available at the beach.

9.3.2 Microbiological quality assessment

Sewage contamination may be identified by a range of microbial, chemical or visual parameters as described in Table 9.5 and Chapter 8. Each gives a different view of the possible source(s) of contamination and should be used appropriately in a staged approach for assessing sewage contamination of bathing beaches. Hence, in addition to identifying which indicators to use, it is also important to identify action levels for the primary indicators selected to assess beaches. A further issue is the number of samples required to make an assessment, taking into account the variability of the beach site under study.

A basic selection of sewage indicators called "primary indicators" is proposed as an essential first step in the evaluation of bathing water. Table 9.13 tabulates primary indicators for marine and fresh water. "Secondary

Table 9.13 A beach categorisation scheme based on the concentrations of primary indicators of sewage contamination in marine and fresh waters

Water source	Indicator(s)	Category	95th percentile
Temperate marine water	Faecal streptococci Enterococci[1]	A	< 10
		B	11–50
		C	51–200
		D	201–1,000
		E	> 1,000
Alternative for tropical marine waters[2]	Sulphite-reducing Clostridia *Clostridium perfringens*	A	< 1
		B	1–10
		C	11–50
		D	51–80
		E	> 80
Temperate fresh water[3]	Faecal streptococci Enterococci[1]	A	< 10
		B	11–50
		C	51–200
		D	201–1,000
		E	> 1,000
	E. coli	A	< 35
		B	36–130
		C	131–500
		D	501–1,000
		E	> 1,000
Optional for tropical fresh water[2]	Sulphite-reducing Clostridia *Clostridium perfringens*	A	< 1
		B	1–10
		C	11–50
		D	51–80
		E	> 80

[1] Source for faecal streptococci/enterococci 95th percentile ranges: WHO, 1998
[2] Based on preliminary data
[3] While studies suggest that there is a differential die-off rate for microbial indicators in marine and fresh waters (see section 9.1.3), current data are not sufficient to derive separate 95th percentiles for freshwater environments; the above faecal streptococci/enterococci percentiles are therefore based on data obtained from marine studies, but may be reconsidered when further freshwater studies have been conducted

indicators" are described for follow-up analysis to assist in the assessment and management of faecal contamination at beaches.

Primary indicators

The minimal non-microbial, primary indicators of faecal contamination in marine environments are sanitary plastics and grease. Although somewhat crude indicators, they have been used as aesthetic health indicators because they are associated with faecal contamination. In freshwaters, sanitary plastics may also act as non-microbial primary indicators, whereas grease will not fulfil such a role.

The primary microbial indicators identified are faecal streptococci and enterococci (temperate marine and freshwaters), *E. coli* (temperate freshwaters) and sulphite reducing clostridia, i.e. *Clostridium perfringens* (temperate and tropical marine and freshwaters). Table 9.13 provides an example of beach categorisation, with "A" representing excellent water quality and "E" designating a beach with unacceptable water quality. A single sample result greater than the unacceptable 95th percentile requires follow-up action, such as a sanitary inspection, to verify that it is a statistical occurrence and not due to a real change in exposure.

Secondary indicators

Secondary indicators aimed at identifying the source of faecal contamination should include sulphite reducing clostridia (*Clostridium perfringens*) in temperate waters. Consideration must be given to the fact that dog excreta from surface run-off may be a source of these organisms, moreover it may be the only significant source other than humans. Other secondary indicators in temperate marine waters include faecal sterols and bacteriophages, such as the F-RNA serogroups I and IV for humans or phages to *Bacteroides fragilis* HSP40. In freshwaters, secondary indicators include faecal sterols and phages as above, but further potential secondary indicators include turbidity and phosphate and ammonium levels.

Measurement of indicators

Although the detail in the available literature varies considerably, the incidence of swimming related illness generally increases with the level of sewage contamination suggested by traditional bacterial indicators. There are few consistent relationships between individual indicator organisms and sewage load, and even fewer consistent relationships between individual indicators and particular pathogens. However, poorer quality water as indicated by total and thermotolerant coliforms, *E. coli*, faecal streptococci and enterococci is consistently associated with increased risk to the health of those using the water for recreational purposes (WHO, 1998).

Various statistical procedures for analysing microbiological indicator counts are discussed in Chapter 8. Regulatory standards are based on the use of these statistical methods. Most regulatory approaches have adopted a percentage compliance approach, in which a given percentage (e.g. 95 per cent) of the sample measurements taken must lie below a specific value in order to meet the standard. This simple percentage does not incorporate within its derivation the probability density function that describes the distribution of indicator organisms at a particular sampling location. The most important weakness in such an approach is that it fails to take account of the overall body of data. Some other approaches, such as use of the geometric mean or percentile values, are less affected by individual data.

The statistic most commonly used as a measure of compliance in the USA has been the geometric mean. By definition a mean is a measure of central tendency. As such, the mean is a statistic around which individual measurements tend to cluster. In the context of water quality monitoring, use of the mean will result in a situation in which the higher indicator organism measurements become obscured by the properties inherent in the calculation of the mean. Use of the geometric mean will further obscure extreme values. The median, another measure of central tendency, has an even greater effect on obscuring the higher levels of individual measurement contained within its derivation.

In contrast, a percentile value may be calculated by using the probability density function that describes the series of measurements taken. In this manner, the percentile value describes the distribution of indicator organism measurements at a particular location. Therefore, inherent in the calculation of the percentile value is the distribution of the entire series of measurements taken, resulting in a more accurate description of indicator organism densities at a particular location. Chapter 8 contains an example of a percentile calculation, using a log normal distribution (see section 8.7.3).

The categorisations in Table 9.13 are based on a minimum of 20 samples of the suggested microbial indicator(s). As the 95th percentile values were derived from limited studies, they are provisional and are meant to serve as a general guideline rather than as a standard. The categorisations should be treated as examples, and individual beaches should be evaluated based on site-specific conditions.

Microbiological categorisation sampling protocol

Figure 9.7 and Box 9.2 illustrate the steps necessary to assign a primary microbiological categorisation to a given beach.

9.3.3 Determination of primary classification

Obtaining a primary classification for a given beach incorporates the results of both the sanitary inspection and the initial microbiological quality assessment described above. Once the appropriate categories for each of these criteria have been determined, a lookup table such as that in Table 9.14 can be used to determine the primary classification for the beach.

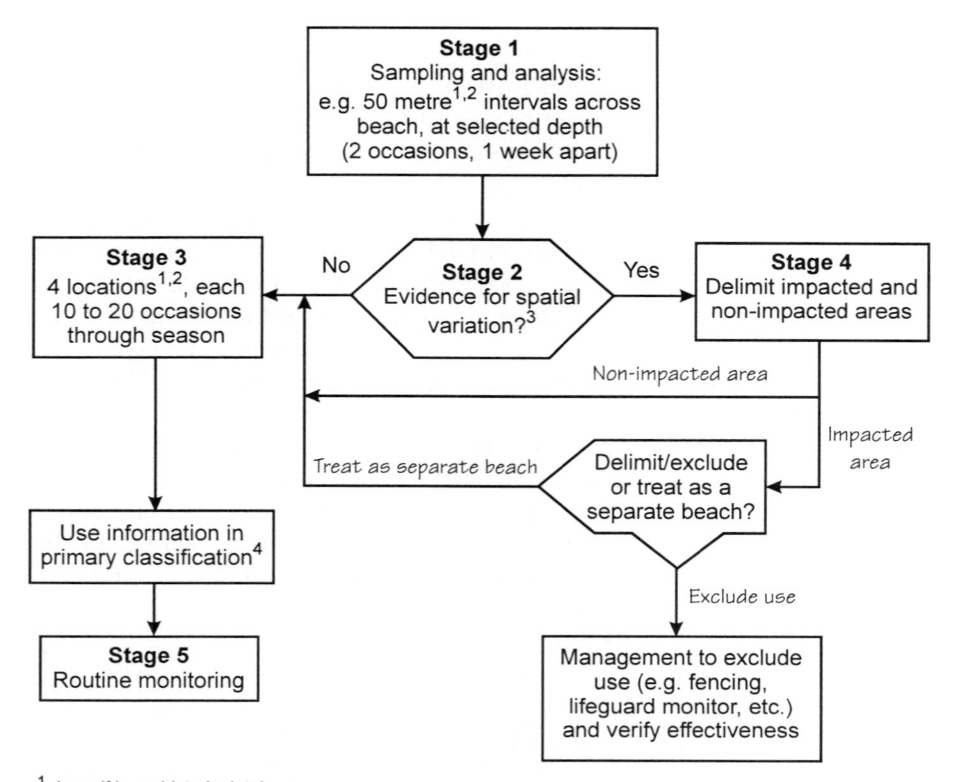

1 Less if large historic database
2 Modified by sanitary inspection
3 For example, across full bandwidth of microbiological categories
4 If variation in quality is recognised then re-classification as described in Section 9.4 may be applicable.

Figure 9.7 An example sampling protocol for primary microbiological categorisation

9.4 Reclassification

Microbiological contamination varies widely and rapidly. In addition, the risks to human health are associated principally with periods of high contamination. Thus:

- where a bathing area is subject to elevated faecal contamination for a limited proportion of the time or over a limited area of the potential bathing areas; and
- where the times of contamination can be predicted in some way; and
- where management interventions can be applied which effectively reduce or prevent exposure at these times,

it is reasonable to modify the beach risk evaluation to take account of the reduction in risk. This approach requires a database that allows an estimation of whether the significant faecal influence is constrained in time and whether

Box 9.2 Example of the practical application of the primary microbiological categorisation protocol

STAGE 1

1. Full width of beach intended for recreational use delimited.
2. Along this full width, collect samples at a selected depth at 50 m intervals on two occasions one week apart at the start of the bathing season. The timing of the sampling should take into account the likely period of maximum contamination from local sewage discharges and bather shedding (i.e. the day after peak numbers of visitors).
3. Concurrently collect sanitary inspection data as described in Chapter 8.

STAGE 2

1. Use Stage 1 data to assess spatial variation.
2. If no significant spatial variation, move to Stage 3.
3. If spatial variation indicated, move to Stage 4.

STAGE 3 (if no spatial variation observed in Stage 2)

1. Select four evenly distributed sampling locations at no greater than 500 m intervals. If the beach is in excess of 2 km in length, include further sampling locations.
2. Conduct microbiological sampling at each of the four locations on 10–20 occasions at equal time intervals throughout one bathing season.
3. At the end of the year, assess Stage 3 data in conjunction with Stage 2 data plus outcomes of sanitary assessments to determine whether there is any significant variation (e.g. in response to rainfall).
4. If significant variation, then assess possibility of reclassification (see Section 9.4). Otherwise, confirm primary classification and proceed to routine monitoring (Stage 5).

STAGE 4 (if spatial variation is found in Stage 2)

1. If spatial variability is exhibited, affected and unaffected zones should be treated as separate bathing areas and each should be classified separately.

Continued

"predictors" can be used to determine when such conditions are likely to occur. In addition, a locally applicable early warning system and subsequent management action that can be deployed in real time, must be determined. Finally, in order for a reclassification to be applied, evidence of the effectiveness of management action is required. Consequently a reclassification should be provisional; although it may be confirmed if the efficacy of management interventions is verified during the initial season of provisional

Box 9.2 Continued

2. Determine the potential source and extent of the affected zone.
3. Delimit the unaffected zone; treat unaffected zone as in Stage 3 with one of the four identified sample locations at the poorer limit of the affected zone.
4. For the affected zone:
   - A monitoring regime for a zone exhibiting spatial variability and likely to be affected by sewage contamination depends on the extent of the zone.
   - It may be that the affected zone has to be managed by exclusion and that no monitoring is required, particularly if the zone is small in extent. Exclusion management action would apply where increased risk is restricted to a specific area. This implies, for example, fencing combined with general and site warning notices or general and site warning notices plus pro-active individual advice (such as from life-guards) not to use areas. The effectiveness of such management would need to be verified.
   - If the affected zone warrants monitoring, then the Stage 3 process must be replicated. In such a case, if the zone is relatively small in area, fewer sample locations may be selected but sampled more frequently to provide a minimum of 20 data points.
5. At the end of the year, all data from a given zone are used to determine the primary classification to be applied.

STAGE 5

In the following year, microbiological monitoring is confined to five samples at each of the four identified locations within an individual zone (zones in excess of 2 km will require further sample locations). The five sampling occasions will be distributed evenly throughout the bathing season. A sanitary inspection should also be conducted. Routine monitoring requirements in subsequent years may vary, depending on the classification of the beach (section 9.5.5).

The individual data sets for the sampling locations will be further analysed to ensure that there is no significant difference between them. Assuming that no such variation is recognised, treat the data from all years as a single statistical body.

reclassification. As the outcome of this process is of significant economic importance, it should be a requirement to ensure independent audit and verification wherever feasible, in order to satisfy the conflicts of interest that may arise.

Note that it may be appropriate to add an additional dimension to the resulting risk assessment to take account of special groups with increased risk, either because of the activity in which they engage or because they seek out areas not used by traditional bathers. Surfers represent such a special

Table 9.14 Primary classification matrix

Sanitary Inspection Category	Microbiological Assessment Category (indicator counts)				
	A	B	C	D	E
Very low	Excellent	Excellent	Good	Good[2]	Fair[2]
Low	Excellent	Good	Good	Fair	Fair[2]
Moderate	Good[3]	Good	Fair	Fair	Poor
High	Good[3]	Fair[3]	Fair	Poor	Very poor
Very high	Fair[3]	Fair[3]	Poor[3]	Very poor	Very poor

1 Reflects susceptibility to faecal influence
2 Implies non-sewage sources of faecal contamination (e.g. livestock) and this should be verified
3 Indicates an unexpected result which requires verification

group. Alternatively, this may require an additional "commentary" element to the classification.

9.4.1 Simple predictive approaches

It is impossible to predict every type of event that may leave an effect on every beach, because the variation is enormous. However, using one key issue that consistently affects bathing water quality, it is possible to delineate the principles that apply when dealing with such events. The objective is to define the conditions under which increased detection of sewage contamination (and, by inference, risk to human health) can be predicted. Exposure to risk at these times may be reduced by direct interventions. If such interventions can be demonstrated to be effective, then upgrading the classification of the beach to reflect the reduced health risk can be justified.

The issue selected to illustrate this predictive approach is rainfall. To provide appropriate information for this process, rainfall data (real time and historic) must be available. The location of existing rain gauges can be surveyed to determine the optimal position from which to predict effects on the beach. In addition, to determine the effect that a rainfall event may have on a bathing water, sources of contamination to the beach must be categorised; primary inputs of concern are CSOs, riverine and storm drains. Examples of the type of information required for each input are given in Box 9.3.

A protocol can be adopted to investigate whether deterioration in water quality at recreational beaches is predictable and hence subject to appropriate management action. The assumption is that a local administration wishes to contend that a beach has experienced water quality deterioration and that this deterioration is predictable. A number of study designs have been adopted and

Box 9.3 Discharge sources associated with rainfall

Generic factors associated with combined sewer overflows (CSOs), riverine and storm drain inputs

Predictive outputs should be evaluated by examining a set of historical data to determine whether the predictor would previously have accurately predicted exposure events. Basic data requirements include: rainfall history, rainfall intensity (a function of amount and duration), sewage flow, location of discharges and definition of the zone of influence. Catchment and population equivalent loadings also need to be defined. Here location and zones of influence (resulting in both inputs and outputs) need to be defined. The zones of influence should lead to the delineation of impacted bathing areas. It is essential to undertake at least one intensive run of monitoring associated with an event or series of events. This monitoring should include a determination of the estimated extent of the impacted area linked to the various baseline data collected. Thus the rainfall intensity leading to a defined impacted area may be determined. If resources do not enable extended feedback monitoring to differentiate between different event intensities, then the predicted worst case zone of impacted area should be defaulted to. These data and their interpretation will provide the predictive base for estimating thresholds for subsequent events.

In some circumstances, a combination of other factors associated with the rainfall event may be used to determine the predictive capacity. These will include climatic and hydrographic conditions — specifically tide current and wind. Such factors could affect the occurrence/non-occurrence of an event, the likely zone of impact, and the duration of the event outcome.

Discharge source: combined sewer overflows

Background information

Combined sewer overflow discharges are derived from localised urban catchments. There are none of the 'softening' effects characteristic of riverine systems, typified by peaks and troughs of contamination. Effects are manifested rapidly. There is a simple, direct relationship between rainfall and discharge. Storage capacity exists on many current systems, and small events may therefore be contained. A widely applied "rule of thumb" is that effects may become obvious when dry weather flow is exceeded threefold. This is already incorporated into many systems. When an event triggers the threshold, the effect is rapid, with a potential for high microbial load and high public health risk.

Utility in prediction

Low rainfall may be accommodated; typically there is a threshold that will trigger an increased risk outcome. The best predictor may not be rainfall itself — it is the actual flow within the system. A relationship between rainfall and flow through the system that will trigger an alert must be determined. While good practice dictates that they should discharge below low water, CSOs may discharge directly onto the beach. Direct measure of the CSO operation forms the process; when they are operating, the risk is real.

Continued

Box 9.3 Continued

Discharge source: riverine

Background information

Rainfall in a catchment affects all its contributing inflows in a complex way over a wide area: delays, complex flow characteristics (including non-"plug" flow) and a series of small plugs may result. Riverine inputs are potentially the sum of multiple discharges from sewage systems, CSOs, storm drains and other industrial and rural sources. Where riverine pollution is dominated by a single pollution source, which may manifest as a plug, rainfall is a likely predictor in a relatively simple relationship. A significant increase in flow after a relatively long low-flow period could lead to sediment remobilisation and associated contamination. All likely influencing factors in a catchment must be categorised and identified (e.g. CSOs, storm drains, likelihood of sediment resuspension and surface run-off from grazing land). Effects of multiple CSOs and storm drain events contributing to a major riverine outcome are very complex to predict. They may lead to delays and staggered loadings, varying in intensity. The resuspension of sediments is related to extended dry periods and river flow. Complex and multiple sources of contamination can result and the health risk is difficult to predict.

Utility in prediction

Outcomes are difficult to predict; generally there is a variable delay in events manifesting themselves. There may be multiple, overlapping sources resulting in an unpredictable duration. Predisposing weather conditions, particularly the first major event after a period of low rainfall or low river flow, should signal a potential risk outcome. In terms of run-off, predictors will include rainfall intensity likely to lead to a threshold effect. Agricultural practices will modify the nature and extent of run-off and, in turn, may vary the threshold.

Discharge source: storm drains

Background information

Storm drain discharges are associated with localised, generally urban sources. In principle storm drains should not be connected to the sewage system, and therefore should not have a high sewage loading. Provided that there is no sewage connection, the likely discharge is generally of low significance to public health. However, the discharge may be associated with high total coliform (and sometimes high thermotolerant coliform) counts, which are a poor predictor of health risk. Generally storm drains discharge directly onto the beach and as a result, if they are connected to the sewage system, there will be an increased health risk.

Utility in prediction

Storm drains respond directly to rainfall. There is no storage capacity and therefore no delay in the outcome of the event. In effect, there is no threshold before a discharge occurs. Thus the system response is almost instant. A flushing effect means that the most significant (albeit generally low) health risk is at the start of the event. There is no simple relationship between the amount of discharge and risk burden; as time progresses the contamination load may be exhausted. The first rainfall releases a discharge contaminant plug, while subsequent rainfall leads to a discharge with little contaminant loading.

could be of use. All assume a sanitary inspection of the types of sources listed in Box 9.3. Although the use of simple predictive approaches requires additional work to plan and implement, such approaches are not highly expensive.

9.4.2 Advanced predictive approaches

More advanced studies have been developed to provide data on: the reasons for short-term elevated microbiological indicator counts; the timing of such elevated analytical results; the time taken for water quality to return to "baseline" conditions; the potential for prediction of water quality change; and the potential for remediation of poor water quality. Although these studies were designed initially for use under percentage compliance based regulatory structures, they are also very valuable tools for the classification approach suggested in this chapter.

Studies of this type from the UK suggest that well founded scientific studies (i.e. "compliance" modelling, budget studies, diffuse source modelling and near-shore modelling) would require between ten thousand and hundreds of thousands of USA dollars, depending on the complexity of the study. Where a full site study is required, the beach authority wishing to claim that prediction of elevated microbiological indicator counts is a feasible management tool for public health maintenance, should plan for and appropriately resource a potentially costly 12 month study.

Compliance modelling

This type of investigation was initially designed to understand the causes of occasional "high values" leading to a failure to comply with percentage compliance based standards. These investigations require reliable microbiological data covering several years and possibly several locations, as well as a set of variables that have been proved to predict microbiological concentrations at the study sites.

Multivariate statistical methods, such as multiple regression, can be applied to the data set to predict faecal indicator concentrations. The success of modelling should be judged on the basis of the explained variance (R^2) of the predictive multivariate model assuming statistical significance. Values for R^2 over 60 per cent for a particular beach year have been achieved. Nevertheless, this approach should not be adopted if there are insufficient sampling periods for each year (e.g. less than 20). In addition, careful control on variable inclusion (and hence multicolinearity) is required in model construction and constant input from a professional statistician is also essential.

The initial modelling study is an exploratory tool. It suggests predictability, which should be confirmed by further sampling of inputs through a budget investigation.

Budget studies
Budget studies can be undertaken if the initial modelling proves the possibility of a relationship with predictable inputs. This type of investigation requires the characterisation of inputs to a bathing water. It is vital that low flow and high flow inputs are measured, together with quantity and quality measurements. Potential sources of pollution include sewage effluent, CSOs and SSOs, rivers, avian inputs, bather loading, septic tanks, industrial discharges, private discharges and lagoon outlets. For these sources, data are required on the type of source and pollution input, frequency of episodic inputs, magnitude of all inputs, (e.g. base flow and episodes, duration of inputs, the flow volume of all inputs), and the microbiological quality of all inputs.

Budget studies provide information that is known to be episodic. Clear evidence that, during specific events, beach microbiological concentrations are commonly dominated by predictable (but non-sewage) sources of faecal indicators would provide local managers with evidence that elevated counts associated with such events would not pose a large risk to public health provided effective management action is taken to limit bather exposure during this time period.

Diffuse source modelling
If riverine inputs to a bathing water are derived from diffuse or non-point source areas, remediation of a beach with poor quality bathing water would require "catchment area" or watershed management. Lumped and distributed models have been applied to predict episodic catchment-derived sources of pollution. The construction of a diffuse source model of the upstream catchment can offer evidence of the contamination being derived from non-sewage sources. Information can therefore be provided by these studies to aid decisions on remediation strategies.

Such modelling requires the definition of sub-catchment units and the implementation of an intensive and targeted data collection exercise to characterise water quality from each characteristic sub-catchment unit. The intensity of agricultural land use and stocking density are of particular importance. Both stochastic multivariate and deterministic modelling have been applied, with good prediction of faecal indicator delivery based on agricultural land use types.

Near-shore hydrodynamic modelling
When the inputs to the beach have been identified and characterised as above, the next question becomes the impact of these constant and episodic inputs at different locations on a specific beach site. One tool applied to this problem is the use of near-shore hydrodynamic modelling. This type of modelling

requires tidal information, water quality dynamics (e.g. T_{90} values for microbiological indicators), wind speed and direction and sampling regimes. Significant data inadequacy exists in the currently available T_{90} values, which describe decay rates. Thus new scientific information is required. In addition to these data, elements such as wave height and sedimentary resuspension may be important predictors of microbiological contamination. However, they are not specifically addressed in current modelling systems.

The near-shore hydrodynamic modelling approach requires complex, finite element modelling. A high level of expertise is necessary to use this approach successfully to predict compliance in shallow near-shore waters. However, such approaches can accommodate both constant and episodic inputs to bathing waters, dynamic change in the near-shore waters, and impact under different tidal states and hydrometeorological conditions.

9.5 Management actions and routine monitoring

Key elements in protecting human health from potential risks associated with recreational or bathing waters are the identification of pollution sources (continuous and intermittent), assessing their impact on the target area and undertaking remedial or management action to reduce their public health significance. Depending on the circumstances, there may be a number of actions that can be taken to reduce public health risk. Such actions would therefore have an impact on the overall classification of the bathing water.

Routine monitoring should be undertaken to determine if the classification status of a beach changes over time. If management actions are shown to be effective and a beach can be reclassified as a result, monitoring requirements may be substantially reduced. Examples of classifications and their associated management and monitoring actions are given in Table 9.15.

9.5.1 Direct action on pollution sources

Direct action should be the principal management action because, if successfully undertaken, it provides a permanent and verifiable reduction of potential health risks. Remedial actions can include: diversion of sewage discharges away from the target area by the construction of long sea outfalls, provision of higher levels of sewage treatment, and increasing storm water retention to reduce frequency of discharge and/or relocation of intermittent discharges. These actions may, however, be outside the control of local communities or regional authorities and an alternative approach of local intervention may be more applicable.

9.5.2 Managing intermittent pollution events

Where there is clear evidence that water quality varies at certain predictable periods, such as following significant rainfall events, it may be possible for

Table 9.15 Examples of classification outcomes and their associated management and monitoring actions

Primary classification	Reclassification	Generic statement for public (non-verifiable, passive action)	Generic management advice[1] (verifiable, active action)	Monitoring requirements[2]
Excellent	–	Excellent beach.	NA	Annual sanitary inspection to ensure no change. Microbiological quality assessment every five years to verify status.
Good	Excellent[3]	This beach is of good quality.	No action needed on health grounds. Action may be warranted for local tourist promotion.	As above.
Fair	Good[4]	Inform public through advice at beach and tourist locations that bathing at location X is discouraged.	Post beach (i.e. bathing discouraged between specified posts). Restrict access (i.e. do not allow car parking). Discourage service industries. Fence area off. Encourage alternatives via car parks, bus stops and service industries.	Annual sanitary inspection to verify no change. Low-level microbiological quality assessment (4 samples on 5 occasions, equally spaced throughout the bathing season). Abnormally high samples need further verification and additional monitoring and possible review of impacted zone. Annual verification of management intervention effectiveness.
Fair	Good[5]	Inform public through advice at beach and tourist locations that bathing is discouraged after periods of heavy rainfall.	Post notice at bathing water. Use lifeguards to warn bathers. Close car parks and service facilities. Stop tourist buses. Encourage use of alternative beaches by providing free transport.	As above.
Poor	Good/fair	This area is of periodic poor quality and bathing is discouraged at certain locations/times.	Advice similar to that for "Fair"	As for "Fair".

Continued

Table 9.15 Continued

Primary classification	Reclassification	Generic statement for public (non-verifiable, passive action)	Generic management advice[1] (verifiable, active action)	Monitoring requirements[2]
Very poor	Not affected by local management	This area may be polluted with (nature of pollution) from (type of source). This may be unpleasant for bathers and presents some risk to human health.	Post generic warning notices similar to the risk statement at access points to the beach. Use posters to inform of alternative locations. Do not allow development of service industries. Make access difficult (e.g. no provision of car parks). Encourage use of alternative bathing areas. Encourage pressure for remedial action.	Annual sanitary inspection to confirm no changes to primary pollution source. Microbiological quality assessment every five years to verify status.

1 The level of action depends on the likely health impact of the event
2 Includes requirements for sanitary inspections and microbiological quality assessments
3 As defined by the conditions of contamination
4 As defined by the area of contamination
5 Increased contamination occurs under certain conditions

local management to undertake verifiable interventions that would reduce public health risks. Interventions would include passive non-verifiable actions, such as advising local residents and tourists not to bathe in the affected zone of the intermittent discharge for a given period following heavy rainfall. Active and verifiable interventions could include posting warnings around the affected zone following a rainfall event, advising bathers not to swim for a given period of time. In addition, advice could be given about the location of alternative bathing waters and transportation could be provided to and from those locations. Lifeguards, if present, could re-enforce the message. More restrictive measures could be the closure of relevant car parks and service industries (but not sanitary facilities).

9.5.3 Management interventions on spatial pollution

It is possible for a bathing water to be only partially affected by a source of human sewage. For example, a riverine input containing sewage from upstream communities may flow across a bathing water causing significant elevation in microbial indicator concentration. Unless direct action can be taken as outlined in section 9.5.1, various alternative options exist for reducing public health exposure. These options can range from the passive provision of information to the general public that bathing at the location was not advised, to actively dissuading bathing, such as by not providing public transportation or car parking near the affected area or by fencing off the area. As suggested in section 9.5.2, the policy of dissuasion should be reinforced by information about alternative bathing areas together with some encouragement, in the form of transport, easier parking or provision of service industries, etc., to entice bathers away from the polluted area.

9.5.4 Management of polluted zones

Where the whole extent of the bathing area is considered to pose a potential health risk and interventions along the lines of those described in section 9.5.1 are not feasible, management actions are needed to reduce the use of the bathing area. As before, information can be given to the public informing them of the water quality problems associated with the bathing water and this can be re-enforced by actions such as making access difficult by controlling car parking facilities, and by closing service industries. Additionally, information regarding alternative bathing waters of a similar nature, but with acceptable water quality, should be provided.

9.5.5 Routine monitoring

Under the classification scheme, routine monitoring always requires an annual sanitary inspection, to confirm that no changes in the primary pollution source(s) have occurred over the course of the year. In addition, microbiological quality assessments should be carried out, although the level of monitoring required for a given beach may depend largely upon its classification, as shown in Table 9.15. Beaches classified as very high or very low quality (i.e. "excellent" or "very poor"), for example, may only need a microbiological quality assessment every few years, to verify that their status has not changed. Mid-level ("good", "fair" and "poor") beaches may require an annual, low-level microbiological quality assessment, with 20 samples being taken at a minimum of four sites on five occasions evenly spaced throughout the bathing season. Beach zones greater than 2 km in length may require additional sampling sites. Further sampling may be necessary if abnormally high microbiological levels are found. If a beach has been reclassified, annual verification of the effectiveness of management interventions would also be required. When results of this routine monitoring suggest that the status of a beach has altered, the classification of the beach should be revised following a process similar to that described in Figure 9.6.

9.6 Evaluation and validation of the proposed approach

A classification scheme of the type proposed in this chapter is of value if it accomplishes one or more of the following goals:

- Contributes to informed personal choice (e.g. individuals, by using the information provided, can and do modify their exposure). This implies inter-location comparability and an informed public.
- Contributes to local risk management (e.g. by excluding or discouraging access to areas or at times of increased risk, thereby reducing overall exposure).
- Assists in making maximum use of the minimum necessary monitoring effort.
- Assists local decision making regarding safety management.
- Encourages incremental improvement and prioritises areas of greatest risk.

In order to evaluate whether the goals have been reached, both field testing and evaluation of the scientific validity of the approach is required. A limited number of intensive studies would be necessary to test the scientific validity of the approach; in recognition of the importance of this, the participants at the Annapolis meeting developed a protocol for such a study. This protocol requires extensive sampling of study sites, as described in the following sections, and should not be confused with the less rigorous microbial assessments necessary for classifying a beach under the scheme described in this chapter.

9.6.1 Validation protocol

Many countries around the world are interested in establishing uniform recreational water monitoring protocols that would provide accurate assessments of water quality in a timely manner. Scientists and public health officials recognise the need for monitoring approaches, such as that proposed in this chapter, that would characterise a bathing water at reasonable cost and within the constraints of limited resources (personnel and equipment and supplies). To establish such protocols, it is important to determine the essential parameters that must be considered in the monitoring programme, e.g. temporal, spatial and environmental considerations. The sampling of a recreational water must be adequate to capture all of these factors to ensure the likelihood that samples portray the water quality at the time they are taken.

The establishment of a robust set of data from multiple, contrasting locations and conditions is essential to determine general sampling requirements that are transferable to most locations world-wide. It is desirable that all parties interested in improved monitoring approaches participate collectively in conducting studies to develop the data for determining the minimum sampling requirements (at least for typical beach environments), in freshwater, estuarine and marine settings. In order to develop such a database, a standard sampling protocol which can be used (and adhered to) by everybody is required, whereby the data derived from each study would be compatible with data from the other sampling studies. The following is a recommended approach to identify the major elements, parameters and conditions to be developed by the sampling protocol that would be applied to beach studies intended to describe the important monitoring features for recreational waters. This protocol should be implemented in conjunction with a sanitary inspection, as described in Chapter 8.

Microbiological parameters

Two microbial indicators of faecal contamination were selected for this sampling study protocol: faecal streptococci or enterococci and sulphite reducing *Clostridium* or *Clostridium perfringens*. The protocol can apply equally to other indicator organisms described in Table 9.5, such as *E. coli* in freshwaters. The indicators proposed in this protocol development were chosen because the methods for their detection and enumeration have been well described and field tested by a number of investigators in numerous recreational water studies as well in other environmental testing. There is a large database that describes the precision, accuracy and coefficients of variation for these methods. These methods were also chosen because they are considered applicable for both marine and freshwater testing.

Temporal study conditions

The studies should be performed at least over the period of a typical bathing season, which can range from several weeks to all year round, depending on latitude and local customs. A three-month sampling period or longer is considered best to obtain a robust set of data to analyse for temporal effects under most circumstances. Under most conditions a minimum of 50 days of sampling is considered a robust study. This should provide satisfactory data to establish important factors or conditions at a study site and that will allow the assessment of important locations for sampling, when to sample, and to establish factors that contribute to microbiological water quality variability. This amount of study data should allow assessment of critical factors that may trigger sampling (e.g. regression, multivariate regression, trends, etc.) when applied to a beach. It should also allow the combination of data from various studies to make the assessments more robust, so that guidance may be derived for dissemination to all persons concerned with public safety at beaches.

Sampling should encompass daily periods and should be conducted at least several times a week. Pollution varies in response to the density of users and the local population who may be discharging to the sewage system (e.g. peak uses may often occur at weekends and holidays). In addition, local events may occur routinely with resultant effects on waters serving recreational areas. The sampling protocol should take account of these factors, so as not to introduce a bias to the data set. Sampling should be carried out hourly over a 12 hour period, for example from 0700 hours to 1900 hours for all sampling locations comprising the beach study site.

Event sampling

Many studies to date have demonstrated that one of the most significant factors leading to increased faecal pollution in recreational waters is rainfall. While the general sampling protocol described above should pick up the effects of rainfall events over a long recreational season, it may not do so for short-term evaluations. For locations subject to rainfall events, the general sampling protocol could, therefore, lead to a lack of data covering these event and their contribution to microbial pollutant loading at a beach. If feasible, it is recommended that at least 20 per cent of the study sampling days should be during and after rainfall events where there is, or there is likely to be, local run-off.

Spatial sampling conditions

It is very important in sampling studies (for establishing uniform monitoring guidelines) to characterise the water at a beach from the swash zone (i.e. the sand area that is covered with waves on an intermittent or occasional basis during the sampling period) out to the most distant locations confining the beach (but at least to chest height), at the depths where exposure is likely to occur, and also along the designated width of the beach (parallel to the shore line). This becomes the designated area of water that a single sampling event, from off-shore sampling periods in a single day, should characterise. The designated area comprises a grid of sampling locations that are sampled for each period.

Sampling grid

The spacing of the length between grid samples running parallel along the beach should be uniform at 20 m and with a minimum of three locations (resulting in a minimum of 60 m total distance). Beaches shorter than this recommended length cannot be considered for incorporation into the sampling validation study.

Sample site distances perpendicular to the shore should be located from the 20 m grid transects. These locations should be:

- Ankle deep (0.15 m from grid transect on shore).
- Knee deep (0.5 m from grid transect).
- Chest deep (1.3 m from grid transect).

Samples should be taken at the following depths:

- Ankle depth sample ~ 0.075 m below the water surface.
- Knee and chest depth samples ~ 0.3 m below the water surface.

Although sand samples are not an absolute requirement for this sample validation study, they are considered to be desirable. Sand samples should be taken from the swash zone and from the top 2 cm of the sand. Enough sand should be taken to enable one portion to be used to establish the dry weight and another portion to be used to elute microbial components for quantification.

Sampling and analysis

A single, discrete sample should be taken from each location at each period. Each sample must be labelled with location, day and time taken, and any other distinguishing characteristics needed to identify the sample. Samples must be iced, packaged and transported by surface or air to the laboratory for processing and analysis. Sample analysis must be initiated within 8–12 hours and all discrete samples must be assayed in triplicate for each dilution (i.e. three dilutions, although this may be reduced to two if, or when, the water becomes well characterised for the presence of indicators under various sampling conditions). Other test or observational variables for each sampling period are:

Physical and chemical variables

- pH (daily).
- Salinity — estuarine (hourly); marine, if no significant riverine influence (daily).
- Turbidity (hourly).
- Water and air temperature (hourly).

Other observations and measurements

- Rainfall — magnitude, duration, time relative to sampling (every 6 hours).
- Wave height (hourly).
- Current direction and speed — fresh and estuarine (hourly).

- Total light or radiation (hourly).
- Tidal state and magnitude (hourly).
- Wind direction and speed relative to beach (hourly).
- Per cent cloud cover (hourly).
- Bather population at each transect point (e.g. by means of photographs) (hourly).
- Animal population — presence and number of horses, donkeys, dogs, shore birds (hourly).
- Boats anchored or moored within 1 km of the beach (hourly).
- Beach debris and sanitation: sanitary plastics, visible grease balls, algae (daily).
- Location of freshwater, storm water, sewage outfall or other intrusion to beach.
- Location of bather facilities (showers, lavatories) and relevance of input from these sources (shower run-off, sewage overflow) to beach.

Database requirements are:

- All raw data should be provided to a computerised database system.
- All data should be entered into a spreadsheet compatible with universal spreadsheet formats.
- All data entered should be validated for accuracy.
- All data should be duplicated in separate computer files for future access.

Analysis of the data should generate the following descriptive statistics:

- Number of samples taken.
- Geometric means per sample, per replicate, etc.
- Standard deviation.
- Quality assurance and quality control results.
- Coefficients of variation, precision and accuracy of methods used by the laboratory.

9.7 References

Bitton, G., Farrah, S.R., Ruskin, R.H., Butner, J. and Chou, Y.J. 1983 Survival of pathogenic and indicator organisms in groundwater. *Groundwater*, **21**, 405–410.

Calderon, R. 1990 United States Environmental Protection Agency, Personal communication.

California State Water Resources Control Board. Undated. *Water Quality Control Plan for Ocean Waters of California,* Californian State Water Recources Control Board, California.

CEPPOL/UNEP (Caribbean Environmental Programme/United Nations Environment Programme) 1991 Report on the CEPPOL seminar on monitoring and control of sanitary quality of bathing and shellfish-growing marine waters in the wider Caribbean. Kingston, Jamaica, 8-12 April 1991.

Technical Report No. 9, UNEP Caribbean Environmental Programme, Jamaica.

Chamberlain, C.E. and Mitchell, R. 1978 A decay model for enteric bacteria in natural waters. In: R. Mitchell [Ed.] *Water Pollution Microbiology*, New York, Wiley, Vol. 2, 325.

Cioglia, L. and Loddo, B. 1962 The process of self-purification in the marine environment III. Resistance of some enteroviruses, *Nouvi Annli Di Igiene E. Microbiologia*, **13,** 11.

(DINAMA) 1998 *Reglamienta del Decreto* 253/79, Agua Clase 2b: "aguas desinadas a recreación por contacto directo con el cuerpo. Uruguay, Ministerio de Vivienda, Ordenamiento Territorial y Medio Ambiente, Direcci.

Divizia, M., Ruscio, V., Degener, A.M. and Pana A. 1998 Hepatitis A virus detection in wastewater by PCR and hybridization. *Microbiologica*, **21**(2), 161–167.

Dufour, A.P. and Ballentine, P. 1986 Ambient water quality criteria for bacteria – 1986 (Bacteriological ambient water quality criteria for marine and fresh recreational waters). EPA A440/5-84-002, United States Environmental Protection Agency, Washington, D.C., 18 pp.

EEC (European Economic Community) 1976 Council directive of 8 December 1975 concerning the quality of bathing water. *Official Journal of the European Communities*, **19,** L 31.

Environmental Agency 1981 *Environmental Laws and Regulations in Japan (III) Water,* Japan Environmental Agency.

Funderburg, S.W., Moore, B.E., Sorber, C.A. and Sagik, B.P. 1978 Survival of poliovirus in model wastewater holding pond. *Progress in Water Technology*, **10**, 619–629.

Grimason, A.M., Smith, H.V., Thitai, W.N., Muiruri, P., Irungu, J., Smith, P.G., Jackson, M.H. and Girdwood, R.W.A. 1992 Occurrence and removal of *Giardia* spp. cysts in Kenyan waste stabilisation pond systems. In: R.C. Thompson, J.A. Reynoldson and A.J. Lymbery [Eds] *Proceedings Giardia—From Molecules to Disease and Beyond,* 6–9 December, 1992, Fremantle, Murdoch University, Perth, 28.

Grimason, A.M., Smith, H.V., Young, G. and Thitai, W.N. 1996a Occurrence and removal of *Ascaris* sp. ova by waste stabilisation ponds in Kenya. *Water Science and Technology*, **33**(7), 75–82.

Grimason, A.M., Wiandt, S., Baleux, B., Thitai, W.N., Bontoux, J. and Smith, H.V. 1996b Occurrence and removal of *Giardia* sp. cysts by Kenyan and French waste stabilisation pond systems. *Water Science and Technology,* **33**(7), 83–89.

Gyurek, L.L. and Finch, G.R. 1998 Modeling water treatment chemical disinfection kinetics. *Journal of Environmental Engineering,* **124**(9), 783–793.

Haas, C.N., Joffe, J., Anmangandla, U., Jacangelo, J.G. and Heath, M. 1996 Water quality and disinfection kinetics. *Journal of the American Water Works Association*, **88**(3), 95–103.

Hanes, N.B. and Fragala, R. 1967 Effect of seawater concentration on the survival of indicator bacteria. *Journal of the Water Pollution Control Federation*, **39**, 97.

INCYTH (Argentina Instituto Nacional de Ciencia y Técnica Hidricas) 1984 Estudio de la factibilidad de la disposición en el mar de los efluentes cloacales de la cuidad de Mar del Plata. Informe Final. Buenos Aires, Secretaría de Recursos Hidrill de Argentina.

Jacangelo, J.G., Adham, S.S. and Lâiné, J.-M. 1995 Mechanism of *Cryptosporidium, Giardia* and MS2 virus removal by MF and UF. *Journal of the American Water Works Association*, **87**(9), 107–212.

JCA (Puerto Rico Junta de Calidad Ambiental) 1983 Reglamento de Estándares de Calidad de Agua, 28 de Febrero de 1983.

Keswick, B.H., Gerba, C.P., Secor, S.L. and Cech, I. 1982 Survival of enteric viruses and indicator bacteria in groundwater. *Journal of Environmental Science and Health*, **A 17**(6), 903–912.

Kramer, H.H., Herwaldt, B.L., Craun, G.F., Calderon, R.L. and Juranek, D.D. 1996 Waterborne diseases: 1993 and 1994. *Journal of the American Waterworks Association*, **88**(3), 66–80.

Leeming, R., Ball, A., Ashbolt, N. and Nichols, P. 1996 Using faecal sterols from humans and animals to distinguish faecal pollution in receiving waters. *Water Research*, **30**(12), 2893–2900.

Long, J. and Ashbolt, N.J. 1994 Microbiological quality of sewage treatment plant effluents. AWT Science and Environment report number **94**/123, Sydney Water Corporation, Sydney, 26 pp.

Madireddi, K., Babcock, R.W., Levine, B., Huo, T.L., Khan, E., Ye, Q.F., Neethling, J.B., Suffet, I.H. and Stenstrom, M.K. 1997 Wastewater reclamation of Lake Arrowhead, California — an overview. *Water and Environmental Research*, **69**(3), 350–362.

McFeters, G.A. and Stuart, D.J. 1974 Comparative survival of indicator bacteria and enteric pathogens in well water. *Applied Microbiology*, **27**, 823–829.

Ministerio del Interior 1976 *Aguas de Balneabilidade*, Portoria No. 536, Brazil.

Ministerio de Salud 1979 *Disposiciones Sanitarias sobre Aguas* – Artculo 69 Ley 05, Colombia.

Ministerio de Salud 1983 Modificaciones a los Artculos 81 y 82 Reglamento de los Ttulos I, II y III de la Ley General de Aguas. Decreto Supremo No 007-83-SA, Peru.

Ministerio de Salud 1986 Higiene comunal, lugares de baño en costas y en masas de aguas interiores, requisitos higiénicos sanitarios, 93–97, La Habana, Cuba.

Ministerio de Salud Pública 1987 Instituto Ecuatoriano de Obras Sanitarias, Proyecto de normas reglamentarias para la aplicación de la Ley, Ecuador.

Morbidity and Mortality Weekly Report 1988 Water related disease outbreaks 1985. Vol. 37., No. SS02; 015, US Centers for Disease Control, Atlanta, GA.

Morbidity and Mortality Weekly Report 1990 Waterborne disease outbreaks 1986–1987. Vol. 39, No. SS01; 001, US Centers for Disease Control, Atlanta, GA.

Morbidity and Mortality Weekly Report 1991 Vol. 40, US Centers for Disease Control, Atlanta, GA.

Morbidity and Mortality Weekly Report 1993 Surveillance for waterborne disease outbreaks. Vol. 42, No. SS05; 001, US Centers for Disease Control, Atlanta, GA.

Omura, T., Onuma, M. and Hashimoto, Y. 1982 Viability and adaptability of *E. Coli* and enterococcus group to salt water with high concentration of sodium chloride. *Water Science and Technology*, **14,** 115–126.

Oppenheimer, J.A., Jacangelo, J.G., Laîné, J.-M. and Hoagland, J.E. 1997 Testing the equivalency of ultraviolet light and chlorine for disinfection of wastewater to reclamation standards. *Water and Environmental Research*, **69**(1), 14–24.

Otaki, M., Yano, K. and Ohgaki, S. 1998 Virus removal in membrane separation process. *Water Science and Technology*, **37**(10), 107–116.

Perezrey, R., Chavez, H. and Baluja, C. 1995 Ozone inactivation of biologically-risky wastewaters. *Ozone-Science and Engineering*, **17**(5), 499–509.

Roberts, P.J.W. 1996 Sea outfalls. In: V.P. Singh and W.H. Hager [Eds] *Environmental Hydraulics.* Kluwer Academic Publishers, Netherlands, 63–110.

Salas, H.J. 1998 *History and Application of Microbiological Water Quality Standards in the Marine Environment.* Centro Panamericano de Ingenieria Sanitaria y Ciencias del Ambiente and Pan American Health Organization, Lima, Peru.

SEDUE (Mexico Secretara de Desarrollo Urbano y Ecologa) 1983 Breviario Jurdico Ecologico, Subsecretaria de Ecologa.

Sieracki, N. 1980 The Effects of Short Exposures of Natural Sunlight on the Decay Rates of Enteric Bacteria and A coliphage in a Simulated Sewage Outfall Microcosm. Master of Science Dissertation, University of Rhode Island.

US EPA 1986 Bacteriological ambient water quality criteria availability. *Federal Register*, **51**(45), 8012.

Venezuela 1978 Reglamento parcial No. 4 de la ley orgínica del ambiente sobre clasificación de las aguas áciouñ.

WHO/UNEP 1977 Health criteria and epidemiological studies related to coastal water pollution. Report of a group of experts jointly convened by WHO and

UNEP, Athens, 1–4 March 1977, Document ICP/RCE 206(5), World Health Organization Regional Office for Europe, Copenhagen.

WHO 1975 Guide and criteria for recreational quality of beaches and coastal waters. Report on a working group Bilthoven, 28 October – 1 November 1974, Document IEURO 3125(1), World Health Organization Regional Office for Europe, Copenhagen.

WHO 1998 *Guidelines for Safe Recreational-water Environments: Coastal and Freshwaters.* Draft for Consultation, World Health Organization, Geneva.

WHO/UNEP 1978 *First Report on Coastal Water Quality Monitoring of Recreational and Shellfish Areas (MED VII).* WHO/EURO document ICE/RCE 205(8).World Health Organization, Copenhagen.

Yates, M.V. and Gerba, C.P. 1998 Microbial considerations in wastewater reclamation and reuse. In: T. Asano, [Ed.] *Wastewater Reclamation and Reuse. Vol. 10*, Water Quality Management Library, Technomic Publishing Co. Inc., Lancaster, Pennsylvania, 437–488.

Chapter 10*

CYANOBACTERIA AND ALGAE

In freshwaters, scum formation by cyanobacterial phytoplankton is of concern to human health. Freshwater algae proliferate quite intensively in eutrophic waters and may contain irritative or toxic substances. Nevertheless, incidents of impairments of human or animal health caused by algae are rarely reported. One example was the closure of a number of bathing sites in Sweden because of mass occurrences of the flagellate *Gonyostomum semen* which causes skin irritations and allergies (Cronberg *et al.,* 1988). Incidents attributed to cyanobacteria are far more numerous and, in most cases, have been caused by species of cyanobacteria that may accumulate to surface scums of extremely high cell density. As a result, the toxins they may contain ("cyanotoxins") reach concentrations likely to cause health effects.

Surface aggregations of planktonic cyanobacteria occur because of their capability to regulate their buoyancy, enabling them to seek water depths with conditions optimal for their growth. Regulation of buoyancy is a slow process, and cells adapted to ambient turbulence may take several days to adapt their buoyancy when conditions change (e.g. turbulence is reduced). Thus, cells or colonies may show excessive buoyancy and accumulate at the water surface. Light winds drive such accumulations to leeward shores and bays, where the resulting scums become thick. In extreme cases, such agglomerations may become very dense, with cells frequently concentrated by a factor of 1,000 or more, eventually reaching in some cases, one million-fold concentrations with a gelatinous consistency. More frequently, surface accumulations are seen as streaks or slimy scums that may look like blue-green paint or jelly. Such situations can change rapidly within hours with changes in the wind direction. Monitoring strategies must take into account this highly dynamic variability of cyanotoxin occurrence in time and space.

Scum formation is influenced by the morphological conditions of the water body, such as the water depth from which cyanobacteria can rise to the surface (i.e. the thickness of the stratum in which they are dispersed) and the length of wind fetch over which surface aggregations can be swept together to form shoreline scums (Figure 10.1). Accumulated scum material may take a

* *This chapter was prepared by I. Chorus and M. Cavalieri*

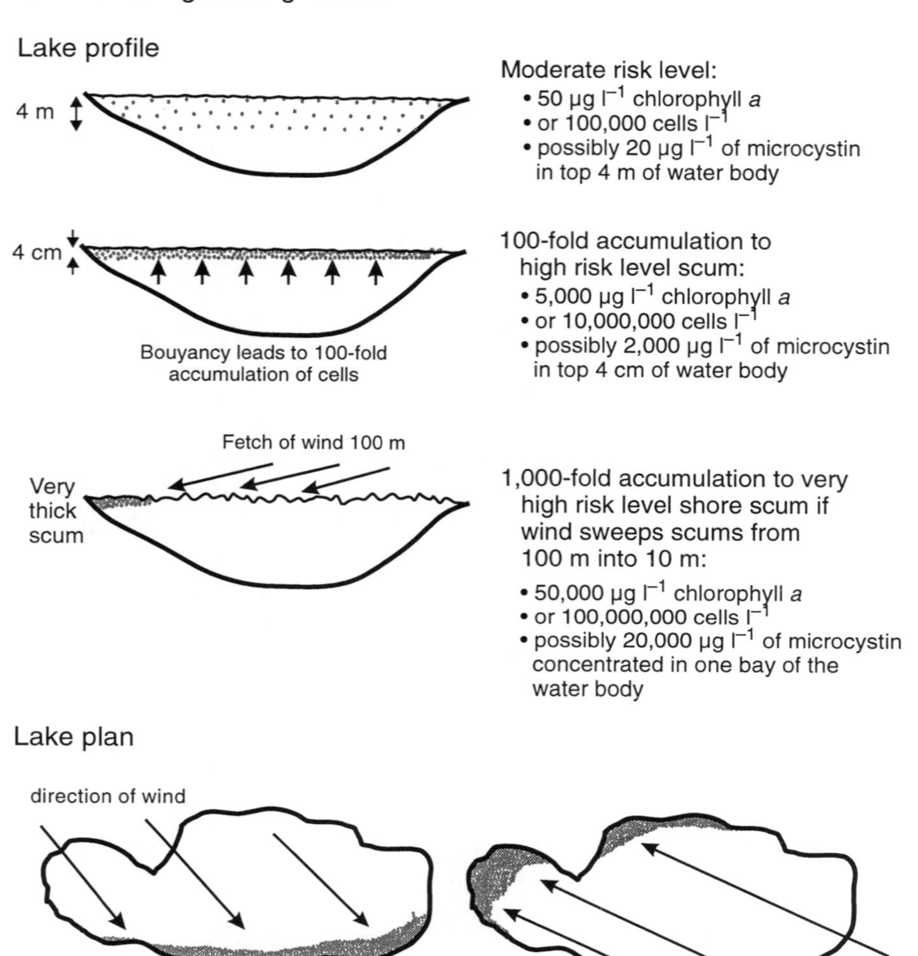

Figure 10.1 Schematic illustration of scum-forming potential changing the cyanotoxin risk from moderate to very high (After Falconer *et al.*, 1999)

long time to disperse especially in shallow bays. Dying and lysing cells within the scum release their contents, including the toxins, into the water. However, toxin dissolved in the water is rapidly diluted and probably also degraded. Cell-bound cyanotoxin concentrations usually are the greater cause for concern in recreational waters.

Most of the problems reported with nuisance and toxin-containing aquatic cyanobacteria in freshwaters have involved planktonic species, i.e. those distributed in the water body or forming surface scums. However, benthic (i.e. bottom-dwelling), species have occasionally surfaced, and been washed ashore where they have caused the death of dogs scavenging upon the material. Benthic cyanobacteria can grow as mats on sediments in shallow water. Some mats become detached and are driven onshore where they result in acutely toxic accumulations (Edwards *et al.*, 1992).

Severe illness due to direct dermal contact with such mats has been reported from tropical marine bathing sites (Kuiper-Goodman *et al.,* 1999). In coastal marine environments many toxic species of dinoflagellates, diatoms, nanoflagellates and cyanobacteria occur, and have led to several forms of human health impacts mainly after consumption of shellfish and fish, i.e. syndromes such as Paralytic Shellfish Poisoning, Diarrhetic Shellfish Poisoning, Amnesic Shellfish Poisoning, Neurotoxic Shellfish Poisoning and Ciguatera. Nevertheless, there exists very little scientific evidence that marine toxic algal blooms cause health problems for recreational users of water. This evidence is reviewed in the WHO *Guidelines for Safe Recreational-Water Environments* (WHO, 1998).

As a result of the lack of evidence for effects of toxic algae on recreational users of marine waters, this chapter is concerned principally with monitoring and assessment of toxic cyanobacteria in freshwaters. Nevertheless, the methods given for cyanobacteria may be employed to assess the development of other planktonic algae. Although various toxic marine algae have been associated occasionally with human health effects, concern for human health centres on toxic cyanobacteria, and therefore these are the subjects of the remainder of this chapter. Further information concerning coastal phytoplankton blooms and associated monitoring strategies are available in Franks (1995), Smayda (1995) and in the UNESCO manual on this subject (UNESCO, 1996).

For lakes, reservoirs and rivers different levels of "alert" (for cyanobacterial cell concentrations and their toxin contents) have been proposed in *Toxic Cyanobacteria in Water* (Chorus and Bartram, 1999) and in the *Guidelines for Safe Recreational-water Environments* (WHO, 1998). These documents present a series of guideline values and situations associated with incremental severity and probability of greater effects in relation to cyanobacterial occurrence (Table 10.1).

10.1 Design of monitoring programmes

Many cyanobacterial species frequently forming mass developments may contain hepatotoxins or neurotoxins, and all of them contain lipopolysaccharides (LPS) in their cell wall. Lipopolysaccharides may be the cause of irritations of the skin, digestive tract, respiratory membranes, eyes and ears that are frequently associated with cyanobacteria. Research in pharmacology and ecotoxicology indicates that cyanobacteria contain a variety of

Table 10.1 Phase 3 monitoring — guidelines for safe practice in managing recreational waters according to three different levels of risk

Level of risk[1]	Health risks	Recommended actions
20,000 cells cyanobacteria per ml or 10 μg l^{-1} chlorophyll *a* with a dominance of cyanobacteria	Short-term adverse health outcomes (e.g. skin irritations and gastro-intestinal illness, probably at low frequency)	Post on-site risk advisory signs Inform relevant authorities
10^5 cells cyanobacteria per ml or 50 μg l^{-1} chlorophyll *a* with a dominance of cyanobacteria	Potential for long-term illness with some species Short-term adverse health outcomes (e.g. skin irritations and gastro-intestinal illness)	Watch for scums Restrict bathing and further investigate hazard Post on-site risk advisory signs Inform relevant authorities
Cyanobacterial scum formation in bathing areas	Potential for lethal acute poisoning Potential for long-term illness with some species Short-term adverse health outcomes (e.g. skin irritations and gastro-intestinal illness)	Immediate action to prevent contact with scums; possible prohibition of swimming and other water-contact activities Public health follow-up investigation Inform relevant authorities

[1] Expressed in relation to cyanobacterial density and given in order of increasing risk

substances not yet identified, but that may have a potential impact on people. The implications of the present state of knowledge for surveillance and management are that any mass development of cyanobacteria may be a potential health hazard. If the cyanobacterial cells contain hepatotoxic microcystins, cause for concern may be higher because of the chronic effects of this potent toxin. Therefore, monitoring should address primarily the occurrence of cyanobacterial mass developments, whereas microcystin analysis may be adequate in specific situations.

Visual inspection of a bathing site is of crucial importance because it shows immediately whether cyanobacteria occur in potentially hazardous densities. However, as scum formation and dispersion may occur within hours and thus too frequently for monitoring, assessment of the risk of cyanobacterial exposure during recreational activities is greatly enhanced by an understanding of the population development of these organisms in a given water body (i.e. through background monitoring of variables that enable their proliferation). Good knowledge of the local growth conditions for cyanobacteria can greatly enhance predictability of bloom formation. As knowledge and understanding of a given site accumulate, regular patterns of cyanobacterial growth may be noticed, and surveillance may as a result be focused upon critical periods. Involving limnological expertise may be very useful, particularly during the development of monitoring programmes and for their periodic assessment for efficacy.

10.1.1 Monitoring strategy for freshwater cyanobacteria

A structured, quantitative investigation approach aims at focusing surveillance efforts upon those sites that are likely to present a risk. It further provides a scheme for immediate assessment and action by discerning three steps of action (Table 10.2).

- Determination of the carrying capacity of the ecosystem for cyanobacteria.
- Site inspection to detect mass developments.
- Quantitative assessment of biomass as a basis for risk assessment when mass developments occur.

Preliminary identification of water bodies potentially harbouring high densities of cyanobacteria is possible on the basis of simple transparency measurements with a Secchi disc. If transparency is high (greater than 2 m) and no significant discolouration of the water can be seen, cyanobacterial densities are unlikely to be high. It should be noted, however, that in large and deep lakes with a large volume from which cells can accumulate to scums, high densities in some parts of the lake cannot be excluded. In water bodies with Secchi disc readings of less than 2 m, the following three steps may be undertaken to assess the risk of toxic cyanobacteria.

Step 1. Determination of the carrying capacity of the ecosystem for cyanobacteria

Algal and cyanobacterial population growth requires phosphorus and nitrogen. The concentrations of these nutrients determine the maximum amount of algae and cyanobacteria that can develop in a given water body, or the "carrying capacity" of an ecosystem for these organisms. The carrying capacity is more often limited by the availability of phosphate but sometimes, particularly in marine ecosystems, it may also be limited by nitrogen. If total phosphorus is not limiting, it may be worthwhile to analyse nitrogen to check whether the carrying capacity may be lower than assumed from the total phosphorus concentrations. Total phosphorus should be measured rather than dissolved phosphate (soluble reactive phosphate (SRP), also known as orthophosphate) because algae and cyanobacteria can store sufficient amounts of phosphate to increase their population 10-fold, even if no dissolved phosphate can be detected. Thus, cell-bound phosphate (which is included when measuring total phosphorus but is missed when measuring only dissolved phosphate) is more meaningful for the assessment of carrying capacity.

Total phosphorus should be assessed several times during the cyanobacterial growth season in order to check temporal variability. If variability of the

Table 10.2 Parameters to be measured or assessed for each of the three phases of the recommended monitoring strategy

Phase or activity	Rationale for monitoring	Variables
Monitoring phases		
Phase 1 (background)	Potential for cyanotoxin problems	Nutrient concentrations (total phosphorus, nitrate and ammonia) Transparency (Secchi disc) Hydrophysical conditions (e.g. flow regime and thermal stratification) Other biological complex interactions
Phase 2 (basic)	Site inspection for indicators of toxic cyanobacteria	Transparency (Secchi disc) Discolouration Scum formation Hydrophysical conditions Temperature Weather conditions (e.g. winds, light) Changes in turbulence (i.e. mixing) Other biological complex interactions
Phase 3 (cyanobacteria)	Qualitative/quantitative assessment of potentially toxic cyanobacterial assemblages	Transparency (Secchi disc) Qualitative microscopic analysis to identify dominant taxa (genus is sufficiently precise) Quantitative microscopic analysis (only as precise as needed for management)[1]
	Determination of cyanobacterial biomass	Chlorophyll *a* analysis (provides an estimate of cyanobacterial biomass in the case of rather monospecific blooms)
Additional activities		
Toxicity	Presence of toxicity	Bioassays
Toxin analysis	Presence of specific toxins (qualitatively and quantitatively)	Chemical analyses

[1] The ratio of the concentration of algal to cyanobacterial cells (or filaments or colonies) in the water sample (as determined by enumeration) may be converted to biomass values

concentrations is low (less than 50 per cent), assessment twice a year may prove sufficient (in subtropical and temperate climates in spring at total overturn and in summer during the main bathing season). If total phosphate concentrations are below 0.01–0.02 mg l^{-1} P, mass developments of cyanobacteria are unlikely and high turbidities, (if present) may have other causes. If total phosphate concentrations are higher, monitoring should move to Step 2 in order to check for the presence of phytoplankton mass developments.

Step 2. Monitoring to detect possible mass developments of phyto-plankton (algae plus cyanobacteria) by visual inspection of bathing sites and immediate actions to prevent health hazards

Monitoring should generally be performed at fortnightly intervals. The areas most likely to be affected should be assessed first, such as the downwind shorelines. Information on changes in wind direction or strength in the preceding 24 hours may be valuable for understanding the movement of surface scums in the water body.

Visual inspection should consider the three following conditions. Each condition may be considered a prerequisite for the following condition.

1. Determine if Secchi transparency is less than 1 m or, in absence of a Secchi disc, if the bottom of the lake cannot be seen at 50 cm depth along the shore line; if so
2. Determine if cyanobacteria are visible as a greenish discolouration of the water or at the water's edge or as green or blue-green streaks on the water surface (note: evenly dispersed greenish discolouration may also be due to algal phytoplankton rather than to cyanobacteria, and microscopic assessment is necessary to determine the causative organism, but surface scums and streaks may be attributed to cyanobacteria); if so
3. Determine if a green or blue-green scum is visible on the water surface in any area.

If cyanobacteria are visible and if their population density exceeds one of the guideline values given in Table 10.1, then action appropriate for the given location should be taken, such as informing responsible authorities, initiating the posting and publication of warning notices, regular monitoring according to Step 3 below, and deciding whether or not to initiate immediate action such as marking or enclosing affected areas (if sufficiently small) and prohibiting access to such areas by water users; restricting access to the water edge that is affected, other than for launching boats; and regulating or restricting access by all recreational users where cyanobacterial blooms cover the general waters.

High nutrient inputs favour the development of cyanobacteria and algae. Therefore, if cyanobacteria are present, inspect the catchment area for signs of sewage outlets, excessive fertilisation close to the shoreline, erosion, or other potential sources of phosphate input. The identification of such sources provides the basis for measures addressing the cause of the problem.

Step 3. Quantitative assessment of cyanobacterial biomass and further actions to prevent health hazards

Upon detection of cyanobacteria at bathing sites, their quantification may be desirable for risk assessment. Two quantitative measures are equally valuable: microscopic cell counts or determination of chlorophyll *a* concentration as a simple measure for algal (including cyanobacterial) density. The choice of methods depends on equipment, expertise and personnel available. If chlorophyll *a* is used and its concentrations remain below 10 $\mu g\ l^{-1}$, hazardous densities of cyanobacteria are unlikely. At higher concentrations, chlorophyll *a* analysis must be supported by qualitative microscopic investigations for dominance of cyanobacteria.

If the results of the measurements exceed the guideline values, immediate management action are suggested (Table 10.1). Furthermore, the detection of cyanobacteria at potentially hazardous concentrations should initiate planning of measures for restoration of bathing water quality. Because of the complexity of factors leading to cyanobacterial proliferation, flexible approaches are important, involving further development of monitoring and of protection measures as information on a given water body is accumulated.

Regular measurements of transparency (Secchi readings) in Step 3 can greatly enhance understanding of the system. If transparency is high (greater than 2 m) and no significant discolouration of the water can be seen, cyanobacterial densities are unlikely to be high. However, samples should be taken to determine the phytoplankton community and the water chemistry, in order to estimate the capacity of the water body for cyanobacterial bloom formation. In deep and stratifying lakes, samples from different levels within the vertical stratification are required because some cyanobacteria accumulate near the thermocline. If transparency is low (less than 1–2 m) and accompanied by a greenish to bluish discolouration, high cyanobacterial densities are likely. In addition, greenish streaks formed by buoyant cyanobacteria (e.g. *Microcystis, Anabaena*) during warm and calm weather may be visible on the water surface. Inspection of downwind areas of the water body is essential when these characteristics are observed, because wind action can readily lead to accumulations of these buoyant organisms. Samples for further analysis (taxonomical and toxicological) should be taken.

Assessment of toxicity or chemical analysis for specific cyanotoxins is not generally recommended for two reasons: (i) the results of toxin analysis provide only a partial basis for risk assessment because only some of the substances causing health outcomes are known and can be analysed, and (ii) the results of an epidemiological study indicate that some health outcomes are not due to known cyanotoxins (Pilotto *et al.*, 1997). Assessment of toxicity using a bioassay may circumvent this problem, but the results are not easily interpreted in terms of human exposure during swimming, particularly with respect to skin reactions. Many cyanotoxin survey studies in different parts of the world have shown that more than half of the field populations investigated did contain toxins, particularly microcystins. Therefore, cyanobacteria are likely to be toxic.

Toxin analysis or toxicity assays may be useful under specific circumstances, because if concentrations of the known cyanotoxins (particularly microcystins) prove to be low, some critical health outcomes (particularly liver intoxication due to microcystins) may be excluded. Warning notices regarding potential irritative effects of cyanobacteria (on the skin, gastrointestinal tract, ears, eyes, respiratory membranes) should nonetheless be posted while the cyanobacterial population density is above the guideline values (Table 10.1). Toxin analysis or toxicity assays may be required in advance of a sports event to improve understanding of the potential risk due to cyanobacteria. If cyanotoxin concentration is low, a local authority may decide to proceed with the event in spite of high cyanobacterial density. Recreational facilities may also choose to invest in monitoring of the known cyanotoxins in order to avoid temporary closure of the facility whenever cyanobacteria proliferate.

10.1.2 Sample site selection

In addition to the general guidance provided in Chapter 2, site selection for cyanobacteria sampling should take account of specific incidents, that are suspected to be associated with exposure to cyanobacteria or algae. Additional factors to be considered in site selection are: the history, if available, of cyanobacterial and algal population development and occurrence of toxins at the water body or coastal strip; local characteristics of the catchment and water body which (may) influence the development and fate of cyanobacterial populations, or even their current location in parts of the water body; and the wider knowledge of the characteristics of cyanobacterial and algal population development and fate and the production and fate of cyanobacterial and algal toxins.

The heterogeneous and dynamic nature of many planktonic population developments may require sampling several sites, such as locations that are prone to accumulation or scum formation, particularly if these are in areas used for recreation; locations at beaches or in the open water body which are used for immersion sports or those involving accidental immersion (such as sailboarding or water-skiing); a central reference site in open, mixed water (experience may indicate if this can be used as a representative site for the main water mass to assess the total population size and thus estimate accumulation or scum formation potential); and decaying accumulations for dissolved toxins. A useful approach for large water bodies is to sample individual zones that differ in some physical, chemical, geomorphological or biological features. Preliminary sampling may help to define these zones, based on their spatial variability and gradients in environmental and plankton properties.

Sampling to assess the total population size may require information on the horizontal distribution of cyanobacteria within the water body, as well as on their distribution with depth. Particularly in thermally stratified water bodies, some cyanobacteria form pronounced population maxima at the depth with optimal light intensity and/or nutrient concentrations. Stratification leads to a water body functioning as two separate masses of water (the epilimnion and the hypolimnion) with different physico-chemical characteristics, with a transitional layer (metalimnion) sandwiched between. Thermal stratification can be determined by measuring vertical profiles of temperature within the water body. In temperate climates, thermal stratification generally occurs seasonally in water bodies of appropriate depth, whereas in tropical climates it often follows diurnal patterns. Thermal stratification has important implications for the distribution of the concentrations of nutrients and the interpretation of phosphorus and nitrogen data. Usually, shallow (2–3 m), wind exposed lakes do not stratify, whereas in temperate climates deeper lakes usually exhibit a stable stratification from spring to autumn. Lakes of intermediate depth (e.g. 5–7 m) may develop transient thermal stratification for a few calm and sunny days; the stratification is then disrupted by the next event of rain or wind. However, even if temperature is uniform throughout the water column, stratification of organisms can occur on calm days. Depth gradients of oxygen concentration and pH are good indicators of such stratification. Depth-integrated samples are more adequate than surface samples for the assessment of population size and nutrient concentrations in such situations. Approaches to optimising depth-integration are discussed by Utkilen *et al.* (1999) (see also section 10.3).

10.1.3 Monitoring frequency

Monitoring at time intervals should aim (i) to give warnings of developing cyanobacterial and algal populations and associated toxin levels; (ii) to provide information on the duration of cyanobacterial and algal populations and toxin levels that exceed guideline values; (iii) to provide information on the decline of cyanobacterial and algal populations and toxins due to natural processes; or (iv) to enable assessment of the persistence or reduction in cyanobacterial populations and toxin levels due to intervention, such as eutrophication control.

In recreational waters where bloom formation is suspected, the frequency of monitoring should be sufficient to provide data to enable an appropriate Alert Levels Framework system to operate (Chorus and Bartram, 1999). For example, monitoring may begin on a fortnightly basis and be increased to twice-weekly whilst alert levels are exceeded, before being reduced again after alert levels and guideline values for cyanobacterial cells and toxins are no longer exceeded. A scheme of suggested frequencies according to the steps of monitoring is presented in Table 10.3.

Table 10.3 Monitoring frequency and parameters for each of the three phases of the recommended monitoring strategy

Phase or activity	Parameters	Frequency
Monitoring phases		
Phase 1 (background)	Transparency (Secchi disc)	At least once a month
	Nutrient concentrations	At least twice a year (spring overturn and summer)
	Hydrophysical conditions (e.g. flow regime, thermal stratification)	At least once a month
Phase 2 (basic)	Transparency (Secchi disc)	Fortnightly
	Discolouration	Fortnightly
	Scum formation	Fortnightly
	Hydrophysical conditions	At least once a month
	Temperature	Possibly continuous
	Weather conditions (e.g. winds, light)	Continuous
	Changes in turbulence (i.e. mixing)	Possibly continuous
Phase 3 (cyanobacteria)	Transparency (Secchi disc)	Twice weekly
	Qualitative microscopic analysis	Twice weekly
	Quantitative microscopic analysis	Twice weekly
	Chlorophyll *a* analysis	Twice weekly
Additional activities		
Toxicity	Bioassays (to confirm the presence of toxicity in cells and/or released into the water)	Possibly at first appearance of situations (as in Phase 3) and in all cases when health problems are suspected or reported
Toxin concentration	Chemical analyses (to confirm the presence of specific toxins both qualitatively and quantitatively)	As above

If a water body prone to cyanobacterial mass development is to be used for water-contact sports on a seasonal basis, or for a single event, monitoring should begin not less than two weeks before the beginning of the season or the event. Monitoring should then continue, with the frequency adjusted to enable decisions to be made about access to the facility throughout the season, or whether to proceed with the event.

10.2 Laboratory and staff requirements

Monitoring for cyanobacterial and algal health hazards makes a range of demands upon analytical resources, some of which are different from those made by other aspects of water quality monitoring. Although a higher level of sophistication will provide more information, cyanobacterial and algal monitoring can be highly effective at a very low level of demand on facilities (Table 10.4).

Background monitoring of physical and chemical variables reflecting bloom-forming potential (such as transparency, nutrient concentrations and hydrophysical conditions) makes limited demands upon analytical resources and capacities and may be readily decentralised. While any capable chemical laboratory can carry out laboratory analysis, some limnological or oceanographic expertise is necessary for planning field work, quality control of data, and interpretation of results.

Basic monitoring for indicators of toxic cyanobacteria focuses mainly on critical site inspection and requires almost no facilities. If performed by local staff with observation skills and increasing experience, regular monitoring and recording of simple variables such as transparency, discolouration and scum formation, provides much information for management.

Monitoring of populations of cyanobacteria and algae requires a microscope and some skill in its use. Health authority staff with experience in microscopy can easily learn to recognise the most important toxin-producing cyanobacteria and algae in the waters under their responsibility, provided occasional training by experts is provided.

10.3 Sampling

Samples taken directly by immersion of a sample bottle or sampling device are termed "grab" samples; they are also known as "spot" or "snap" samples. For sampling cyanobacteria, grab samples are often taken from the surface. Composite or integrated samples consist of several sub-samples collected separately (e.g. from different parts of the water body) and then mixed together. They are taken for quantitative, representative samples when the variables to be assessed are unevenly distributed (but information on distribution is not required), for example, when assessing the total content of a substance in a water body (e.g. total phosphorus potentially available for phytoplankton growth) or the total population of an organism (e.g. taking into account the horizontal or vertical variations in distribution of cyanobacterial populations due to the presence of physico-chemical gradients). If knowledge of the precise distribution is required, each sub-sample can be evaluated individually. Composite samples may be:

- *Depth-integrated.* These are most commonly made up of two or more equal sub-samples collected with a sampler at predetermined depth intervals from the surface to just above the bottom. Selection of depths for subsampling must be adequate to account for stratification of temperature, substances and organisms in the water body (Utkilen *et al.,* 1999).

Table 10.4 Monitoring approaches, their requirements and options for their organisation

Monitoring type	Parameters of interest	Analytical demands	Who	Where	Notes
Background[1]	Nutrient concentrations (i.e. total phosphorus, nitrate and ammonia); flow regime; thermal stratification; transparency	Low, basic (i.e. photometer, boat, depth sampler and Secchi disc)	Environmental officers or consultants with limnological expertise	Local, regional	Readily incorporated into water resource monitoring
Basic[2]	Transparency; discolouration; scum formation	Minimal (i.e. Secchi disc and regular site inspection by trained staff)	Environmental or health officers	Local	Very high return in relation to input
Cyanobacteria	Dominant taxa – quantity (often determination of genus is sufficiently precise; quantification only as needed for management)	Low, basic (i.e. microscope; photometer is useful)	Environmental or health staff; consultants with limnological expertise	Local, regional	Specific training is required, but quite easily achieved; very high return in relation to input
Toxicity	Toxicity	Low, but skilled (i.e. biotests)	Toxicologists	Central	Demands on skills rather high
Toxin concentration	Toxin content	High[3]	Analytically skilled staff	Central	High return in relation to effort; enables de-warning if bloom proves to be non-toxic

1 Potential for cyanotoxin

2 Site inspection for indicators of toxic cyanobacteria

3 Methods with lower demands are currently under development, for example ELISA

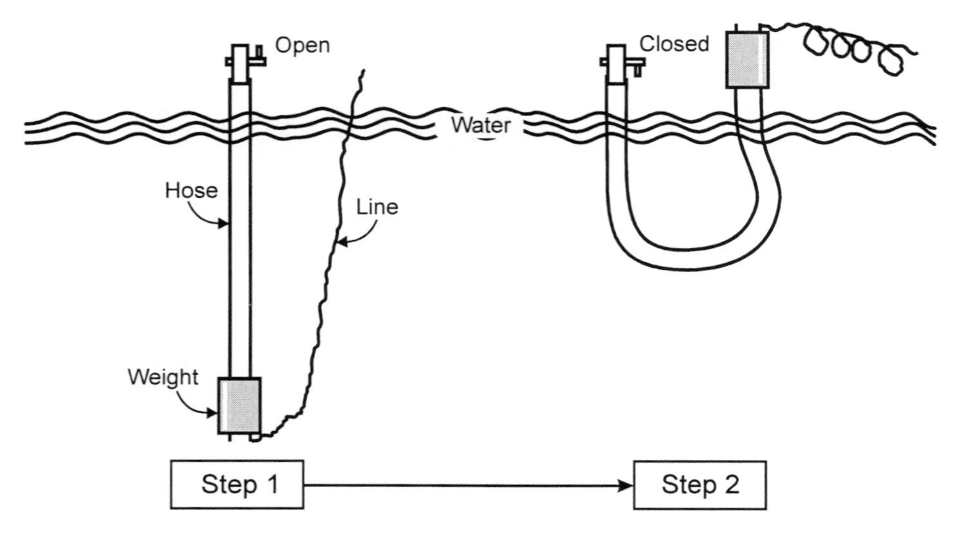

Figure 10.2 A hosepipe sampler for deep waters

Continuous depth integration can be obtained with a tube sampler or a pumping system (see Figures 10.2–10.4).

- *Area-integrated.* These are made by combining a series of samples taken at various sampling points spatially distributed in the water body (usually all at one depth or at predetermined depth intervals).
- *Time-integrated.* These are made by mixing equal volumes of water collected at a sampling station at regular time intervals.
- *Discharge-integrated.* These are integrated over time intervals adapted to the discharge at regular intervals over the period of interest. A common arrangement is to sample every 3 hours over a 24-hour period. The composite sample is then made by mixing portions of the individual samples that are proportional to the rate of discharge at the time the sample was taken.

A depth sampler, sometimes called a grab sampler or bottle (Van Dorn or Niskin type), is designed in such a way that it can retrieve a sample from any predetermined depth (Venrick, 1978). It consists of a tube that can be closed at its ends by spring-loaded flaps that are triggered by dropping a weight (called a messenger) down the lowering rope. A sample obtained in this way can be used for all chemical analyses except dissolved oxygen. This common sampler is relatively inexpensive, robust and can be deployed from almost any vessel. It gives samples for quantitative analysis.

The hosepipe sampler is a piece of flexible plastic piping of several metres in length, weighted at the bottom, and provides a simple mechanism for

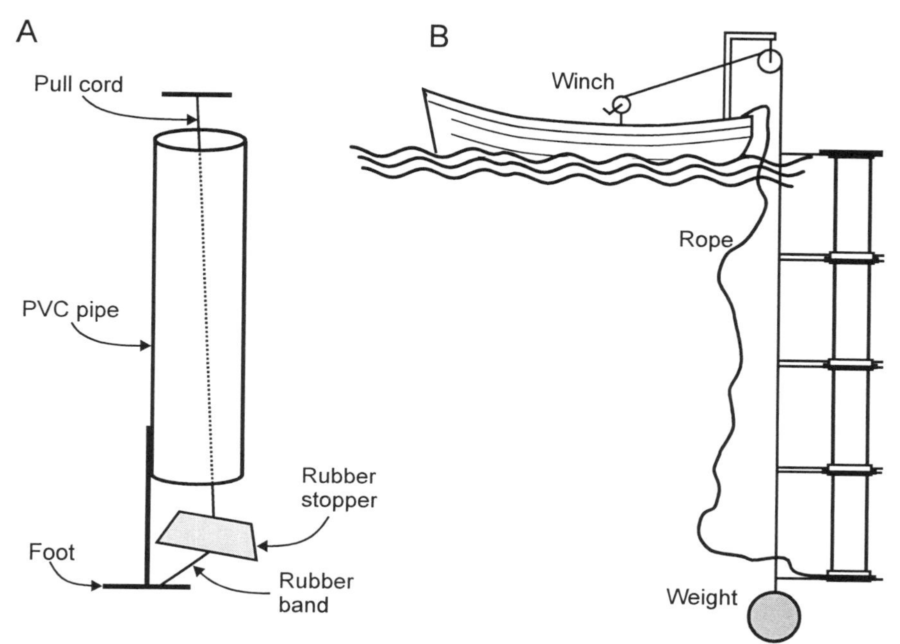

Figure 10.3 **A**. The pipe sampler: a simple device for depth integrated samples from shallow water bodies; **B**. The segmented tube sampler (After Sutherland *et al*., 1992)

collecting and integrating a water sample from the surface to the required depth in a lake. The hosepipe is lowered with its upper end open, trapping a water column as it is lowered (Figure 10.2, Step 1). The upper end is closed before hauling up the lower (open) end by means of an attached rope (Figure 10.2, Step 2). The total length of this tube can be up to 30–35 m and it is suitable for relatively calm waters. Another device suitable for taking depth-integrated samples for shallow water columns (less than 5 m deep) or surface waters of deeper water bodies is a simple pipe sampler (Figure 10.3A). The segmented tube sampler (Lindahl, 1986; Sutherland *et al.,* 1992) is a similar alternative for relatively shallow and calm waters. It consists of lengths (1–3 m each) of PVC (polyvinylchloride) pipe linked with valves, with a total length of up to 20 m (Figure 10.3B).

Integrated samples can also be obtained using a battery-operated water pump (electric diaphragm pumps are effective) and flexible plastic piping (Figure 10.4) which is operated at a steady pumping rate while the water inlet is drawn upwards between the desired depths at a uniform speed. This apparatus may also be operated to sample many litres of water from the same depth or to filter, for example through a plankton net, large quantities of water from a fixed depth for qualitative and for quantitative (volumes can be

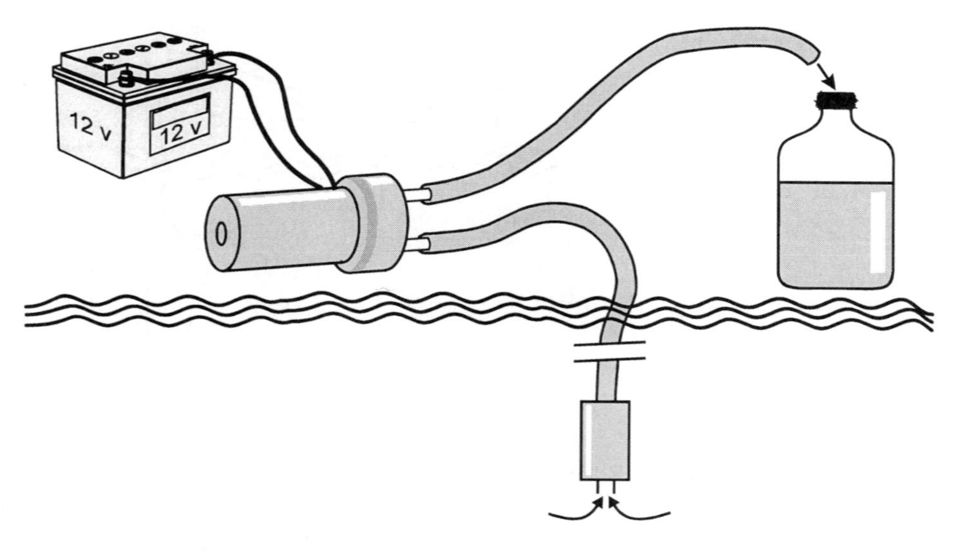

Figure 10.4 A pump sampling system

measured) analyses. General discussions are given in Beers (1978) and Powlik *et al.* (1991) for peristaltic pumps, and Voltolina (1993) and Taggart and Legget (1984) describe diaphragm pumps.

Sampling scums

Scums usually occur near shorelines at low water depths and therefore working with a grab sampler or a plankton net may be difficult. Sampling scums is carried out more easily with a wide-necked plastic or glass container. When sampling scums their heterogeneous density must be taken into account. Two different approaches may be developed. The first aims to assess the maximum density of cyanobacteria and/or highest toxin level by taking a sample where the scum is thickest (move the bottle mouth along the surface to collect the dense mats of buoyant cyanobacteria). The second approach aims to simulate conditions where shallow waters are mixed by bathers and playing children (agitate the scum before submerging the bottle). Both types of approach may be used for comparison.

Plankton net sampling

Sampling with a plankton net (Tangen, 1978) is mainly performed when large quantities of cell material are required (e.g. for toxicity testing) or when only qualitative analysis of phytoplankton is necessary. Sampling with a net causes a bias according to the size and shape of the organisms, i.e. the finest mesh size of 10 µm will miss small cells, such as unicells of *Microcystis* spp.

and picoplankton. Furthermore, filamentous cyanobacteria may be under-represented because some filaments may slip through the net.

The sampling depth is dependent on the taxa of algae and cyanobacteria present. Floating cells (*Microcystis, Anabaena*) are harvested within the upper few metres, whereas sampling of well mixed or stratified water bodies showing a depth distribution of cyanobacteria (e.g. *Planktothrix*) may include deeper water layers. For sampling the water column (or parts of it) lower the plankton net (25–50 μm mesh) to the desired depth, wait until the rope is taught, and then draw it back slowly to the surface. The net should be drawn very slowly out of the water to allow the water to run through it because large nets are sometimes heavily loaded with water when the pores are clogged with plankton. Rinsing plankton off the netting can be assisted by shaking the net slightly while raising it out of the water slowly.

For sampling surface blooms horizontal net hauls are more appropriate in order to filter floating cells. The plankton net should be moved parallel to the water surface. It can also be towed behind a boat moving slowly.

A disadvantage of collecting material with a plankton net is that the water volume filtered through the net cannot be determined precisely. Calculations based on the area of the net opening and length and distance hauled are not recommended because they overestimate strongly the amount of water actually filtered (due to clogging of the pores, only a fraction of the water volume will actually have passed through the net).

10.3.1 Determination of phosphorus and nitrogen

Water samples are collected with a clean sampler, a water pump or directly with the sample bottle. In shallow, unstratified lakes the sampling depth is less important than in deep, stratified lakes, where at least one sample from the epilimnion, one from the metalimnion and one from the hypolimnion should be taken. If this is not possible, a single surface sample will provide useful information, but only an incomplete picture of the growth conditions for cyanobacteria and algae.

A 100 ml sample container (see section 10.3.4) should be filled, immediately closed and stored cool. If analysis aims at differentiating between the different forms of phosphorus and nitrogen, the storage time should be as short as possible, and no more than 24 hours if the samples can be stored cool. Risks during extended storage involve transformations between dissolved and particulate fractions as well as between nitrate and ammonia. Preferably, the samples should be filtered in the field using membrane filters (0.45 μm pore diameter) pre-washed with a few millilitres of sample, and the filtered fractions should be stored separately. Alternatively samples should be filtered within 4 hours after sampling.

During filtration of samples for analysis of the dissolved nutrient fractions it is particularly important to avoid contamination. Filtering devices may be contaminated with higher concentrations from previous samples, especially if nutrient-enriched deep water layers had been filtered previously. This is especially important for phosphate, because it tends to adsorb to materials. Rinsing the equipment with double-distilled water between localities or with the sample to be filtered is recommended. It is helpful to take samples with low concentrations first (e.g. usually surface water) and to move on to samples in which higher concentrations are expected (e.g. deep waters).

10.3.2 Quantitative and qualitative determination of cyanobacteria and algae

For microscopic determination of cyanobacteria and algae, and for their microscopic quantification grab samples of 50–200 ml are put into a brown glass bottle and fixed immediately with Lugol's iodine solution or formaldehyde solution. Lugol's solution renders cells heavier, thus facilitating enumeration. Addition of 1–2 ml of Lugol's iodine per 100 ml of sample results in a 1 per cent final concentration (note: very hypertrophic waters may require more preservative). Formaldehyde should be avoided because it presents health risks to the user (it is a potent allergen), or it should be used only under conditions of excellent ventilation in the laboratory and at the microscope. It has the advantage of not causing discolouration of the sample, but the disadvantage of not enhancing settling in counting chambers as effectively as Lugol's solution. Additional fresh, unpreserved samples are useful for microscopic identification because the iodine in the Lugol's solution covers the characteristic colour of the cyanobacteria. Such samples need not be quantitative and may be collected with a plankton net (10 μm mesh) or as a grab sample at a site with high cyanobacteria and algal density. Unpreserved samples for identification may be stored for several hours without appreciable deterioration if kept cool during transportation to the laboratory.

If biomass is to be quantified by chemical analysis of chlorophyll *a* concentration, samples of 1 litre (or less if cell densities are high) of water are taken and filtered as soon as possible. Filtration in the field involves problems of filter transport, which is possible either on ice in an icebox, or submerged in ethanol used for extraction. Direct sunlight must be avoided during filtration and transport.

Materials and method

- ✓ Brown (or white) glass bottles: 50–200 ml, preferably pre-stocked with a few drops of Lugol's iodine solution.
- ✓ Brown glass bottles or dark plastic bottles of 1 litre (for chlorophyll *a*)
- ✓ Lugol's solution (Willén, 1962).

For chlorophyll *a* analysis, apparatus for field or laboratory filtration of the water samples includes:

- ✓ Electric vacuum pump (if filtration is to be performed in the field, a system using a 12V power supply or hand vacuum pump is necessary).
- ✓ Filtration device.
- ✓ Glass fibre filters (average pore size 0.7 µm, filter diameter 47 mm).
- ✓ Either ice and icebox or ethanol and glass vessels for filter transport.

Lugol's solution

Dissolve 20 g potassium iodide into 200 ml of distilled water, mix and add 10 g of sublimated iodine (the solution must not be supersaturated with iodine because this can result in crystal formation with consequent interference in counting). Supersaturation can be tested by diluting 1 ml of stock solution to 100 ml with distilled water to give concentrations similar to those in preserved samples. If iodine crystals appear after standing, more potassium iodide (approximately 5 g) should be added and the test repeated. If no crystals appear, 20 ml of glacial acetic acid must be added. Store stock solution in a dark bottle and use within one year.

10.3.3 Cyanotoxin analysis

For toxicity testing, a large amount of cell material may be collected (without determining the water volume from which it originates) with a plankton net as described in section 10.3. If the concentration of cell-bound toxin is to be related to the water volume from which the cells were collected, the best approach is to filter a defined volume as described by Lawton *et al.* (1994) through a 0.45 µm mesh membrane filter. The volume chosen should be sufficient to provide a pronounced greenish layer of material on the filter without clogging the filter. Care must be taken to stir the sample to disperse the cells evenly every time immediately before a sample is poured onto the filter. The filtrate may be used for the analysis of toxins dissolved in the water.

10.3.4 Sample containers

The laboratory that conducts the analyses should ideally provide containers and bottles for the transport of samples. These should be pre-labelled and well-arranged in suitable containers (if cooling is not necessary, soft drink crates, with subdivisions for each bottle, are cheap and very practical). For routine sampling of the same sites, it is advisable always to use the same bottle for each site and each variable. This avoids cross-contamination, which is a particular concern for phosphorus analyses. For most samples glass containers are appropriate, but often plastic containers (which are more stable than glass) can be used instead. The following containers are recommended for the transport of cyanobacteria and related samples:

- *Phosphorus analysis.* 100 ml glass bottles pre-washed and stored with sulphuric acid (4.5 mol l^{-1}) or hydrochloric acid (2 mol l^{-1}) until use, rinse well before use.
- *Nitrate, ammonia and total nitrogen.* Glass or polyethylene bottles (100 ml).
- *Microscopic identification of cyanobacteria.* Wide-mouth polyethylene bottles (100 ml) are appropriate for studying living material in a fresh grab or net sample.

- *Microscopic identification and quantification of cyanobacteria.* Brown glass bottles (100 ml) already containing preservative. Clear (plastic or glass) bottles may be used if the samples can be stored in the dark.
- *Chlorophyll a analysis.* 1 litre brown (plastic or glass) bottles are recommended to avoid degradation of chlorophyll by sunlight. Clear (plastic or glass) bottles may be used if the samples can be stored in the dark. If filtration for chlorophyll *a* analysis is performed in the field, moist filters are best transported either on ice, by folding them with the cell layer inside and wrapping them in aluminium foil, or by immersing them in ethanol immediately after filtration. For the latter, 50 ml wide-mouth brown glass bottles or 10 ml tightly sealing test tubes may be used.

For toxin analysis, cell material is either collected with a plankton net or by filtration. Material enriched with a plankton net is best transported in wide-mouth plastic containers, in which it may be frozen for subsequent freeze-drying. Material on filters is best transported either dry, if filters can be dried rapidly without direct sunlight, or on ice by folding them with the cell layer inside and wrapping them in aluminium foil. Samples for analysis of dissolved cyanotoxins are acquired by filtration either in the laboratory or in the field. They are transported (filtered or unfiltered) in 1 litre plastic containers.

10.3.5 Sample transport and preservation

Samples must be labelled clearly with sampling site (station), date and time of sampling and depth of sampling. Preservation of samples, filtered volumes and any irregularity should be noted in the field record. In general, samples should be stored cool and dark in a storage box (if necessary, with solid coolant) during sampling and transportation. Storage of samples between 2 °C and 5 °C may preserve many types of samples, but checks should be made to confirm this with each sample type. Preferably, the samples should be analysed immediately after sampling. If a storage time longer than 12 hours is necessary, quick-freezing of samples to –20 °C is recommended. Samples to be filtered must be filtered before freezing.

For the analysis of ammonia, storage is particularly critical. Samples can be cooled in a refrigerator but should be analysed within three hours of collection. Preservation for longer periods is not recommended. Filtration of samples should be avoided. It is nearly impossible to obtain filters free of ammonia and filtration may also evaporate the ammonia contained in the sample.

If samples for chlorophyll *a* are filtered in the field, filters must be either transported frozen, or submerged in ethanol. The ethanol should be boiling (at 75 °C) when put onto the filters. If this is not possible in the field, transporting filters submerged in cold ethanol and heating the ethanol later in the laboratory may be preferable to the risk of degradation occurring on filters transported dry but poorly cooled. The suitability of this approach should be checked for any given situation.

Samples for microscopic enumeration preserved with Lugol's iodine at the time of collection (see section 10.3.2) are relatively stable and no special storage is required, although they should be protected from extreme temperatures and strong light. Nevertheless, it is recommended that samples are examined and counted within a few weeks because some species of phytoplankton are sensitive to prolonged storage and Lugol's iodine solution and may degrade if stored for many months or even years. Unpreserved samples for microscopic quantification require immediate attention, either by addition of preservative or by following alternative counting methods that do not use preserved cells. Where unpreserved samples cannot be analysed immediately they should be stored in the dark with the temperature kept stable, at about +4 °C.

10.4 On-site analysis

A significant advantage of on-site testing is that tests are carried out on fresh samples which have not been contaminated or the characteristics of which have not otherwise changed as a result of storage. Some analyses such as temperature and transparency can only be carried out in the field.

10.4.1 Transparency

The transparency of water is strongly influenced by turbidity due to particles such as phytoplankton (cyanobacteria and algae) or suspended silt. It can be measured easily in the field. If transparency is low (less than 1–2 m) and accompanied by greenish to bluish discolouration, streaks, or even scums, high cyanobacterial densities are likely. Such results may initiate immediate inspection of further downwind sites and the collection of samples for cyanobacterial analysis.

Transparency can be obtained approximately with a Secchi disc (see Figure 10.5). The disc is made of rigid plastic or metal, but the details of its design are variable. It may be 20 or 30 cm or even larger in diameter and is usually painted white. Alternatively, it may be painted with black and white quadrants. The disc is suspended on a light rope or chain so that it remains horizontal when it is lowered into the water. The suspension rope is graduated at intervals of 0.1 and 1 m from the level of the disc itself and usually the rope does not need to be more than 5 m in length. A weight fastened below the disc helps to keep the suspension rope vertical while the measurement is being made. Transparency estimated in this way (submersible photometers are also available for these measurements) is taken to be the mean of the depths at which the disc disappears when viewed from the shaded side of the boat and

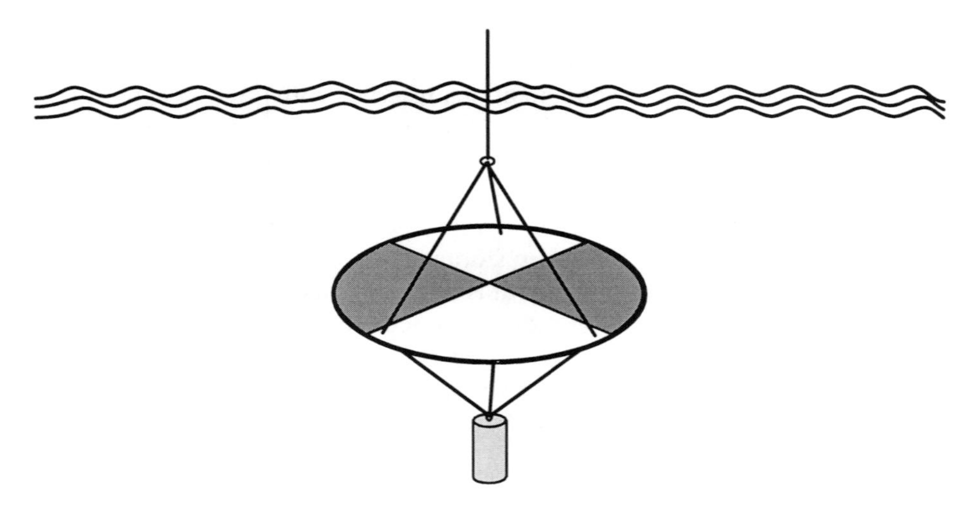

Figure 10.5 The Secchi disc for measuring transparency

at which it reappears upon raising after it has been lowered beyond visibility. Observation of the disc through a tube (painted black inside) with a transparent pane at the lower end and held just below the surface improves precision, particularly if the water surface is perturbed (Wetzel and Likens, 1990).

If cyanobacteria occur as floating streaks or mats on the water it is difficult to obtain representative transparency data. Depending on the measuring site, values can vary from 0 to greater than 2 m. It may be useful to determine transparency in areas without floating cells as well as within scums. The Secchi disc has to be lowered very carefully to prevent destroying the formation of accumulated cyanobacterial cells, and before taking the Secchi disc measurement the surface scums should be given time to return to their original water coverage again.

An improvised transparency determination may be recommended by local authorities for users of recreational sites known to be affected frequently by cyanobacterial mass developments. If bathers cannot see their feet while standing knee-deep in the water because of greenish turbidity, bathing should be avoided.

10.4.2 Temperature

Temperature is best measured *in situ* with a probe because water samples gradually reach the same temperature as the surrounding air. If this is not possible, it may be measured with a thermometer or probe in a water sample of at least 1 litre, immediately after taking the sample. Graduations of 0.1 °C are appropriate.

Procedure

1. If measuring in a sample, immerse the thermometer in the water until the temperature reading is constant. Record the reading to the nearest 0.1 °C.
2. If using a probe, lower the probe to the required depth. Hold it at that depth until the reading on the meter is constant. If a complete profile of temperatures is to be taken, measurements should be made at 1 m intervals from the surface to the bottom. Near the surface, or in areas of large thermal discontinuities, measurements should be made at intervals of less than 1 m.

10.4.3 *In situ* fluorometric analysis of chlorophyll *a* and remote sensing

Submersible fluorometers exist which can provide fine scale profiles vertically and horizontally. This is valuable, particularly for monitoring large water bodies and coastal areas for highly variable patterns of chlorophyll concentration (algal and/or cyanobacterial biomass). This approach has been used successfully to monitor variability in phytoplankton biomass and species composition, as well as surface temperature, salinity and nutrient concentrations, in the Baltic Sea. Fully automated analyser systems are installed on three passenger ferries. The system allows high frequency sampling with a spatial resolution of about 100 m and a temporal resolution of 1–3 days. The project uses, especially during the cyanobacterial bloom period, satellite images to detect the extent of the algal surface accumulations. The data are used to provide information on the Baltic Sea phytoplankton on the Internet at http://www.fimr.fi. Another example of the use of a flow-through system deployed on ferries has been reported from Japan (Harashima *et al.,* 1997).

Real-time data for chlorophyll *a* distribution and concentrations, and potentially for cyanobacterial phycobiliprotein pigments in freshwaters, can be generated by the remote sensing of water optical properties by high resolution airborne scanners (Cracknell *et al.*, 1990; Jupp *et al.*, 1994). However, flight times may be infrequent and data collection depends on factors such as cloud cover. Nevertheless, the remote sensing of cyanobacterial populations as a contribution to water body management has excellent potential.

Visible satellite imagery provides a synoptic perspective but, with few exceptions, does not yet have the ability to discriminate between different phytoplankton taxa nor is it effective during inclement weather. New techniques like the use of Advanced Very High Resolution Radiometers (AVHRR) on board the polar orbiting National Oceanic Atmospheric Administration (NOAA) satellites have been applied to monitor large-scale algal blooms, for example in North America (Gower, 1997) and the Baltic Sea (Kahru, 1997).

Table 10.5 Differentiation of phosphorus fractions

Fraction	Definition
1. Soluble reactive phosphorus	Filtered sample
2. Dissolved organic phosphate	Digested filtered sample
3. Particulate phosphorus	Total phosphorus less the dissolved organic phosphate fraction (i.e. 4 minus 2)
4. Total phosphorus	Digested unfiltered sample

10.5 Determination of nutrients in the laboratory

Whereas for phosphorus, the method given in section 10.5.1 is widely accepted and easy to perform with common laboratory equipment, several methods may be considered appropriate for nitrate, depending on available equipment. The method given in section 10.5.2 demands the least equipment but, due to an evaporation step, reproducibility may be poorer than for the method given in section 10.5.3. Both methods use hazardous chemicals which require appropriate safety protection and hazardous waste collection. Ion chromatography, if available, evades this problem and may be the preferred option.

10.5.1 Phosphorus

Some widely accepted digestion methods for dissolving particles to release all of the phosphate achieve this aim only incompletely. The procedure of Koroleff (1983a) for determining total phosphate has proved to be simple and efficient and is the basis of the ISO/FDIS protocol 6878 (ISO, 1998). This method is applicable to many types of water including seawater, in a concentration range of 5–800 $\mu g\ l^{-1}$ of P, or higher if the samples are diluted. Differentiation by the fractions shown in Table 10.5 is possible through filtration. Further information is available in Wetzel and Likens (1990) and APHA/AWWA/WPCF (1995).

Principle

Digestion or mineralisation of organophosphorus compounds to SRP (also known as orthophosphate) is performed in tightly sealed screw-cap vessels with persulphate, under pressure and heat in an autoclave (in the absence of which good results have also been obtained with a household pressure cooker), or simply by gentle boiling. Polyphosphates and many organophosphorus compounds may also be hydrolysed with sulphuric acid to molybdate-reactive orthophosphate. Many organophosphorus compounds are converted to SRP by mineralisation with persulphate. Orthophosphate ions are reacted with an acid solution containing molybdate and antimony ions to form an antimony phosphomolybdate complex. The complex is reduced with ascorbic acid to form a strongly coloured molybdenum blue complex. The absorbance of this complex is measured to determine the concentration of SRP present. An overview of the procedure, necessary equipment and chemicals is provided below (see also ISO, 1998).

Reagents

Only reagents of recognised analytical grade, and only distilled water having a phosphate content that is negligible compared with the smallest concentration to be determined in the samples, should be used.

- ✓ Sulphuric acid (1.84 g ml^{-1}) (Caution, eye protection and protective clothing are necessary).
- ✓ Sulphuric acid, solution, $c(H_2SO_4)$ = 9 mol l^{-1}. Add 500 ± 5 ml of water to a 2 litre beaker. Cautiously add, with continuous stirring, 500 ± 5 ml of sulphuric acid (1.84 g ml^{-1}) and mix well. (Caution, eye protection and protective clothing are necessary).
- ✓ Sulphuric acid, solution, $c(H_2SO_4)$ = 4.5 mol l^{-1}. Add 500 ± 5 ml of water to a 2 litre beaker. Cautiously add, with continuous stirring, 500 ± 5 ml of sulphuric acid solution (9 mol l^{-1}) and mix well. (Caution, eye protection and protective clothing are necessary)
- ✓ Sulphuric acid, solution, $c(H_2SO_4)$ = 2 mol l^{-1}. Add 300 ± 3 ml of water to a 1 litre beaker. Cautiously add, with continuous stirring and cooling, 110 ± 2 ml of sulphuric acid solution (9 mol l^{-1}). Dilute to 500 ml ± 2 ml with water and mix well. (Caution, eye protection and protective clothing are necessary)
- ✓ Ascorbic acid, 100 g l^{-1} solution. Dissolve 10 g ± 0.5 g of ascorbic acid in 100 ml ± 5 ml of water. The solution is stable for two weeks if stored in an amber glass bottle in a refrigerator and can be used as long as it remains colourless.
- ✓ Sodium hydroxide, solution, c(NaOH) = 2 mol l^{-1}. Dissolve 80 g of sodium hydroxide pellets in water, cool and dilute to 1 litre with water. (Caution, eye protection and protective clothing are necessary).
- ✓ Acid molybdate, solution. Add the molybdate solution (I) to 300 ml ± 5 ml of sulphuric acid 9 mol l^{-1} with continuous stirring. Add the tartrate solution (II) and mix well. (Caution, eye protection and protective clothing are necessary).
- ✓ Molybdate solution (I). Dissolve 13 g ± 0.5 g of ammonium heptamolybdate tetrahydrate [$(NH_4)_6Mo_7O_{24}4H_2O$] in 100 ml ± 5 ml of water.
- ✓ Tartrate solution (II). Dissolve 0.35 g antimony potassium tartrate hemihydrate [$K(SbO)C_4H_4O_6½H_2O$] in 100 ml ± 5 ml of water. The reagent is stable for at least two months in an amber glass bottle.
- ✓ Sodium thiosulphate pentahydrate, 12 g l^{-1} solution. Dissolve 1.2 g sodium thiosulphate pentahydrate ($Na_2S_2O_35H_2O$) in 100 ml water. Add about 50 mg of anhydrous sodium carbonate (Na_2CO_3) as preservative. This reagent is stable for at least four weeks if stored in an amber glass bottle.
- ✓ Potassium peroxodisulphate solution. Add 5 g potassium peroxodisulphate (K_2 S_2O_8) to 100 ml water, stir to dissolve. The solution is stable for at least two weeks, if the supersaturated solution is stored in an amber borosilicate bottle, protected from direct sunlight.
- ✓ Soluble reactive phosphate (orthophosphate), stock standard solution corresponding to 50 mg of P per litre. Dry a few grams of potassium dihydrogenphosphate to constant mass at 105 °C. Dissolve 0.2197 g of KH_2PO_4 in about 800 ml of water in a 1,000 ml volumetric flask. Add 10 ml of 4.5 mol l^{-1} sulphuric acid and make up to the mark with water. The solution is stable for at least three months if stored in a well stoppered glass bottle. Refrigeration to about 4 °C is recommended.
- ✓ Soluble reactive phosphate (orthophosphate), standard solution corresponding to 2 mg of P per litre. Pipette 20 ml of SRP stock standard solution into a 500 ml volumetric flask.

Make up to the mark with water. Prepare this solution each day it is required. One millilitre of this standard solution contains 2 μg of P.

Apparatus

- ✓ Ordinary laboratory apparatus and filter assembly with membrane filters, 40–50 mm diameter with 0.45 μm pore size.
- ✓ Pre-cleaned glass bottles for filtered samples.
- ✓ Spectrometer, suitable for measuring absorbance in the visible and near infrared regions. Capable of accepting optical cells with pathlengths from 1 cm to 5 cm. The most sensitive wavelength is 880 nm, but if a loss of sensitivity is acceptable, absorbance can be measured at the second maximum of 680–700 nm. The detection limit of the method is lower if a spectrometer capable of accepting 10 cm pathlength optical cells is available.
- ✓ Autoclave (or pressure cooker): used for digestion of samples at 115–120 °C.
- ✓ Borosilicate flasks, 100 ml, with glass stoppers tightly fastened by metal clips (heat resistant polypropylene bottles or conical flasks (screw capped) are also suitable). Before use, clean the bottles or flasks by adding about 50 ml of water and 2 ml of 1.84 gm l^{-1} sulphuric acid. Place in an autoclave for 30 minutes at 115–120 °C, cool and rinse with distilled water. Repeat the procedure several times and store filled with distilled water and covered.

Before use, all glassware should be washed with 2 mol l^{-1} hydrochloric acid at 45–50 °C and rinsed thoroughly with water. Do not use detergents containing phosphate. Preferably, the glassware should be used only for the determination of phosphorus. After use it should be cleaned as above and kept covered until use. Glassware used for the colour development stage should be rinsed occasionally with sodium hydroxide solution to remove deposits of the coloured complex that has a tendency to stick (as a thin film) on the walls of glassware.

Procedure

If filtration is necessary for the determination of total soluble phosphorus and/or dissolved phosphate, filter the sample within 4 hours after sampling. If the sample was cooled, bring it to room temperature before filtration. Wash a 0.45 μm membrane filter to ensure it is free of phosphate by passing through it 200 ml of water, previously heated at 30–40 °C. Filter the sample discarding approximately the first 10 ml of filtrate and collecting 5–40 ml depending on the concentrations expected. The filtration time should not exceed 10 minutes. If necessary a larger diameter filter should be used. Add 1 ml of 4.5 mol l^{-1} sulphuric acid per 100 ml of test sample. The acidity should be about pH 1, if not, adjust with NaOH 2 mol l^{-1} or H_2SO_4 2 mol l^{-1}. Store in a cool dark place until analysis is possible.

The mineralisation method using potassium peroxodisulphate is described here. This method will not be efficient in the presence of large quantities of organic matter. In this case oxidation with nitric acid–sulphuric acid is necessary. This latter procedure must be carried out in an efficient fume cupboard.

1. Pipette up to a maximum of 40 ml of the test sample (appropriately prepared) into a 100 ml conical flask. If necessary, dilute with water to about 40 ml.
2. Add 4 ml of potassium peroxodisulphate solution and boil gently for 30 minutes. Periodically, add sufficient water so that the volume remains between 25 ml and 35 ml.
3. Cool, adjust pH to between 3 and 10, transfer to a 50 ml volumetric flask, and dilute with water to about 40 ml.

Thirty minutes is usually sufficient to mineralise phosphorus compounds; some polyphosphoric acids need up to 90 minutes for hydrolysis. Alternatively mineralise for 30 minutes in an autoclave at between 115 °C and 120 °C. Most ordinary kitchen pressure cookers are adequate if laboratory equipment is not available. Any arsenate present will cause interference. If arsenic is known or suspected to be present in the sample, eliminate the interference by treating the solution with sodium thiosulphate solution immediately after the mineralisation step. In

Table 10.6 Selection of appropriate volumes for test portions in relation to the concentration of soluble reactive phosphorus

SRP concentration range (mg l^{-1})	Volume of test portion (ml)	Thickness of optical cell (cm)
0.0 to 0.2	40	4 or 5
0.0 to 0.8	40	1
0.0 to 1.6	20	1
0.0 to 3.2	10	1
0.0 to 6.4	5	1

the case of seawater mineralised in an autoclave, free chlorine must be removed by boiling before the arsenate is reduced by thiosulphate. Iron concentrations above 600 mg l^{-1} (e.g. in mining lakes) and sulphide (detectable by its smell) will also interfere.

4. Test portion after the mineralisation or filtration step. The maximum volume of test portion to be used is 40 ml. This is suitable for the determination of SRP concentrations of up to 0.8 mg l^{-1} when using an optical cell of 1 cm pathlength to measure the absorbance of the coloured complex formed by the reaction with acid molybdate reagent. Smaller test portions may be used as appropriate in order to accommodate higher phosphate concentrations (Table 10.6). Phosphate concentrations at the lower end of the calibration ranges are best determined by measuring absorbance in optical cells of 4, 5 or 10 cm pathlength.

   Carry out a blank test in parallel with the determination, by the same procedure, using the same quantities of all the reagents as in the determination, but using the appropriate volume of water instead of the test portion.
5. Calibration solutions. (A1) To prepare the set of calibration solutions transfer, by means of a pipette, 1, 2, 3, 4, 5, 6, 7, 8, 9 and 10 ml of the SRP standard solution to 50 ml volumetric flasks. Dilute with water to about 40 ml. These solutions represent SRP concentrations from 0.04 mg l^{-1} to 0.4 mg l^{-1}. If total phosphate or total soluble phosphate is being determined, proceed according to the mineralisation method chosen. Then proceed to colour development. Proceed accordingly for other ranges of phosphate concentration (Table 10.6). Typically, the test portion volume will be in the range of 5–10 ml.
6. Colour development . (A2) Add to each 50 ml flask, while swirling, 1 ml of ascorbic acid 100 g l^{-1} and, after 30 seconds, 2 ml of acid molybdate solution I. Make up to the mark with water and mix well.
7. Spectrometric measurements. (A3) Measure the absorbance of each solution at 880 nm after 10–30 min, or if loss of sensitivity can be accepted at 700 nm. Use water in the reference cell.
8. Plotting the calibration graph. (A4) Plot a graph of absorbance against the phosphorus content (in mg l^{-1}) of the calibration solutions. The relationship between absorbance and concentration is linear. Determine the reciprocal of the slope of the graph. Check the graph from time to time, especially if new packages of chemicals are used. Run a calibration solution with each series of samples.
9. Determination. (B1) Colour development — proceed as in (A2) using the test portion appropriately processed. (B2) Spectrometric measurements — proceed as in (A3).

Expression of results

The concentration of total phosphorus expressed in mg l^{-1} is given by the equation:

$$P_{tot} = \frac{(A - A_0)V_{max}}{fV_s}$$

where:

- A is the absorbance of the test portion
- A_0 is the absorbance of the blank test
- f is the slope of the calibration graph (e4), in litres per milligram
- V_{max} is the reference volume of the test portion (50 ml)
- V_s is the actual volume, in ml, of the test portion.

The test report should contain complete sample identification, reference to the method used, the results obtained and any further details likely to influence the results.

10.5.2 Spectrometric method for nitrate using sulphosalicylic acid

This method does not require sophisticated equipment and is suitable for surface and potable water samples (ISO, 1998). The method may be used up to nitrate-nitrogen concentrations of 0.2 mg l^{-1} using the maximum test portion volume of 25 ml, and can be expanded by using smaller test portions. The limit of detection lies within 0.003 and 0.013 mg l^{-1}, using cells of path-length 40 mm and a 25 ml test portion volume. A nitrate-nitrogen concentration of 0.2 mg l^{-1} gives an absorbance of about 0.68 units, using a 25 ml test portion and cells of 40 mm pathlength. The main interferences are chloride, SRP, magnesium and manganese (II). Interference problems can be avoided with other spectrometric methods such as ISO 7890-1 and 7890-2 (ISO, 1986a,b).

Reagents

- ✓ Sulphuric acid ≈18 mol l^{-1} = 1.84 g ml^{-1} (Caution, eye protection and protective clothing are necessary).
- ✓ Glacial acetic acid ≈ 17 mol l^{-1} = 1.05 g ml^{-1} (Caution, eye protection and protective clothing are necessary).
- ✓ Alkali solution = 200 g l^{-1}. Dissolve with care 200 g ± 2 g of sodium hydroxide pellets in about 800 ml of water. Add 50 g ± 0.5 g of EDTA-Na_2 and dissolve. Cool to room temperature and make up to 1 litre with water in a measuring cylinder. Store in polyethylene bottle. This reagent is stable indefinitely. (Caution, eye protection and protective clothing are necessary).
- ✓ Sodium azide solution = 0.5 g l^{-1}. Dissolve with care 0.05 g ± 0.005 g of sodium azide in about 90 ml of water and dilute to 100 ml with water in a measuring cylinder. Store in a glass bottle. This reagent is stable indefinitely. (Caution: this reagent is very toxic if swallowed. Contact between the solid reagent and acid liberates very toxic gas).
- ✓ Sodium salicylate solution 10 g l^{-1}. Dissolve 1 g ± 0.1 g of sodium salicylate in 100 ml ± l ml of water. Store in a glass polyethylene bottle. Prepare this solution freshly on each day of operation.
- ✓ Nitrate, stock standard solution 1,000 mg l^{-1}. Dissolve 7.215 g ± 0.001 g of potassium nitrate (previously dried at 105 °C for at least 2 h) in about 750 ml of water. Transfer to a 1 litre volumetric flask and make up to one litre mark with water. Store the solution in a glass bottle for not more than two months.

- ✓ Nitrate, standard solution 100 mg l^{-1}. Pipette 50 ml of the stock standard solution into a 500 ml volumetric flask and make up to the 500 ml mark with water. Store the solution in a glass bottle for not more than one month.
- ✓ Nitrate, working standard solution 1 mg l^{-1}. Into a 500 ml volumetric flask, pipette 5 ml of standard nitrate solution. Make up to 500 ml mark with water. Prepare the solution freshly on each occasion of use.

Apparatus

Standard laboratory apparatus plus:

- ✓ Spectrometer, capable of operating at 415 nm with cells of 40 or 50 mm pathlength.
- ✓ Evaporating dishes, about 50-ml capacity. If the dishes are new or not in regular use, they should be rinsed first with water and taken through the procedure as in the colour development step (see below).
- ✓ Water bath, boiling, capable of accepting at least six of the evaporating dishes.
- ✓ Water bath, capable of thermostatic regulation to 25 °C ± 0.5 °C.

Procedure

Warning: This procedure involves the use of concentrated sulphuric acid, acetic acid, sodium hydroxide and sodium azide solutions. Eye protection and protective clothing are essential when using these reagents. They must never be pipetted by mouth.

The maximum test portion volume, which can be used for the determination of nitrate concentrations up to 0.2 mg l^{-1}, is 25 ml. Use smaller test portions as appropriate in order to accommodate higher nitrate concentrations. Because surface water samples contain suspended matter, allow them to settle, centrifuge them or filter them through a washed glass fibre filter before taking the test portion. Neutralise samples with a pH value greater than 8 with acetic acid before taking the test portion.

Carry out a blank test in parallel with the determination, using 5 ml ± 0.05 ml of water instead of the test portion and designate the measured absorbance A_b.

1. Calibration. To prepare the set of calibration solutions add, to a series of clean evaporating dishes, using a burette, 1, 2, 3, 4 and 5 ml respectively of the working nitrate standard solution, corresponding to nitrate amounts of 1, 2, 3, 4 and 5 µg in the respective dishes.
2. Colour development. Add 0.5 ml ± 0.005 ml of sodium azide solution and 0.2 ml ± 0.002 ml of acetic acid. Wait for at least 5 minutes, and then evaporate the mixture to dryness in the boiling water bath. Add 1 ml ± 0.01 ml of sodium salicylate solution, mix well and evaporate the mixture to dryness again. Remove the dish from the water bath and allow the dish to cool to room temperature.

   Add 1 ml ± 0.01 ml of sulphuric acid and dissolve the residue in the dish by gentle agitation. Allow the mixture to stand for about 10 minutes. Then add 10 ml ± 0.1 ml of water followed by 10 ml ± 0.1 ml of alkali solution.

   Transfer the mixture to a 25-ml volumetric flask but do not make up to the 25 ml mark. Place the flask in the water bath at 25 °C ± 0.5 °C for 10 min ± 2 min. Then remove the flask and make up to the 25 ml mark with water.
3. Spectrometric measurements. Measure the absorbance of the solution at 415 nm in cells of pathlength 40 or 50 mm against distilled water as a reference. Designate the absorbance measured as A_s.
4. Plotting the calibration graph. Subtract the absorbance of the blank solution from the absorbances of each of the calibration solutions and plot a calibration graph.
5. Determination. Pipette the selected test portion of volume *V*, such that the aliquot contains a mass of nitrate-nitrogen between 1 µg and 5 µg, into a small evaporating dish. Then proceed as in the preceding "Colour development" and "Spectrometric measurements" steps.

Table 10.7 The effect of other substances on the results obtained with the spectrometric method for nitrate using sulphosalicylic acid

Other substance	Amount of other substance in a 25 ml test portion (μg)	Effect of other substance in a 25 ml test portion	
		m(N) = 0.00 μg (μg N)	m(N) = 5.00 μg (μg N)
Sodium chloride	10,000	+ 0.03	– 0.73
Sodium chloride	2,000	+ 0.01	– 0.16
Sodium sulphate	10,000	+ 0.04	– 0.16
Sodium hydrogen carbonate	10,000	– 0.02	– 0.52
Sodium hydrogen carbonate	2,000	– 0.03	– 0.18
Calcium chloride	5,000	+ 0.23	+ 0.38
Calcium chloride	2,500	+ 0.02	– 0.14
Iron (III) sulphate	20	+ 0.08	– 0.02
Manganese (II) sulphate	20	+ 0.92	+ 0.99
Manganese (II) sulphate	5	+ 0.05	+ 0.13
Zinc sulphate	20	– 0.02	+ 0.07
Copper sulphate	20	+ 0.03	+ 0.19
Ammonium chloride	500	– 0.12	– 0.17

6. Correction for test portion absorption. If absorption by the test portion at the analytical wavelength is known, or suspected, to interfere (as may arise with highly coloured samples), carry out the operations given in the preceding "Colour development" and "Spectrometric measurements" steps, on the duplicate test portion but omitting the addition of sodium salicylate solution. Designate the absorbance measured be A_c.

Expression of results
Calculate the absorbance due to nitrate in the test portion, A_t, from the equation:

$$A_t = A_s - A_b$$

or, when a correction for sample absorption has been made, from the equation:

$$A_t = A_s - A_b - A_c$$

In both equations A_s, A_b and A_c refer to the sample, blank and correction absorbances respectively (see relevant sections above). Read off from the calibration graph, the mass of nitrate, *m(N)* in micrograms, corresponding to the absorbance value A_t.

The nitrate content in the sample, in mg l^{-1}, is given by the formula: $m(N)/V$ where V is the volume of the test portion (in ml). The effect of other substances on this method is provided in Table 10.7.

10.5.3 Spectrometric method for nitrate by reduction of nitrate to nitrite

Nitrates are reduced to nitrites almost quantitatively by amalgamated granulated cadmium. Separate methods for nitrate reduction and for nitrite determination are presented below.

Determination of nitrate: scope, field application and principle

This method is based on the reduction of nitrate ions to nitrite. Because it determines the sum of nitrite and nitrate ions, a separate determination of nitrite must be conducted, and its concentration subtracted from the sum of nitrate and nitrite. At concentrations higher than about 20 μM NO_3–N, calibration factors for a low and high range must be established. The reduction is carried out at a pH of about 8.5. Ammonium chloride buffer is used to control the pH and to complex the liberated cadmium ions (Carlberg, 1972).

Reagents

- ✓ Sulphanilamide (SAN) (Caution: possibly hazardous, use fume cupboard, solution must not be discharged to a public sewer). Dissolve 10 g SAN, $NH_2C_6H_4SO_2NH_2$, in a mixture of 100 ml concentrated hydrochloric acid, HCl (36 per cent) and 600 ml bi-distilled water (BdW). After cooling dilute the solution to 1 litre. At room temperature, when stored in glass bottles, the reagent is stable for several months.
- ✓ Naphtylamine solution (NED) (Caution: possibly hazardous, use fume cupboard, solution must not be discharged to a public sewer). Dissolve 1 g N-(1-naphthyl)-ethylene-diamine dihydrochloride, $C_{10}H_7NHCH_2NH_22HCl$, in BdW and dilute to 1 litre. The solution should be stored in a tightly closed dark bottle, with 3–4 drops of saturated $HgCl_2$ solution in a refrigerator (Kirkwood, 1992). The solution contains 10 μmoles ml^{-1}.
- ✓ Buffer solutions:

  25 per cent stock buffer. Dissolve 250 g ammonium chloride, NH_4Cl in BdW and 25 ml concentrated ammonium hydroxide 25 per cent. Dilute to 1 litre.

  2.5 per cent work buffer (WB). Dilute 100 ml of stock buffer with BdW to 1 litre.

  Wash buffer solution (WbS). Dilute 20 ml of 2.5 per cent WB with BdW to 1 litre.
- ✓ Hydrochloric acid 2M. Dilute 165 ml of concentrated commercial HCl (37 per cent) with BdW to 1 litre.
- ✓ Mercuric chloride solution 1 per cent (Caution, highly toxic). Dissolve 5 g mercuric chloride, $HgCl_2$, in 500 ml BdW.
- ✓ Synthetic seawater (SSW). Dissolve 36 g sodium chloride, NaCl, 12 g magnesium sulphate heptahydrate, $MgSO_47H_2O$, and 0.25 g sodium carbonate, $NaHCO_3$, with BdW and dilute to 1 litre. For analytical purposes this is equivalent to a salinity of 40 psu. For calibration work the SSW may be diluted to the desired salinity.
- ✓ Nitrate standard stock solution (NO_3-N SSS). Dissolve 0.25278 g potassium nitrate, KNO_3 (molecular weight 101.11) dried at 110 °C to constant weight, in BdW and dilute to 250 ml. Store in a tightly closed dark bottle with 2–3 drops of a saturated $HgCl_2$ solution in a refrigerator (Kirkwood, 1992). The solution contains 10 μmoles ml^{-1}.
- ✓ Cadmium coarse pulver. Sieve commercially available granulated cadmium and retain and use the fraction between 35 and 40 mesh, i.e. around 0.5 to 0.42 mm. (Caution: Cadmium is a poisonous metal. It should be handled with great care. All operations on the dry metal, particularly the granules, must be carried out in a well-ventilated area, e.g. a fume cupboard. Never inhale the dust. Cadmium must be treated as hazardous waste).
- ✓ Amalgamated cadmium. The required amount of cadmium metal is about 35 g per reduction column (RC) (Figure 10.6). The sieved granules are rinsed from the oxides by washing with 2M HCl. Then they are washed with plenty of water to eliminate all HCl. All the washed metal is transferred to a round-bottom flask that is filled with 1 per cent $HgCl_2$ solution. The flask is closed with a glass stopper. After this step all contact between air

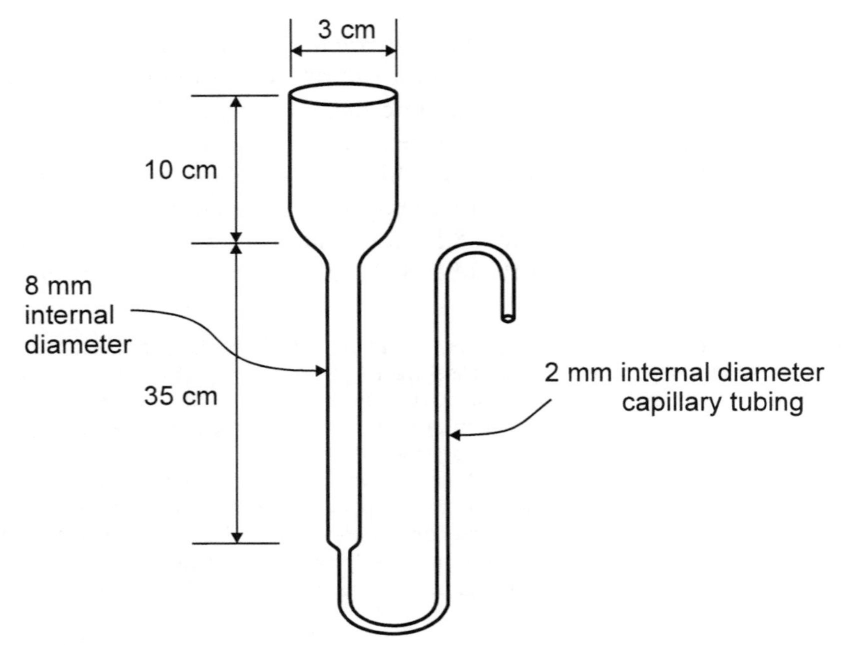

Figure 10.6 A reduction column

and the metal should be avoided. The flask is rotated for 90 minutes in a horizontal position or shaken with suitable equipment. Finally the flask is opened and the turbid sublimate solution rinsed out with BdW. (Caution: the used $HgCl_2$ solution must not be discharged to a public sewer). When a suitable volume of $HgCl_2$ is collected, 25 ml concentrated HCl is added per litre and then precipitated with hydrogen sulphide, H_2S, or sodium sulphide, Na_2S. The liquid is filtered and the precipitate stored, discarding the clear filtrate.

Apparatus and equipment
- ✓ Test tubes with glass or plastic stoppers, graduated or marked at 25 ml volume.
- ✓ Automatic syringe pipettes of 1 ml.
- ✓ 25 ml automatic pipette.
- ✓ 500 ml round-bottom flask.
- ✓ Reduction columns (Figure 10.6).
- ✓ Spectrometer, with 1, 5 and 10 cm pathlength cells.

Preparation of the reduction columns
A small ball of thin copper wire is placed at the bottom of the RC and above the wire a small ball of glass wool. The RC is filled entirely with water. The metal granules are poured into the RC, with a small plastic spoon (making sure that no cavities are formed in the column) and filled to about 1 cm below the reservoir. The amalgamated metal is activated by passing through about 150 ml of Wash buffer Solution (WbS) containing about 100 µM NO_3-N then rinsed thoroughly. The RC is packed with WbS only, before being used for analysis. A newly prepared RC reduces nitrate with an efficiency of 95–100 per cent.

Table 10.8 Preparation of working standard solutions of nitrate by dilution with bi-distilled water or synthetic seawater

Volume of solution D[1] (ml)	Total volume (ml)	Resultant concentration (μM NO_3^--N)
25.0	500	10.00
25.0	1,000	5.00
5.0	1,000	1.00

[1] Solution "D" is prepared by diluting 5 ml of an SSS in 250 ml, giving a solution which contains 0.20 μmoles NO_3^--N per ml

Calibration

There is a significant salinity effect in the calibration for nitrate measurements by manual methods using Hg–Cd reduction columns. Freshly amalgamated RC show a salinity effect of less than 10 per cent while the same RC, after several weeks use, shows a higher discrepancy (up to 30 per cent) when calibration against Working Standard Solution (WSS) made up from BdW and compared with standards in SSW of 35 psu. Working Standard Solution should therefore be made from SSW or the magnitude of the salinity effect should be recorded frequently, whereafter proper correction should be applied to the data. A series of WSS is prepared from the NO_3-N SSS by dilutions with BdW (or SSW) using volumetric flasks (Table 10.8).

Triplicates of WSS and the blank samples with BdW are analysed as described below. Each RC should be calibrated using blanks and calibration solutions. The linear regression of the absorbances measured in the spectrometer against the concentrations of the WSS (including absorbances of the blank samples, concentration equals 0) gives the calibration factor (cf). Using 5 cm cells, a cf of approximately 4.3 should be obtained.

Analysis

Pour 25 ml of sample into the reservoir. Immediately add 1 ml of WB using an automatic syringe pipette, followed by another 25 ml of the sample. Let this pass through the amalgamated metal. Collect drops in a test tube: the first 25 ml are discarded (use them as washer); the second 25 ml are the nitrite sample. The turbidity reference samples are unnecessary. The RC is now ready to receive the next sample. After every analytical batch, the RC must be flushed with WbS. It should never be left to dry. The concentration of the nitrate in the samples is calculated by multiplying their absorbances (A_s) by the cf:

$$\text{concentration } NO_3\text{-N} = \text{cf} \times A_s \text{ [μM]}$$

Control of the reduction efficiency

The reduction efficiency of each RC must be controlled from time to time, preferably for every analytical batch. Duplicates of WSS for nitrite must be analysed, followed by WSS for nitrate of the same concentration:

$$\text{Reduction efficiency (\%)} = \frac{\text{Absorbance of the nitrate WSS} \times 100}{\text{Absorbance of the nitrite WSS}}$$

If the reduction efficiency decreases below 85 per cent, empty the RC, wash the filings quickly with 2M HCl and rinse well with water. Dry the filings, sieve and reamalgamate as described above.

Determination of nitrite: scope, field application and principle

This method is specific for nitrite ions and is applicable to all types of marine waters. It is not appreciably affected by salinity, small changes in reagent concentrations, or by temperature (Grasshoff, 1983). Using 5 cm cells, the detection limit is about 0.02 μM and shows linearity up to about 10 μM. The determination is based on the reaction of nitrite ions with sulphanilamide with the formation of a diazonium compound which, coupled to a second aromatic amine, forms a coloured azo dye.

Reagents

- ✓ Sulphanilamide (SAN) — as in nitrate determination.
- ✓ Naphtylamine solution (NED) — as in nitrate determination.
- ✓ Nitrite standard stock solution (NO_2-N SSS). Dissolve 0.17250 g sodium nitrite, $NaNO_2$ (molecular weight: 69.00), dried at 110 °C to constant weight, in BdW and dilute to 250 ml. Store in a tightly closed dark bottle with 2–3 drops of a saturated $HgCl_2$ solution in a refrigerator (Kirkwood, 1992). The solution contains 10 μM ml^{-1}. (Note: aged solid reagent, even if it is of analytical grade, may contain less than 100 per cent $NaNO_2$ because it is unstable in air, and therefore should not be used for the preparation of SSS).

Apparatus and equipment

- ✓ Test tubes with glass or plastic stoppers, graduated or marked at 25-ml volume.
- ✓ Automatic syringe pipettes of 1 ml.
- ✓ 25 ml automatic pipette.
- ✓ 500 ml round-bottom flask.
- ✓ Reduction columns (Figure 10.6).
- ✓ Spectrometer, with 1, 5 and 10-cm pathlength cells.

Sampling

Nitrite is an intermediate compound in the simplified redox sequence from ammonia to nitrate, and therefore it cannot be preserved properly. Avoid filtration of samples. If samples are slightly turbid or have a visible natural colouration and contain no other disturbing substances, (such as samples taken from nearshore areas), analyse them together with turbidity blanks, but do not filter.

Calibration

Perform calibration in solutions made with BdW. A series of WSS from the NO_2-N SSS is prepared by dilution (Table 10.9). From each of the WSS above, transfer 25-ml triplicates into the tubes. In addition, prepare with BdW one set of triplicates of "blank samples", adding reagents to all tubes as described later. The linear regression of the absorbances against the concentrations of the WSS (including absorbances of the blank samples, concentration equals 0) gives the calibration factor, cf.

Analysis

Transfer 25 ml of sample into the test tubes. If the turbidity has to be determined, transfer 25 ml of sample to a second tube. Pipette 0.5 ml of SAN into the tubes, mix well, and, after 2 minutes but not exceeding 8 minutes, add 0.5 ml of NED into one of the tubes. Stopper and shake. After 8 minutes read at 543 nm in a 5-cm pathlength cell, using the tube without NED as a reference. Colour is stable for two hours.

$$\text{Concentration } NO_2\text{-N} = cf \times A_s \ [\mu M]$$

Table 10.9 Preparation of working standard solutions for nitrite by dilution with bi-distilled water

Volume of solution D[1] (ml)	Total volume (ml)	Resultant concentration ($\mu M\ NO_2^-$-N)
5.0	250	1.00
5.0	500	0.50
1.0	200	0.25
1.0	500	0.10
1.0	1,000	0.05

[1] Solution "D" is prepared by diluting 5.0 ml of a SSS in 1 litre, giving a solution which contains 0.05 μmoles NO_2^--N per ml

10.5.4 Determination of ammonium

Scope, field of application and principle

This method is specific for ammonium and applicable to all kinds of natural waters (Koroleff, 1983b). Ammonium refers to the sum of ammonia and ammonium ions, because the original proportion of each in a water sample is pH dependent. The detection limit is about 0.10 μM (in 5 cm pathlength cells). The Lambert-Beer's law is followed up to about 40 μM. Interferences from amino acids and urea can be neglected. To compensate for the influence of salinity on the developed colour, a correction factor has to be applied (see below).

Reagents

Most of the reagents are caustic and very toxic, therefore mouth pipetting should not be used.

- ✓ Ammonium-free water (AFW). If the ammonium blank concentrations are higher than 0.3 μM, the water should be treated. Therefore, water that has been passed through an acidic deionisation cation exchange resin should be used. Alternatively, 2 ml of concentrated sulphuric acid (96 per cent) and 1 g of potassium peroxodisulphate, $K_2S_2O_8$, can be added per litre. The solution should be boiled for 10 minutes (without the condenser) to remove ammonium, and then distilled to give a residue of 150 ml. Ammonium-free water should be stored in a tightly sealed plastic container with thick walls.
- ✓ Citrate buffer solution. Dissolve 67 g of trisodium citrate dihydrate, $Na_3C_6H_5O_72H_2O$, 34 g boric acid, H_3BO_3, 19 g citric acid dihydrate, $C_6H_8O_72H_2O$, and 30 g sodium hydroxide, NaOH, in AFW and dilute to 1 litre. The solution is stable and should be stored in a well-stoppered glass bottle at room temperature.
- ✓ Reagent A, phenol-nitroprusside solution. Dissolve 35 g phenol, C_6H_5OH and 0.4 g of sodium nitroprusside dihydrate, $Na_2Fe(CN)5NO2H_2O$, in AFW and dilute to 1 litre. Store in a tightly closed bottle in a refrigerator. The solution is stable for several months. (Caution, highly toxic and must be treated as hazardous waste).
- ✓ Reagent B, hypochlorite solution. Dissolve 4 g of the sodium salt of the dichloroisocyanuric acid and 15 g NaOH in AFW and dilute up to 1 litre. Store in a tightly closed bottle in a refrigerator. The solution is stable for several months.

Table 10.10 Preparation of working standard solutions for ammonia by dilution with ammonium-free water

Volume of solution D[1] (ml)	Total volume (ml)	Resultant concentration ($\mu M\ NH_3$-N)
20.0	250	4.0
5.0	250	1.0
5.0	500	0.5
2.0	500	0.2

[1] Solution "D" is prepared by diluting 5.0 ml of a SSS to 1 litre, giving a solution which contains 0.05 μmoles NH_3-N per ml

- ✓ Ammonium Standard Stock solution (NH_3-N SSS). Dissolve 0.13373 g ammonium chloride, NH_4Cl, (molecular weight: 13.49), dried at 110 °C to constant weight, in AFW and dilute to 250 ml. Store in a tightly closed bottle with some drops of chloroform, in a refrigerator. The solution contains 10 μM ml^{-1}.

Apparatus and equipment

- ✓ 25 ml test tubes with ground glass stoppers.
- ✓ Automatic syringe pipettes of 1 ml and 2 ml.
- ✓ 25 ml automatic pipettes.
- ✓ Spectrometer with cells of 1, 5 or 10 cm pathlength.

Analytical procedures

Test tubes should be carefully cleaned according to the following procedure: every tube is filled with about 25 ml water and reagents are added as described later. All ammonium contained in the tubes (dissolved in the water or adhered to the glass walls) will react. Tubes are then rinsed with AFW and stored filled with AFW. The tubes should be kept stoppered when not used. They should not be washed between the different sets of calibrations or analyses, merely rinsed with AFW (Caution: serious contamination from the air can result from smoking).

Calibration

In order to avoid disturbances from variations in pH and salinity in the samples, the calibration can be carried out in two ways. In areas where the salinity variations are small, WSS are diluted with ammonium-free seawater (AFSW) (i.e. surface seawater preferably collected shortly after a plankton bloom) (Table 10.10). For work in estuaries, where the brackish water displays large salinity variations, a calibration in AFW, followed by corrections for the salinity (Table 10.11) of each sample, is preferred. A series of WSS from the NH_3-N SSS is prepared by dilution with AFW or AFSW, using volumetric flasks.

From each of the WSS above, 25 ml triplicates are transferred to the test tubes. In addition, two sets of "blank samples" are prepared, also in triplicate, but with AFW only. To all the tubes, the reagents are added as described below in "Analysis of the samples", but to one of the blank sets a double volume of reagents is added. The blank samples here correct for the absorbance caused by the residual ammonium impurities in the AFW. The second set of blanks, those with double volume of reagents, corrects for the ammonium impurities in the reagents only. The linear regression of absorbances measured against the concentrations of the WSS (including absorbances of the first set of blanks, concentration equals

Table 10.11 Salinity correction factors (cf) for ammonia analysis

	Salinity (psu)									
	<8	11	14	17	20	23	27	30	33	36
pH	0.8	10.6	10.5	10.4	10.3	10.2	10.0	9.95	9.90	9.80
cf	1.00	1.01	1.02	1.03	1.04	1.05	1.06	1.07	1.08	1.09

Source: Koroloff, 1983b

0) gives the cf. The product of cf and the absorbance of the second sets of blanks will be a constant (K), which is deduced from the results obtained with the samples and may vary from analysis to analysis because of influences such as age of solutions and contaminants from chemicals or air. Using a 5 cm cell, the current cf is approximately 11.

Analysis of the samples

1. With the automatic pipette, dispense 25 ml of the sample into the test tubes.
2. With the automatic syringe pipette, add 1.5 ml of citrate buffer and mix (a vortex mixer works very well).
3. Add 0.7 ml of reagent A and using the automatic syringe pipette mix well.
4. Add 0.7 ml of reagent B and with the automatic syringe pipette mix well.
5. Stopper the test tubes, shake and keep in the dark for at least 8 hours until colour has developed. The absorbance will become constant during a maximum of 48 hours.
6. Read at 630 nm in a cell with suitable pathlength. As a reference, AFW is used.

$$\text{Concentration } NH_4^+\text{-N} = cf \times A_S - K\ [\mu M]$$

If the WSS was diluted with AFW and the samples are seawater, the results must be corrected using the correction factors given Table 10.11, depending on the salinity of the sample.

Alternative methods

A manual spectrometric method is given in ISO (1984a) where a blue compound, formed by reaction of ammonium with salycilate and hypochlorite ions in the presence of sodium nitrosopentacyanoferrate (III) is analysed at a limit of detection of 0.003–0.008 mg l^{-1}. An automated procedure is given by ISO (1986c) and distillation and titration method in ISO (1984b). Further details are also given in Wetzel and Likens (1990) and APHA/AWWA/WPCF (1995).

10.6 Algal and cyanobacterial identification and quantification

Approaches to the determination of cyanobacterial and algal taxa, numbers and biomass present in a sample are not yet internationally harmonised. The methods used are very variable and can be undertaken at very different levels of sophistication. Rapid and simple methods addressing the composition of a sample at the level of genera (rather than species) are often sufficient for a preliminary assessment of potential hazard and for initial management decisions. Further investigation may be necessary in order to address quantitative questions of whether cyanobacteria are present above a threshold density. More detailed taxonomic resolution and biomass analyses will be required if population development or toxin content is to be predicted. Distinction between these approaches is important because managers must decide how available staff hours can be most effectively invested. In many cases, the priority will be evaluation of a larger number of samples at a lower level of precision. Furthermore, investing time in regular intralaboratory calibrations encompassing these steps is likely to be more effective than investing time in counting protocols that reduce error, e.g. from 20 per cent to 10 per cent, but at quadrupled effort. The choice of methods also requires informed consideration of sources of variability and error at each stage of the monitoring process, from sampling to counting.

10.6.1 Identification

Microscopic examination of a bloom sample is very useful, even when quantification is not being carried out. The information obtained regarding the cyanobacteria detected can provide an instant alert that harmful cyanotoxins may be present. This information can determine the choice of bioassay or analytical technique appropriate for determining toxin levels. Most cyanobacteria can be distinguished readily from other phytoplankton and particles under the microscope at 200–400 times magnification by their morphological features.

Cyanobacterial and algal taxonomy, following the established botanical code, differentiates by genera and species. However, this differentiation is subject to some uncertainty, and organisms classified as belonging to the same species may nonetheless have substantial genetic differences, for example, with respect to microcystin production, and these cannot be differentiated microscopically. The distinction of genera is very important for assessing potential toxicity, but microcystin content varies extremely at the level of genotypes or strains, rather than at the level of species. This is one reason why identification to the taxonomic level of genera (e.g. *Microcystis, Planktothrix, Aphanizomenon* and *Anabaena*) is frequently sufficient. It may be preferable to give only the genus name especially if differentiation between species by microscopy is uncertain on the basis of current taxonomic knowledge, lack of locally available expertise, or lack of characteristic features of the specimens to be identified. This must be emphasised because "good identification practice" has frequently been misunderstood to require determination down to the species level, and this has led to numerous published misidentifications of species. Practitioners in health authorities with some experience in using a microscope can easily learn to recognise the major cyanobacterial genera and some prominent species that occur in the region they are monitoring. Such efforts should not be deterred by the pitfalls of current scientific work in cyanobacterial taxonomy which targets differentiation to the species level.

More precise identification of the dominant organisms down to species level may be useful for a more accurate estimate of toxin content. For example, *Planktothrix agardhii* and *Planktothrix rubescens* have both been shown to contain microcystins, but each species contains different analogues of the toxin with different toxicity.

For establishing cyanobacterial identification in a laboratory, initial consultation and later occasional co-operation with experts on cyanobacterial identification is helpful. Training courses for beginners should focus on the genera and species relevant in the region to be monitored. Experts can assist in deriving an initial list of these taxa and the criteria for their identification. In the course of further monitoring experience, experts should be consulted periodically for quality control and for updating such a list. Helpful publications for determination of genera and species are presented in Box 10.1.

10.6.2 Quantification by direct counting methods

Microscopic enumeration of cyanobacterial cells, filaments or colonies has the advantage of assessing directly the potentially toxic organisms. Little equipment is required other than a microscope. The method may be rather time-consuming, ranging from 10 minutes to 4 hours per sample depending upon the accuracy required and the number of species to be differentiated. Precise and widely accepted counting procedures are time consuming and require a moderate level of expertise, but serve as a basis to assess performance of simplified methods developed to suit the expertise and requirements of sampling programmes tailored to the assessment of toxic cyanobacteria. A summary of methods is provided in Table 10.12. Detailed information on sampling and on counting marine toxic phytoplankton is given in UNESCO (1995, 1996), and for marine cyanobacteria in Falconer (1993).

Sample concentration by sedimentation or centrifugation

Direct counting of preserved cells is typically carried out by Utermöhl's counting techniques (Utermöhl, 1958) using a counting chamber and inverted microscope. Cells from a sample preserved in Lugol's iodine are allowed to sediment onto the glass bottom of the chamber where they can be counted.

Counting chambers and sedimentation tubes are commercially available or can be constructed by the investigator. The most commonly used chambers have a diameter of 2.5 cm and a height of about 0.5–2 cm and thus contain 2–10 ml of sample. These can fit easily on the inverted microscope stage. If larger volumes of water have to be sedimented (for example, when cell density is low) then the height of the tube has to be increased. These extended

Table 10.12 A summary of methods for the quantification of algae

Method	Volume (ml)	Sensitivity (cells per litre)	Preparation time (minutes)
Compound microscope			
Sedgewick-Rafter Cell (counting cell)	1	1,000	15
Palmer-Malloney Cell (counting cell)	0.1	10,000	15
Drops on slide		5,000–10,000	1
Inverted microscope			
Utermöhl (sedimentation chamber)	2–50	20–500[1]	2–24[2]
Epifluorescence microscopy			
Counting on filters (fluorochrome: Calco Flour)	1–100	10–1,000	15

1 Cells per ml
2 Hours

Source: UNESCO, 1996

tubes, however, are too tall to fit on the inverted microscope stage and the light would have to pass through a considerable thickness of liquid before reaching the sedimented specimens. This can be overcome using a tube in two sections, which allows the supernatant to be removed after settling without disturbing the sedimented cells on the bottom glass. The amount of sedimented water required depends on the density of cells, on the counting technique (fields or transects) and on the magnification being used.

Samples for sedimentation must be equilibrated to room temperature before they are placed in the settling chamber, to prevent air-bubbles from developing. The water sample must be gently inverted several times to ensure even mixing of the particles before being poured into the sedimentation chamber. The chamber must be placed on a horizontal surface to settle making sure that the content is not disturbed or exposed to temperature changes or direct sunshine. Sedimentation times vary depending on the height of the sedimentation tube and the preservative used. Various sedimentation times have been recommended in the literature (Lund *et al.*, 1958). Samples preserved in Lugol's iodine should be allowed at least 3–4 hours per centimetre height of liquid to settle. For samples preserved in neutralised formaldehyde, twice this time is required. Buoyant cyanobacterial cells (e.g. of *Microcystis* spp.) occasionally do not settle unless their gas vesicles are destroyed by applying hydrostatic pressure. Once the samples have settled, phytoplankton density can be determined by counting the organisms on the bottom of the chamber.

If an inverted microscope is not available, and samples with low cyanobacterial density need to be counted, it is possible to concentrate samples

Box 10.1 Sources of information for identification of cyanobacteria and algae

Anagnostidis, K. and Komárek, J. 1988 Modern approach to the classification system of cyanophytes. *Archives Hydrobiology* Supplement **80** (Algol. Studies 50–53), 327–472.

Balech, E. 1995 *The genus Alexandrium Halim (Dinoflagellata)*. Sherkin Island Marine Station, Ireland.

Bourelly P. 1968 Les algues d'eau douce, T. II. *Les algues d'eau douce. Initiation à la systématique. Tome III. Les algues jaunes et brunes*. Boubée, Paris, 438 pp.

Bourelly P. 1970 Les algues d'eau douce, Tome III. *Les algues d'eau douce. Initiation à la systématique. Tome III. Les algues bleues et rouges. Les Eugléniens, Peridiniens et Cryptomonadines*. Boubée, Paris, 512 pp.

Bourelly, P. 1972 Les algues d'eau douce. T.I. *Les algues d'eau douce. Initiation à la systématique. Tome III. Les algues vertes*. Boubée, Paris, 572 pp.

Carr, N.G. and Whitton, B.A. 1973 *The Biology of the Blue-Green Algae*. Botanical Monographs. 9, Blackwell, Oxford, 676 pp.

Carr, N.G. and Whitton, B.A. 1982 The biology of Cyanobacteria. *Botanical Monographs* **17,** Blackwell, Oxford, 688 pp.

Fay, P. and Vanbaalen, C. 1987 *The Cyanobacteria*. Elsevier, Amsterdam, 543 pp.

Fogg, G.E., Stewart, W.D.P., Fay, P. and Walsby, A.E. 1973 *The Blue-Green Algae*. Academic Press, London, 459 pp.

Hasle, G.R. and Fryxell, G.A. 1995 Taxonomy of Diatoms. In: G.M. Hallegraeff, D.M. Anderson and A.D. Cembella [Eds] *Manual on Harmful Marine Microalgae*. IOC Manuals and Guides No. 33, UNESCO, United Nations Educational, Scientific and Cultural Organization, Paris, 339–364.

Komárek, J. and Anagnostidis, K. 1985 Modern approach to the classification system of cyanophytes. I. Introduction. *Archives Hydrobiology* Supplement **71** (Algol. Studies 38/39), 291–302.

Komárek, J. and Anagnostidis, K. 1986 Modern approach to the classification system of cyanophytes. *Archives Hydrobiology* Supplement **73** (Algol. Studies 43), 157–226.

Skulberg, O.M., Carmichael, W.W., Codd, G.A. and Skulberg, R. 1994 Taxonomy of toxic cyanophyceae (Cyanobacteria). In: I.R. Falconer [Ed.] *Algal Toxins in Seafood and Drinking Water*. Academic Press, London, 145–164.

Starmach, K. 1966 *Cyanophyta. Flora slodkowodna Polski 2*. Polska Akademia, Warszawa, 807 pp.

Taylor, F.J.R., Fukuyo, Y. and Larsen, J. 1995 In: G.M. Hallegraeff, D.M. Anderson and A.D. Cembella [Eds] *Manual on Harmful Marine Microalgae*. IOC Manuals and Guides No. 33, UNESCO, United Nations Educational, Scientific and Cultural Organization, Paris, 283–317.

Tomas, C.R. 1993 *Marine Phytoplankton: A Guide to Naked Flagellates and Coccolithophorids*. Academic Press, Inc., New York.

Tomas, C.R. 1996 *Identifying Marine Diatoms and Dinoflagellates*. Academic Press, Inc., New York.

sufficiently to enable a drop (of defined volume by using a micropipette) to be counted under a standard microscope. A 100 ml measuring cylinder can be used to sediment the sample, allowing 4 hours per centimetre of sedimentation height. The supernatant can then be carefully abstracted down to the bottom 5 ml. The sample is thus concentrated by a factor of 20. Gentle centrifugation may be applied for further concentration.

Where sedimentation is not possible, centrifugation (360×g for 15 minutes using 10–20 ml sample) can offer a rapid and convenient method of concentrating a sample (Ballantine, 1953). Centrifugation may be aided by addition of a precipitating agent, such as potassium aluminium sulphate (1 per cent solution) added at 0.05 ml per 10 ml sample. Fixation with Lugol's solution enhances susceptibility to separation by centrifugation. However, buoyant cells may still be difficult to pellet and may require disruption of gas vesicles prior to centrifugation by applying sudden hydrostatic pressure (Walsby, 1994), for example in a well sealed syringe or by banging a cork into the bottle very tightly.

Counting cyanobacteria and algae

When counting cyanobacteria the units to be counted must be defined. The majority of planktonic cyanobacteria are present as filamentous or colonial forms consisting of a large number of cells which are often difficult to distinguish separately. The accuracy of quantitative determination depends on the number of counted objects (e.g. cells or colonies); the relative error is approximately indirectly proportional to the square root of the number of objects counted. The number of colonies, not the number of cells, is decisive for accurate enumeration. However, the number of colonies is often not very high even in water containing a heavy bloom where only several dozen colonies may be present in a 100 ml sample. Both filaments and colonies can differ greatly in the number of cells present, hence results given as number of colonies, for example stating that 1 ml of sample contains an average of 2.43 colonies of *Microcystis aeruginosa,* gives little information on the quantity of cyanobacteria.

Typically unicellular species are counted as cells per millilitre and filamentous species can be counted as the number of filaments, with an average number of cells per filament quoted (often the cells per filament in the first 30 filaments encountered are counted and averaged), or they can be measured as total filament length by estimating the extension of each filament within a counting grid placed in the ocular of the microscope. The latter is more precise if filament length is highly variable. For colonial species, disintegration of the colonies and subsequent counting of the individual cells is preferable to counting colonies and estimating colony size. Colonies sometimes disintegrate after several days when fixed with Lugol's iodine solution.

For more stable colonies, disintegration can be achieved by ultrasonication. This often separates cells very effectively and, in cases where colonies do not totally break down into single cells, their size may be reduced sufficiently to allow single cells to be counted. Sometimes, this is not successful and it is necessary to estimate the geometric volume of individual colonies. If colonies are relatively uniform in size, the average number of cells per colony may be determined and then the colonies may be counted. Generally, the use of values for numbers of cells per colony published in the literature is not recommended because the size of colonies varies greatly.

There are several methods for counting organisms. Most approaches aim at counting only a defined part of the sample and calculating back to the volume of the entire sample. The most common methods are: total surface counting, counting in transects and counting in fields. Counting the total chamber bottom may be very time consuming. It is usually only appropriate for very large counting units (cells, colonies, and filaments) at low magnification. Counting cells in transects from one edge of the chamber to the other, passing through the central point of the chamber, is more efficient. Some inverted microscopes are equipped with special oculars that enable the transect width to be adjusted as required. However, in many cases, horizontal or vertical sides of a simple counting grid can be used to indicate the margin of the transect. Back-calculating to 1 ml of sample can be done by measuring the area of the transects and of the chamber bottom, together with the volume of the counting chamber.

Cyanobacteria and algae occurring in randomly selected fields may be counted. When changing the position of the chamber to find the next field, it is preferable to avoid looking through the microscope to ensure random choice of fields. Microscopic field area covered by a counting grid is usually considered as one field. However, if no counting grid is available, the total spherical field can be considered. Back-calculating to 1 ml of sample can be done by measuring the area of the field and of the chamber bottom together with the volume of the counting chamber.

The density of different species in one sample can vary and there can also be several orders of magnitude difference between the sizes of the species, and therefore it is necessary to select the counting method that is adequate for the sample. Total chamber surface counting with low magnification (100×) may be useful for large species whereas transect or field counting with higher (200× to 400×) magnification is used for smaller or unicellular cyanobacteria and algae. Accurate enumeration using transects or fields assumes even distribution of cyanobacteria and algae on the bottom of the chamber surface after sedimentation. Due to the inevitable convection currents in the sedimentation chamber, that are very difficult to avoid, cells very rarely settle evenly on the surface of the bottom glass — they are almost always more dense either in the middle or around the circumference of the chamber. In some cases, density also varies between opposite edges of the bottom glass. The misestimation that arises from uneven distribution can be minimised by transect counting or by taking a fairly even distribution of randomly selected field. Counting four perpendicular diameters can minimise this error. The relation between counting time and accuracy is best if about 100 counting units (cells, colonies, and filaments) are settled in one transect. This may be achieved by diluting or concentrating samples so that the number of units of the important species lies in this range.

Specimens occurring exactly on the margin of the counting area (transect or field) present the common problem of whether to count them or not. For transect counting, those specimens that lie across the left margin are ignored, while those that cross the right margin are included. In field counting, two predetermined sides of the grid are included, the other two are ignored.

There are different recommendations in the published literature concerning how many specimens per species should be counted for reliable results. Mass developments of phytoplankton populations are generally characterised by dominance of 1–3 species. It is unusual for more than six to eight species to contribute to the majority of biomass. Therefore, it is suggested that 400–800 specimens in each sample are counted, leading to a maximum error of the total count of 7–10 per cent. In this situation 10–20 per cent of the error is accounted for by the few dominant species, 20–60 per cent is accounted for by the subdominant species and the rest of the species can be considered as insufficiently counted. If only cyanobacteria are to be counted, and only one or two species are present, counting up to the precision level of 20 per cent (by counting 400 individual units per species) can be accomplished within less than one hour.

Other counting chambers (e.g. Sedgewick-Rafter or haemacytometers) are available for use with a standard microscope. Samples might require prior concentration or dilution. It can also be useful to monitor samples under high magnification with oil-immersion (1,000×) to check the sample for the presence of very small forms that may be overlooked during normal counting.

The use of mechanical or electronic counters for recording cell counts can shorten counting time considerably, especially if only a few species are counted.

Simplifications

One alternative method, which has been found to be useful, is syringe filtration. This method is considerably less time-consuming because it does not depend on lengthy sedimentation times. Water samples (10 ml) are filtered through a membrane filter disc (13 mm) contained in a filtration device. The filter with the captured phytoplankton is dried at room temperature, and then placed on a drop of immersion oil on a microscope slide. A further drop of immersion oil is placed onto the surface of the filter, which makes the filter transparent. The sample is observed under a standard microscope (200× or 400×) without using a micro-cover slip. All cells on the surface of the filter are counted and the number of cells per litre can be calculated.

For optimising the relationship between the time spent and the information gained, various simplifications are possible. No method of enumeration is definitive, and personal creativity as well as understanding of potential pitfalls may compensate for lack of the ideal equipment or time. For each method applied (for improvisations as well as for "benchmark methods") it is crucial to check for reproducibility and comparability of the method established in the laboratory (parallel counts should not deviate by more than 20 per cent). Furthermore, clear statements of the units in which the results are given are of critical, but often unrecognised, importance. Unfortunately, many reported results are unclear about the size of units quoted, i.e. "one colony", or the size of "one filament". Such terminology varies between laboratories and makes it impossible to compare results.

For estimation of error, UNESCO (1996) gives the following equation (see Table 10.12): at a 95 per cent level of confidence, the relative limits of expected concentrations = $\pm(2 \times 100\%) / (n^{0.5}/n)$. For example, if in a sample volume of 10 ml only 50 cells of species "x" are counted, the result is 5,000 cells per litre. Assuming a deviation of 28 at counting 50, this results in ± 2,800 cells per litre. If the sample was concentrated 10 fold, so that 500 cells were counted, this would result in a higher accuracy of 5,000 ± 900 cells per litre. As a result of the extremely dynamic changes of cyanobacterial density in many water bodies (often amounting to more than 10 fold within a few hours) the precision obtained in counting 50 cells may, in many cases, be quite sufficient for estimating the potential risk involved.

10.6.3 Determination of cyanobacterial biomass

Cell size can vary considerably within species and by a factor of 10 to 100 or more between species; and toxin concentration relates more closely to the amount of dry matter in a sample than to the number of cells. Hence, cell numbers are often not an ideal measure of population size or potential toxicity. This can be overcome by determining biomass. Two approaches are available: estimation from cell counts and average cell volumes, and estimation from chemical analysis of pigment content (chlorophyll *a*).

Cyanobacterial and algal counts and cell volumes

Biovolume can be obtained from cell counts by determining the average cell volume for each species or unit counted, and then multiplying this by the cell number present in the sample to give the total volume of each species. The specific weight of plankton cells is almost 1 mg mm^{-3} and therefore biovolume corresponds quite closely to biomass. Average volumes can be determined by assuming idealised geometric bodies for each species (e.g. spheres for *Microcystis* cells, cylinders for filaments), measuring the relevant geometric dimensions of 10–30 cells (depending upon variability) of each species, and calculating the corresponding mean volume of the respective geometric body.

Simplification for biomass estimates

If the deviation of numbers of dominant species counted in two perpendicular transects is less than 20 per cent between both transects, it is not necessary to count further transects. If the standard deviation of cell dimensions measured on 10 cells is less than 20 per cent, it is not necessary to measure further cells.

If a set of samples from the same water body and only slightly differing sites (e.g. vertical or horizontal profiles) is to be analysed, enumerate all samples, but measure cell dimensions only from one. Check others by visual estimate for deviations of cell dimensions and conduct measurements only if deviations are suspected.

Chlorophyll a *analysis*

The concentration of chlorophyll *a* may be used as a sensitive approximation of algal biomass and as an alternative to counting and measuring biovolumes. Chlorophyll *a* concentrations of mixed phytoplankton populations give an overestimation of the biomass of the cyanobacteria and algae of interest. This degree of overestimation can be assessed by a brief microscopic estimate of the share of cyanobacteria and algae in relation to other phytoplankton biomass. Nevertheless during cyanobacterial mass developments chiefly consisting of one species, chlorophyll *a* may be a good measure of biomass. The method requires relatively simple laboratory equipment and is considerably less time-consuming than microscopic enumeration. A useful analytical protocol is given in ISO (1992). This method involves an extraction procedure with hot ethanol to inactivate chlorophyllase and accelerate the lysis of pigments.

Apparatus

- ✓ Spectrometer, for use in the visible range up to 800–900 nm, with a resolution of 1 nm, a bandwidth of 2 nm or less, sensitivity less or equal to 0.001 absorbance units and with optical cells of pathlength between 1 cm and 5 cm.
- ✓ Vacuum filtration device, filter holder with clamp.
- ✓ Vacuum water pump or electric vacuum pump (in the laboratory).
- ✓ Glass fibre filters free of organic binder (average pore size 0.7 µm, 47 mm diameter).
- ✓ Filters for filtration of the extracts (average pore size 0.7 µm, 25 mm diameter), or as an alternative a centrifuge (possibly refrigerated), with an acceleration of 6,000 g and a swinging rotor suitable for extraction tubes.

- ✓ Extraction vessels, e.g. wide-necked amber glass vials with polytetrafluorethylene (PTFE) lined screw caps, typically of 30 ml to 50 ml capacity and suitable for centrifugation at 6,000g.
- ✓ Water bath, adjustable to 75 °C ± 1 °C with a rack for extraction vessels.

Filtration

1. Samples must be shaken before filtration in order to mix thoroughly. Filter a measured volume of the sample (normally between 0.1 and 2 litres, depending on the concentration of algae and cyanobacteria). Pour into the filter cup, drop by drop, recording the volume, to avoid filter clogging.
2. Filter continuously and do not allow the filter to dry during filtration of a single sample. Vacuum pressure during filtration should not exceed 0.5 atmospheres.
3. The vacuum pressure should be reduced just before the filters become dry, in order to leave a thin layer of water and avoid rupture of the algal or cyanobacterial cells.
4. Some analysts recommend adding 0.2 ml of magnesium carbonate suspension (1 per cent (w/v) $MgCO_3$, shaken before use) to the final few millilitres of water in the filter cup. Avoid touching the filter with fingers. It preferable to use forceps. Direct sunshine must also be avoided because chlorophyll degrades rapidly.
5. Filters must either be frozen immediately (see below) or covered with hot ethanol to avoid pigment degradation.
6. Filters must be folded so that the cell layer is protected from rubbing off onto packaging materials. Wrapping folded filters in aluminium foil is a practical solution because it protects the filter and enables labelling on the foil.

Procedure

If extraction is not performed immediately, filters should be placed in individual labelled bags or Petri dishes and stored at –20 °C in darkness. Samples are readily transported in this form.

Extract the filters with a total volume of 10–40 ml (the volume to be used depends on the size of the photometer cuvette) of boiling 90 per cent ethanol (v/v) at 75 °C and leave overnight (24 hours) at +4 °C or in darkness at approximately 20 °C for 24–48 hours. Ethanol containing a denaturant is used successfully in many laboratories. However, denaturants vary and it is prudent to use ethanol without a denaturant or to run comparative analyses to assess the effect of the denaturant. A comparative determination with 90 per cent pure ethanol is recommended.

Homogenisation either by ultrasonication or with a tissue grinder, may be performed to disrupt cells and enhance extraction, after having poured part of the boiling ethanol onto the filter and having used the rest to rinse the apparatus. However, homogenisation is not likely to be essential for extraction of cyanobacteria.

Clarification of the slurry

1. Centrifuge the ethanol and the filter for 15 minutes at 6,000 g. This should result in a clear supernatant.
2. Carefully decant the clear supernatant with a Pasteur pipette into a calibrated flask with stopper. Fill to the mark with ethanol, stopper and mix. This is the extract volume V_e in millilitres.
   As an alternative, filter the slurry through a filter (see Apparatus section above) into a calibrated flask with a stopper. Wash the extraction vessel with ethanol and transfer quantitatively into the calibrated flask. Fill to the mark with ethanol, stopper and mix. This is the extract volume V_e.
3. Store the flasks in darkness and proceed promptly to the measurement step.

Measurement

Blank the spectrometer with the same ethanol at each wavelength before reading sample.

1. Transfer the clear extract into the cuvette using a pipette, either a) leaving sufficient volume in the cuvette for the addition of HCl (if it is preferred to proceed by adding HCl directly into the cuvette) or b) leaving a sufficient volume of extract in the flask for a second measurement after acidification of the extract left in the flask.
2. Record the absorbance at 750 nm (750a) and 665 nm (665a) against a reference cell filled with ethanol. The absorbance at 665 nm should fall between 0.01 and 0.8 units. This may be achieved by choosing a suitable volume of water to be filtered, extractant volume, dilution, or pathlength, etc. To start with, take 0.5 litre of sample, 20 ml of ethanol and a 5 cm cuvette.
3. Proceed with the acidification step, either a) adding HCl directly into the cuvette, or b) acidifying the extract left in the flask. In either case, add 0.01 ml of HCl 3 mol l^{-1} per 10 ml of extract volume and agitate gently for 1 minute.
4. Record absorbance at 750 nm (750b) and 665 nm (665b) after between 5 minutes and 30 minutes.

Calculation and expression of results

1. Calculate absorbance of the extract before acidification: $663_a - 750_a = A_a$
   Calculate absorbance of the extract after acidification: $665_b - 750_b = A_b$
2. Calculate chlorophyll *a* concentration (Chla) in mg m^{-3}

$$\text{Chla} = 29.6 \times (A_a - A_b) \times {}^{V_e}\!/_{V_s} \times d$$

where:

- V_e is the volume, in millilitres, of the extract
- V_s is the volume, in litres, of the filtered water sample
- d is the pathlength, in centimetres, of the cuvette

3. Phaeopigment concentration (Phaeo) in mg m^{-3} may be calculated to indicate the portion of inactive cyanobacterial and algal biomass:

$$\text{Phaeo} = 20.8 \times A_b \times {}^{V_e}\!/_{V_s} \times d - \text{Chla}$$

Note: The ratio of chlorophyll *a* to phaeophytin *a* gives an indication of the effectiveness of sample preservation, as well as of the condition of the cyanobacterial algal population. When samples are concentrated by filtration for the purposes of analysis, the cells die. Consequently, the chlorophyll immediately starts to degrade to phaeopigments. If filters are not extracted rapidly with hot ethanol, or frozen, chlorophyll *a* concentrations start to reduce. Occasionally, other factors disturb this method, resulting in very low or even negative values for chlorophyll *a*. If this occurs, the following calculation should be made:

$$\text{Chlorophyll } a + \text{Phaeophytin } a = \frac{12.2(663a) \times V_e}{V_s \times 1} \quad \text{mg m}^{-1}$$

This should result in a similar value as for the sum of the concentrations of both pigments determined separately, as above.

10.7 Detection of toxins and toxicity

Laboratory methods used to evaluate toxins can vary greatly in their degree of sophistication and the information they provide. Relatively simple, low cost methods can be employed which rapidly evaluate the potential hazard and allow management decisions to be taken. In contrast, highly sophisticated analytical techniques determine precisely the identity and quantity of cyanotoxins. Information obtained from simple rapid screening methods, such as microscopic examination can be used to make an informed decision on the type of bioassay or physicochemical technique that will be adequate. Currently, there is no single method that can be adopted that will provide adequate monitoring for all cyanotoxins in the different sample types that might have to be evaluated. The increasing variety and number of individual cyanotoxins being discovered make the goal of very specific and sensitive analytical methods that would detect all relevant toxins increasingly complex and ultimately unachievable (Yoo *et al.,* 1995).

In conclusion, it is strongly suggested that a monitoring programme of toxin concentrations should not be adopted as a matter of course but only when specifically indicated as discussed in section 10.1. For most recreational sites, monitoring the development of blooms rather than toxins is a more rational approach. A comprehensive review of methods and approaches is given in Lawton *et al.* (1999) and for marine algal and cyanobacterial toxins in UNESCO (1995).

10.8 Elements of good practice

- Monitoring of recreational water-use areas should be sufficient to identify the risk of blooms, taking into account actual or potential accumulation of toxic cyanobacteria and algae.
- Sampling points should be located to represent different water masses (stratified waters, waters coming from river mouths, etc.) in the investigation area and the sources of nutrients (discharges, upwellings, etc.). Possible transport mechanisms of toxic phytoplankton should be considered, possible physical forcings should be identified and sampling schemes arranged accordingly.
- In areas of high risk, sampling for algae should be carried out at least weekly. During development of blooms, sampling should be intensified to daily.
- Monitoring of toxicity (using bioassays, chemical or immunological procedures) is only justified where reason exists to suspect that hazards to human health may be significant. In such cases, long-term information on phytoplankton populations (toxic, harmful and others) should be collected where appropriate.
- Analysis of toxins should only be undertaken where standard, replicable and reliable analyses can be performed.
- Where conditions are such that monitoring is considered essential, temperature, salinity (in marine coastal areas), dissolved oxygen, transparency, presence of surface water stratification, phytoplankton biomass (chlorophyll), surface current circulation (transport of algae) and meteorological patterns (such as seasonal rainfall, storms and special wind regimes) should be considered.

10.9 References

APHA/AWWA/WPCF 1995 *Standard Methods for the Examination of Water and Wastewater*. 19th edition, American Public Health Association, Washington.

Ballantine, D. 1953 Comparison of different methods of estimating nanoplankton. *Journal of the Marine Biology Association of the United Kingdom,* **32,** 129–147.

Beers, J.R. 1978 Pump sampling. In: A. Sournia [Ed.] *Phytoplankton Manual*. United Nations Educational, Scientific and Cultural Organization, Paris. 41–49.

Carlberg, S. R. 1972 New Baltic Manual. Cooperative Research Report, *ICES,* **29**(A), 62–72.

Chorus, I. and Bartram, J. [Eds] 1999 *Toxic Cyanobacteria in Water: a Guide to Public Health Significance, Monitoring and Management*. E & FN Spon, London.

Cracknell, A.P., Wilson, C.C., Omar, D.N., Mort, A. and Codd, G.A. 1990 Toxic algal blooms in lochs and reservoirs in 1988 and 1989. In: *Proceedings of the NERC Symposium on Airborne Remote Sensing.* Natural Environment Research Council, Swindon, UK, 203–210.

Cronberg G, Lindmark, G and Björk, S. 1988 Mass development of the flagellate *Gonyostomum semen* (Raphidophyta) in Swedish forest lakes - an effect of acidification? *Hydrobiologia,* **161**, 217–236.

Edwards, C., Beattie, K.A., Scrimgeour, C.M. and Codd, G.A. 1992 Identification of anatoxin-a in benthic cyanobacteria (blue-green algae) and in associated dog poisonings at Loch Insh, Scotland. *Toxicon,* **30**(10), 1165–1175.

Falconer, I.R. 1993 Measurement of toxins from blue-green algae in water and foodstuffs. In: I.R. Falconer [Ed.] *Algal Toxins in Seafood and Drinking Water.* Academic Press, London, 165–175.

Falconer, I., Bartram, J., Chorus, I., Kuiper-Goodman, T., Utkilen, H., Burch, M. and Codd, G.A. 1999 Safe levels and safe practices. In: I. Chorus and J. Bartram [Eds] *Toxic Cyanobacteria in Water: a Guide to Public Health Significance, Monitoring and Management*. E & FN Spon, London.

Franks, P.J.S. 1995 Sampling Techniques and Strategies for Coastal Phytoplankton Blooms. In: G.M. Hallegraeff, D.M. Anderson and A.D. Cembella [Eds] *Manual on Harmful Marine Microalgae.* UNESCO, IOC Manuals and Guides No.**33**, United Nations Educational, Scientific and Cultural Organization, Paris, 25–43.

Gower, J.F.R. 1997 Bright Plankton Blooms on the West Coast of North America observed with AVHHR imagery. In: M. Kahru and C.W. Brown [Eds] *Monitoring Algal Blooms: New Techniques for Detecting Large–Scale Environmental Change*. Springer, Berlin, Hieldelberg, New York, 25–40.

Grasshoff, K. 1983 Determination of nitrite. In: K. Grasshoff *et al.* [Eds] *Methods of Seawater Analysis*. Verlag Chemie, Weinheim, Deerfield Beach, Florida, Basel, 143–150.

Harashima, A., Tsuda, R., Tamaka, Y., Kimoto, T., Tatsuta, H. and Furusawa, K. 1997 Monitoring Algal Blooms and Related Biogeochemical changes with a flow-through system deployed on ferries in the adjacent Seas of Japan. In: M. Kahru and C.W. Brown [Eds] *Monitoring Algal Blooms: New Techniques for Detecting Large–Scale Environmental Change*. Springer, Berlin, Hieldelberg, New York, 85–112.

ISO 1984a *Water Quality—Determination of Ammonium. Part 1. Manual Spectrometric Method.* ISO 7150-1, International Organization for Standardization, Geneva.

ISO 1984b *Water Quality—Determination of Ammonium.* ISO 5664, International Organization for Standardization, Geneva.

ISO 1986a *Water Quality—Determination of Nitrate. Part 1. 2,6-Dimethylphenol Spectrometric Method.* ISO 7890-1, International Organization for Standardization, Geneva.

ISO 1986b *Water Quality—Determination of Nitrate. Part 2. 4-Fluorophenol Spectrometric Method after Distillation.* ISO 7890-2, International Organization for Standardization, Geneva.

ISO 1986c *Water Quality—Determination of Ammonium. Part 2.* ISO 7150-2, International Organization for Standardization, Geneva.

ISO 1988 *Water Quality—Determination of Nitrite. Part 3.* ISO 7980-3, International Organization for Standardization, Geneva.

ISO 1992 *Water Quality—Measurement of Biochemical Parameters. Spectrometric Determination of the Chlorophyll-a Concentrations.* ISO 10260. International Organization for Standardization, Geneva.

ISO 1998 *Water Quality—Spectrometric Determination of Phosphorus using Ammonium Molybdate*. ISO/FDIS 6878. International Organization for Standardization, Geneva.

Jupp, D.L.B., Kirk, J.T.O. and Harris, G.P. 1994 Detection, identification and mapping of cyanobacteria using remote sensing to measure the optical quality of turbid inland waters. *Australian Journal of Marine and Freshwater Research,* **45,** 801–828.

Kahru, M. 1997 Using Satellites to Monitor Large Scale Environmental Change: A Case Study of Cyanobacteria Blooms in the Baltic Sea. In: M. Kahru and C.W. Brown [Eds] *Monitoring Algal Blooms: New Techniques for Detecting Large–Scale Environmental Change.* Springer, 43–57.

Kirkwood, D. S. 1992 Stability of solutions of nutrient salt during storage. *Marine Chemistry,* **38,** 151–164.

Koroleff, F. 1983a Determination of phosphorus. In: K. Grasshoff, M. Ehrhardt and K. Kremling [Eds] *Methods of Seawater Analysis.* Verlag Chemie, Weinheim; Deerfield Beach, Florida, Basel, 125–139.

Koroleff, F. 1983b Determination of ammonia. In: K. Grasshoff, M. Ehrhardt and K. Kremling [Eds] *Methods of Seawater Analysis.* Verlag Chemie, Weinheim; Deerfield Beach, Florida, Basel, 175–180.

Kuiper-Goodman, T., Falconer, I. and Fitzgerald, J. 1999 Human Health Aspects. In: I. Chorus and J. Bartram [Eds] *Toxic Cyanobacteria in Water: a Guide to Public Health Significance, Monitoring and Management.* E & F N Spon, London.

Lawton, L., Edwards, C., Codd, G.A. 1994 Extraction and high–performance liquid chromatographic method for the determination of microcystins in raw and treated waters. *Analyst,* **119,** 1525–1530.

Lawton, L., Harada, K.-I., Kondo, F. 1999 Laboratory analysis of cyanotoxins. In: I. Chorus, and J. Bartram [Eds] *Toxic Cyanobacteria in Water: a Guide to Public Health Significance, Monitoring and Management.* E & F N Spon, London.

Lindahl, O. 1986 A dividable hose for phytoplankton sampling. *ICES, C.M.*, L, 26, annex 3.

Lund, J.W.G., Kipling, C. and LeCren, E.D. 1958 The inverted microscope method of estimating algal numbers and the statistical basis of enumeration by counting. *Hydrobiologia,* **11,** 143–170.

Pilotto, L.S., Douglas, R.M., Burch, M.D., Cameron, S., Beers, M., Rouch, G.R., Robinson, P., Kirk, M., Cowie, C.T, Hardiman, S., Moore, C. and Attewell, R.G. 1997 Health effects of recreational exposure to cyanobacteria (blue-greens) during recreational water-related activities. *Australia and New Zealand Journal of Public Health*, **21,** 562–566.

Powlik, J.J., St. John, M.A. and Blake, R.W. 1991 A retrospective of plankton pumping systems with notes on the comparative efficiency of towed nets. *Journal of Plankton Research,* **13,** 901–912.

Smayda, T.J. 1995 Environmental Monitoring. In: G.M. Hallegraeff, D.M. Anderson and A.D. Cembella [Eds] *Manual on Harmful Marine Microalgae.* IOC Manuals and Guides No. 33, United Nations Educational, Scientific and Cultural Organization, Paris, 405–431.

Sutherland, T.F., Leonard, C. and Taylor, F.J.R. 1992 A segmented pipe sampler for integrated profiling of the upper water column. *Journal of Plankton Research*, **14,** 915–923.

Taggart, C.T. and Legget, W.C. 1984 Efficiency of large–volume plankton pumps, and evaluation of a design suitable for deployment from small boats. *Canadian Journal of Fisheries and Aquatic Science.* **41,** 1428–1435.

Tangen, K. 1978 Nets. In: A. Sournia [Ed.] *Phytoplankton Manual.* United Nations Educational, Scientific and Cultural Organization, Paris, 50–58.

UNESCO 1995 Manual on Harmful Marine Microalgae. In: G.M. Hallegraeff, D.M. Anderson and A.D. Cembella [Eds] *IOC Manuals and Guides No. 33,* United Nations Educational, Scientific and Cultural Organization, Paris, 177–211.

UNESCO 1996 *Design and Implementation of Some Harmful Monitoring Systems.* IOC Technical Series No. 44, United Nations Educational, Scientific and Cultural Organization, Paris.

Utkilen, H., Fastner, J. and Bartram, J. 1999 In: I. Chorus and J. Bartram [Eds] *Toxic Cyanobacteria in Water: a Guide to Public Health Significance, Monitoring and Management.* E & F N Spon, London.

Utermöhl, H. 1958 Zur Vervollkommnung der quantitative Phytoplankton–Methodik. *Mitt. Int. Verein. Theor. Angew. Limnol,* **5,** 567–596.

Venrick, E.L. 1978 Water Bottles In: A. Sournia [Ed.] *Phytoplankton Manual.* United Nations Educational, Scientific and Cultural Organization, Paris, 33–40.

Voltolina, D. 1993 The origin of recurrent blooms of *Gymnodinium sanguineum* Hirasaka in a shallow coastal lagoon. *Journal of Experimental Marine Biology and Ecology,* **168,** 217–222.

Walsby, A.E. 1994 Gas vesicles. *Annual Review of Microbiology,* **58,** 94–144.

Wetzel, R.G and Likens, G.E. 1990 *Limnological analyses.* Springer-Verlag, New York.

Willén, T. 1962 Studies on the phytoplankton of some lakes connected with or recently isolated from the Baltic. *Oikos,* **13,** 145–152.

WHO 1998 *Guidelines for Safe Recreational-Water Environments: Coastal and Fresh-waters.* Draft for Consultation. World Health Organization, Geneva.

Yoo, S., Carmichael, W., Hoehn, R. and Hrudey, S. 1995 *Cyanobacterial Blue–Green Algal Toxins: A Resource Guide.* AWWA Research Foundation, 229 pp.

Chapter 11*

OTHER BIOLOGICAL, PHYSICAL AND CHEMICAL HAZARDS

In addition to those described in Chapters 3, 7, 8 and 10 there are a number of other diverse biological, physical and chemical hazards that could be encountered in the recreational water environment. Many of these are local in nature and should be addressed in monitoring programmes where they are known or suspected to be locally important. The characteristics of the hazard and the local conditions define the appropriate remedial measures. It is important that standards, monitoring and implementation enable preventative and remedial actions in this time frame that will prevent health effects arising from such hazards. The WHO *Guidelines for Safe Recreational-water Environments* (WHO, 1998) emphasise the importance of identification of circumstances that will support a continuously safe environment for recreation. This includes awareness of biological hazards such as those discussed in this chapter.

The following sections provide a summary of the assessment and control of some biological, physical and chemical hazards encountered in recreational waters. The on-site visit form should be adapted to take account of locally-occurring hazards and any special features of a particular recreational-use area. Inspections should be carried out annually.

During an environmental health assessment, biological risks such as the presence of disease-causing, poisonous or venomous animals or plants, and physical hazards such as extreme water temperatures, should be noted. Effective ways of informing the public and, where possible, protecting them from such hazards should be recommended.

11.1 Biological hazards

11.1.1 Health hazards

Injuries from dangerous aquatic organisms may be sustained in a number of ways, for example accidental encounters with venomous sessile or floating organisms when bathing or treading on stingrays, weeverfish or sea urchins.

* *This chapter was prepared by K. Pond, G. Rees, B. Menne and W. Robertson*

Table 11.1 Relative risk to humans posed by selected groups of organisms

Attacks and poisonings by dangerous organisms	Mild discomfort	Requires further medical attention	Requires emergency medical attention
Non-venomous organisms			
Sharks		✓	✓
Barracudas		✓	
Needlefish		✓	✓
Groupers		✓	
Piranhas		✓	
Conger eels		✓	
Moray eels		✓	
Electric fish		✓	
Giant clams		✓	
Seals & sea lions		✓	
Hippopotami		✓	✓
Crocodiles		✓	✓
Venomous invertebrates			
Sponges	✓	✓	
Hydroids	✓	✓	
Portuguese man of war	✓	✓	✓
Jellyfish	✓	✓	
Box-jellyfish		✓	✓
Hard corals	✓	✓	
Sea anemones	✓	✓	
Blue-ringed octopus			✓
Cone shells		✓	✓
Bristle-worms	✓	✓	
Crown of thorns	✓	✓	
Sea urchins (most)	✓		
Flower sea urchin		✓	✓
Venomous vertebrates			
Stingrays		✓	✓
Catfish	✓	✓	
Weeverfish	✓	✓	
Stonefish		✓	✓
Surgeonfish	✓	✓	
Snakes		✓	✓

Source: WHO (1998)

Unnecessary handling, or provocation of venomous organisms during seashore exploration, or invading the territory of animals when swimming may also lead to injury. Table 11.1 lists the relative risk to humans posed by venomous and non-venomous organisms which may be encountered in recreational bathing areas.

11.1.2 Monitoring and assessment

Routine monitoring for biological hazards is justified only where they exist. Monitoring should respond to local conditions; for example where there is a known hazard, such as jellyfish, it is important to identify the source and the nature of the hazard (Raupp *et al.*, 1996).

11.1.3 Remedial measures

Many serious incidents can be avoided through an increase in public education and awareness. Surveillance systems should be in place to provide warnings to the public in areas where it is known that sharks, jellyfish and other hazardous organisms are common (see examples given in Box 11.1) (Fenner, 1998). It is important to identify and to assess the risks from various aquatic organisms in a given region. Where the health outcomes are known to be mild, remedial measures should be based primarily upon raising public awareness and providing information to the public. This may be done in a number of ways (see Chapter 6) and may require only simple messages, such as advice to wear suitable footwear whilst exploring the intertidal area or to avoid handling marine or freshwater organisms. Where the health outcomes are known to be more severe it may be necessary to declare exclusion zones in bathing areas or to restrict bathing where or when appropriate.

11.2 Microbiological hazards

11.2.1 Leptospirosis

Leptospirosis (Weil's disease or haemorrhagic jaundice) is usually characterised by the sudden onset of fever and chills, severe headache, muscular pain, abdominal pain, nausea and conjunctivitis. The causative bacterium is of the genus *Leptospira*. Other symptoms may include aseptic meningitis, conjunctival haemorrhage, rash, jaundice and cough with bloodstained sputum. The organism enters the body either through abraded skin or by contact with mucous membranes. The incubation period is 10–12 days (range 3–30 days) and symptoms persist for approximately one week. Prolonged mental health symptoms may occur after leptospirosis but the relationship is not well documented.

Monitoring and assessment

Water becomes contaminated with leptospires from the urine of infected domestic animals and rodents. Therefore, the common faecal indicator organisms cannot be relied upon as indicators of the presence of the *Leptospira*. The detection of pathogenic leptospires in water is difficult. They are relatively slow growing during enrichment and do not compete well against other more rapid growing organisms. Culture reactions and serology are required to

Box 11.1 Remedial measures to deal with jellyfish in bathing areas in Barcelona, Spain

Jellyfish are commonly found in coastal environments, although their normal environment is approximately 50 miles away from the coastlines in oceanic waters. The factors that govern their presence at the coastline are still unclear. Although dry weather conditions (dry years) are considered to contribute to jellyfish along the coastline, their presence should be considered a natural phenomenon not linked to pollution. Their presence usually indicates a flow of oceanic waters towards the coastline.

In order to alert bathers to the hazards posed by jellyfish an information leaflet has been produced for the public, including illustrations of the most common species, and a code of behaviour with recommendations in the event of encountering jellyfish. It also includes telephone numbers that sailors or fishermen can call if a large number of jellyfish are seen moving in the direction of the coast, so that preventative measures can be undertaken. The Jellyfish Expertise Centre, Institute of Marine Sciences, Barcelona has developed an Internet web page with practical information and recommendations. Jellyfish can be prevented from reaching the beach by removing them from the water using Pelican boats that are used to eliminate undesirable floating pollution on the water. More than 400 jellyfish have been removed in a single day by such means. Preventative measures that have been applied in other regions include the installation of nets and bubble screens. Unfortunately, jellyfish tend to clog nets and to break into pieces that continue to sting. If a jellyfish is detected at the beach, megaphones or loudspeakers alert bathers, safety warning flags can be changed to red to ensure that people stay out of the water. Additionally, the Red Cross and lifeguards can be prepared to deal with bather queries as well as to provide First Aid to those who have been stung. More than 40 stings have been attended in a single day when warnings have been ignored. Several municipalities within Barcelona now have small boats with shallow draughts that can remove jellyfish from bathing areas nearer to the beach.

A jellyfish stranded on the beach should not be handled because it can still sting — its venom-filled pouch remains active. Therefore, it is also important to remove any jellyfish on the beach, although it is necessary to be aware that small broken pieces may still be active while remaining moist. Lightweight protective clothing, such as a lycra suit, or a layer of petroleum jelly spread on unprotected skin, have been shown to protect swimmers against stings. Severe allergic reactions are uncommon unless a person has a history of allergy (atopy or asthma) or has been stung previously or has heart disease. However, cases of toxicity and allergic contact dermatitis and leukocytoclastic vasculitis have been reported following a jellyfish sting.

Source: Maria Figueras, Unitat de Biologia i Microbiologia, Facultat de Medicina i Sciences de la Salut, Sant Llorenc 21 43201 Reus.

distinguish pathogenic and saprophytic strains. Routine monitoring is not recommended but recreational waters should be examined for leptospires when they are suspected to be the source of an outbreak of leptospirosis. For example, a recent outbreak of leptospirosis following a triathlon race in Springfield, USA, clearly illustrated the need for monitoring, particularly where large numbers of people are at risk.

Remedial measures, interpretation and reporting

The risk of leptospirosis can be reduced by preventing direct animal access to swimming areas, by treating farm animal wastes prior to discharge and by informing users about the risks of swimming in water that is accessible to domestic and wild animals. Outbreaks of the disease are not common and the risk of leptospirosis associated with swimming areas is low. Outbreaks associated with salt water have never been reported. As a precautionary measure, domestic and wild animals should not have access to swimming areas.

11.2.2 Schistosomes

Among human parasitic diseases, schistosomiasis (sometimes called bilharziasis) ranks second behind malaria in terms of socio-economic impact and health consequences in tropical and subtropical areas. The disease is endemic in 74 developing countries and world-wide some 200 million people are infected.

The major forms of schistosomiasis are caused by five species of water-borne flatworms, or blood flukes, called schistosomes. Humans become infected after contact with water containing the infective stage of the parasite. Intestinal schistosomiasis caused by *Schistosoma mansoni* occurs in the Eastern Mediterranean, Sub Saharan Africa, the Caribbean and South America. Oriental or Asiatic intestinal schistosomiasis, caused by the *S. japonicum* (including *S. mekongi* in the Mekong River basin) group of parasites, is endemic in South East Asia and in the Western Pacific region. Another species of *S. intercalatum* has been reported from 10 African countries. Urinary schistosomiasis, caused by *S. haematobium*, is endemic in Africa and the Eastern Mediterranean. Each of the five species may give rise to acute or chronic disease with widely differing symptoms and clinical signs.

Monitoring and assessment

A computerised global database for schistosomiasis has been established by WHO. The database includes information on epidemiology, control activities, people responsible for control, and water resources for each endemic country.

Schistosomiasis and water are inextricably linked and the high prevalence of schistosomiasis in many parts of the world is closely related to human contact with natural water bodies. Some water contact is occupational and to some extent necessary, but most transmission of schistosomiasis occurs during water contact for domestic and recreational purposes. In endemic areas used by local populations or tourists, monitoring programmes should be implemented. Health education, information and communication are therefore important in a strategy to control morbidity. The objectives of health education are to help people understand that their own behaviour, principally water use practices and indiscriminate urination and defecation, as well as failure to use available screening services or to comply with medical treatment, is a key factor in transmission.

In the short term, where the prevalence of schistosomiasis is high, population-based chemotherapy can reduce the prevalence, severity and morbidity of the disease. Long-term operational and budgetary planning should be made for diagnostic facilities and re-treatment schedules, as well as treatment throughout the health care system and for transmission control. In areas where tourism is important to the economy, molluscides (chemicals that kill the aquatic intermediate host snails) may be applied at specific locations. However, molluscicides are expensive and have impacts on other aquatic life.

11.3 Sun, heat and cold

11.3.1 Health risks

Prolonged recreational use of water can lead to exposure to extreme cold or excessive solar radiation. Staying on the adjacent land area can also lead to enhanced exposure to ultraviolet (UV) radiation, because of the reflection of the sun's rays from the surface of the water. Children are often more at risk because they tend not to use sunglasses and because they spend long periods of time going in and out of the water without a protective sunscreen. Skin cancers and cataracts are important health concerns — the United Nations Environment Programme (UNEP) has estimated that over 2×10^6 non-melanoma skin cancers and 200,000 malignant melanomas occur globally each year. It has been reported by WHO that over half of the world's blind population lose their eyesight because of cataract (WHO, 1993b). It is believed that in about 20 per cent of cataract sufferers disease is triggered by short wave ultraviolet light. Direct exposure to UV radiation has both harmful and beneficial effects on humans.

A number of epidemiological studies have implicated solar radiation as a cause of skin cancer in fair-skinned humans (IARC, 1992) and severe sunburn in children has been shown as a risk factor for malignant melanoma (Katsambas and Nicolaidou, 1996; Weinstock, 1996).

Severe heat stress is also a potential risk for anyone exposed to high temperatures. The most common clinical syndromes are heat cramps, heat exhaustion and heat stroke. Recreational water users commonly expose themselves to prolonged periods of high temperatures, which is often exacerbated if they are undergoing physical exercise.

Exposure to extreme low temperatures, such as often experienced by swimmers in the open sea can also present health risks. When the body temperature falls there is a sense of confusion, a reduction in swimming capability (coupled with an overestimation of swimming capability), a possible loss of consciousness, and death by hypothermic cardiac arrest or drowning.

11.3.2 Remedial measures

Health education campaigns play an important part in the prevention of health effects due to sun, heat and cold exposure. In the UK, for example, 89 per cent of health authorities were found by Sabri and Harvey (1996) to be implementing primary prevention programmes in an attempt to meet the Government's "Health of the Nation" target of stabilising incidence of skin cancer by the year 2005. However, relatively few of these prevention programmes have been evaluated (Melia *et al.,* 1994). Data from Australia have shown that the "Slip! Slop! Slap!" campaign that was initiated in 1980, and its follow-up campaign "Sunsmart" in 1988, were effective in increasing awareness and self-reported sun-protection behaviour (Harvey, 1995). Evaluation of other health education campaigns in various countries have reported improved public knowledge about the dangers of exposure to sunshine but no significant change in sun-protection behaviour (Cameron and McGuire, 1990). Where a corresponding change in behaviour was reported, it was due to the use of sunscreens (Hughes *et al.,* 1993). To avoid the adverse effects of exposure to UV radiation, the correct use of sunscreens and protective clothing should be advocated, taking into consideration the UV index. Currently, however, there is no direct experimental evidence that sunscreens are effective in reducing skin cancer incidence (Sabri and Harvey, 1996). Melia *et al.* (1994) concluded that the benefits of education about sun protection are as yet unproven but, if organised effectively, education is likely in the long term to reduce the risk of most skin cancers and photo-aging. They noted that local initiatives require a multidisciplinary approach to ensure co-operation between general practitioners, dermatologists, pathologists and health promotion officers. Such initiatives should be supported by a national programme promoting sun protection and awareness together with national monitoring of changes in knowledge and behaviour.

Integrated coastal management (ICM) plans may help with remedial measures by displaying information in prominent places at the recreational bathing area, developing campaigns and high profile media events. At the local level, ICM has a large role to play in information dissemination.

11.4 Physical and chemical hazards

11.4.1 Health hazards

Local factors, such as agricultural or, industrial activity, can have a strong influence on the aspects of physical and chemical water quality. Therefore, before standards can be set, it is essential to understand the general characteristics of the water body of interest, together with the effect of local environmental conditions, the processes affecting the concentrations of the physical and chemical variables, and the factors that may modify the toxicity of these variables. It may, therefore, be more appropriate to identify water quality standards on a local basis rather than to adopt national standards for physicochemical aspects of recreational water quality. In determining the likely hazards of physicochemical variables, it is important to evaluate the degree of exposure that recreational users will encounter (the use of wet suits for example, will prolong immersion in cold climates).

11.4.2 Monitoring and assessment

During an inspection for chemical hazards, attention should be paid to the presence of industrial effluent disposal facilities, such as outfalls, sewers and rivers, tributaries, streams or ditches. Adjacent activities and facilities, such as intensive agriculture, electricity generating stations, dredging operations, naval bases, shipyards and terminals, should also be identified and their impact should be assessed. The assessment should also consider the impact of physical characteristics of the local beach and of the meteorological conditions on the dispersion and dilution of contaminants in the recreational-use area. During an inspection it may be necessary to collect representative water samples to confirm the presence of specific chemical contaminants, to establish their magnitude and variability, to identify the source(s) and to evaluate human exposures and health affects. Chemical analyses should only be undertaken where standard, replicable and reliable methods are available. In some cases, simple physical tests, such as pH, turbidity or colour, can be measured on-site and used as surrogates for general chemical contamination. Routine monitoring for specific chemicals should only be considered when the inspection indicates a significant hazard to human health.

In assessing local problems, initial screening for risks associated with ingestion may be undertaken by applying the WHO *Guidelines for Drinking-Water Quality* (WHO, 1993a, 1998), with an appropriate correction factor. Although the Guidelines for Drinking-Water Quality are not, generally, based upon short-term exposures, for the purpose of assessing the risk associated with occasional use of recreational waters, public health authorities can use a reference value of 100 times the value Guideline (for other than acute adverse effects) for an initial screening assessment of recreational water pollution with chemicals with known health effects arising from long-term exposure. It should be emphasised that the exceedence of such a reference value does not necessarily imply a risk, but indicate that public health authorities should evaluate the situation.

If, on consideration, it seems probable that contamination is occurring and recreational users are exposed to significant quantities, chemical analysis will be required to support a quantitative risk assessment. Care should be taken in designing the sampling programme to account, for example, for temporal variations and the effects of water currents. If resources are limited and the situation is complex, samples should be taken first at the point considered to give rise to the worst conditions. Only if this gives cause for concern should there be a need for more extensive sampling.

It is important when evaluating physicochemical hazards that the risks are not overestimated, in relation to risks from other hazards, such as drowning or microbiological contamination, which will be almost invariably much greater.

In areas that are used or proposed for bathing, it is suggested that physicochemical variables such as pH, salinity, aesthetics, clarity, turbidity, colour, oil and grease, inorganic and organic chemicals are considered. Analytical methods and the minimum sample volumes to be taken for these variables are fully detailed in Bartram and Ballance (1996).

11.5 Elements of good practice

- Monitoring for other locally important hazards is justified only when it is suspected that hazards to human health may be significant. Such occurrences may be highly localised.
- Analyses should only be undertaken where standard, replicable and reliable methods are available for known variables.
- Approaches to the assessment of the significance of locally important hazards depend on the type of hazard and should take account of the magnitude and frequency of the hazards, severity and occurrence of health effects and other local factors.

11.6 References

Bartram, J. and Ballance, R. [Eds] 1996 *Water Quality Monitoring. A Practical Guide to the Design and Implementation of Freshwater Quality Studies and Monitoring Programmes*. E & FN Spon, London.

Cameron I.H. and McGuire, C. 1990 Are you dying to get a suntan? The pre- and post-campaign survey results. *Health Education Journal* **49,** 166–170.

Fenner, P.J. 1998 Dangers in the ocean: the traveller and marine envenenation. I. Jellyfish. *Journal of Travel Medicine* **5,** 135–141.

Harvey, I. 1995 *Prevention of Skin Cancer: A Review of Available Strategies*. Public Health Care Evaluation Unit, Department of Social Medicine, University of Bristol, Bristol, UK, 31 pp.

Hughes, B.R., Altman, D.G. and Newton, J.A. 1993 Melanoma and skin cancer: evaluation of a health education programme for secondary schools. *British Journal of Dermatology,* **128,** 112–117.

IARC 1992 *Solar and ultraviolet radiation*. IARC monograph on the evaluation of carcinogenic risk to humans. International Association for Cancer Research, Lyon, 55 pp.

Katsambas, A. and Nicolaidou, E. 1996 Cutaneous malignant melanoma and sun exposure. Recent developments in epidemiology. *Archives of Dermatology,* **132**(4), 144–150.

Melia, J., Ellman, R. and Chamberlain, J. 1994 Meeting the health of the nation target for skin cancer: prevention and monitoring trends. *Journal of Public Health Medicine*, **16**(2), 225–232.

Raupp, U., Milde, P., Goerz, G., Plewig, G., Burnett, J. and Heeger, T. 1996 Case study of jellyfish injury. *Haurtarzt*, **47,** 47–52.

Sabri, K. and Harvey, I. 1996 The prevention of skin cancer: is there a gap between evidence and practice? *Public Health,* **110** (6), 347–350.

Weinstock, M.A. 1996 Controversies in the role of sunlight in the pathogenesis of cutaneous melanoma. *Photochem-Photobiol*., **63**(4), 406–410.

WHO 1993a *Guidelines for Drinking-Water Quality.* Second edition. Volume 1 Recommendations. World Health Organization, Geneva, 188 pp.

WHO 1993b Ultraviolet radiation can seriously damage your health. Press Release WHO/102, 17 December, World Health Organization, Geneva, 2 pp.

WHO 1998 *Guidelines for Safe Recreational Water Environments: Coastal and Fresh Waters*. Draft for consultation. World Health Organization, Geneva.

Chapter 12 *

AESTHETIC ASPECTS

A clean beach is one of the most important characteristics of a waterside resort sought by visitors (Oldridge, 1992; Morgan *et al.*, 1993). Accumulations of coastal debris raise a number of concerns: risks to marine wildlife, potential human health hazards and threats to the economy of coastal communities especially in tourist areas. In extreme cases people may avoid visiting an area if it is littered with potentially hazardous and unaesthetic items such as sanitary and medical waste. Beach quality can be viewed from two perspectives:

- It is the responsibility of the receiving area to ensure clean beaches and water.
- It is the responsibility of the user to behave in an appropriate manner and to avoid spoiling the beach with litter.

Aesthetics does not deal with a health burden directly but affects well being and health gain. The effects of aesthetic issues on the amenity value of marine and riverine environments have been defined by the World Health Organization (WHO) as: loss of tourist days; resultant damage to leisure/tourism infrastructure; damage to commercial activities dependent on tourism; damage to fishery activities and fishery-dependent activities; and damage to the local, national and international image of a resort (Philipp, 1993). Such effects were experienced in New Jersey, USA in 1987 and Long Island, USA in 1988 where the reporting of medical waste, such as syringes, vials and plastic catheters, along the coastline resulted in an estimated loss of between 37 and 121 million user days at the beach and between US$ 1.3×10^9 and US$ 5.4×10^9 in tourism-related expenditure (Valle-Levinson and Swanson, 1991).

The robustness of scientific techniques used in litter analysis is of varying quality and methodologies involved for any beach aesthetic programme must be comparable, have a quantitative basis and, more importantly, be easily understood by the end user. The reduction of beach litter for visual, olfactory and health reasons should be a paramount aim for society. Ideally litter should be cut off at source, but in reality this has been found difficult. Fundamental to this aim are universal education programmes.

* *This chapter was prepared by A.T. Williams, K. Pond, R. Philipp*

Box 12.1 Public perception of microbiological quality and aesthetic aspects

A study was carried out in 1987 in the UK to investigate the public perceptions of beach and sea pollution with particular reference to the perception of bathing water quality. Samples were taken from two holiday resorts. On the basis of pre-existing microbiological evidence, the two resorts were chosen so they would have contrasting levels of measured sea pollution.

Interviewers were instructed to select respondents of a wide variety of ages and apparent social classes, recruiting approximately equivalent numbers of men and women on or near the beach in each of the two resorts. The interview schedule was designed to elicit the public's perception of beach and sea pollution, their perception of the quality of bathing water and their reporting of any of a list of symptoms. Respondents were also asked about their, and their children's (if applicable), swimming and other water-related activities. Sampling took place over an eight-week sampling period during the summer months in 1987. All interviews took place on the beach or in the immediate surroundings.

- The microbiological results for Resort 1 indicated higher levels of microbiological contamination than at Resort 2.
- The sea at Resort 1 was more likely to be seen as discoloured, dirty, cloudy, having film, oil or slime than at Resort 2.
- The frequency of reported debris in both the sea and beach was significantly greater at Resort 1.
- There was a higher incidence of discarded food or drink containers reported on the beach than in the sea.

Swimmers and non-swimmers at Resort 1 showed a significantly different percentage of holidaymakers reporting symptoms of illness such as stomach upsets, nausea, diarrhoea or headaches compared with holidaymakers at Resort 2.

Source: University of Surrey, 1987

12.1 Beach litter visual triggers

The presence of clear water does not guarantee that the water is uncontaminated and free from pathogens but the presence of certain items on a beach may however, imply poor microbiological water quality (University of Surrey, 1987) (Box 12.1). Equally, beaches free from litter do not imply that the sanitary quality of the sand is good (Mendes *et al.*, 1997). The general public usually infer that a highly littered beach has poor water quality and it is logical to assume that people prefer to visit a clean beach rather than a dirty beach (Rees and Pond, 1995a). It has been reported by WHO, that "*Good health and well-being require a clean and harmonious environment in which physical, psychological, social and aesthetic factors are all given due importance*" (WHO, 1989 p. 5).

Marine litter is defined as "*solid materials of human origin that are discarded at sea or reach the sea through waterways or domestic or industrial outfall*" (NAS, 1975 p. 104). However, the question remains whether a single item of sanitary waste on a beach necessarily means that a beach is dirty or, alternatively, how many condoms, sanitary towels or metal cans it takes to make a beach aesthetically displeasing. Aesthetics is defined by Collins Concise Dictionary (1995 p. 19) as relating to "*(a) pure beauty rather than to other consideration, (b) relating to good taste*". It relates to personal preferences, which in turn encompass things perceptible by the senses (sight, smell, taste, touch and hearing), gender, socio-economic status, psychological profile, climate, "sense of well being", age, culture, and whether the observer or user is local or a tourist (Dinius, 1981; Williams, 1986; Oldridge, 1992; Morgan *et al.*, 1993; Bonaiuto *et al.*, 1996). Certain aspects of aesthetic pollution have a greater impact on the public than others and it has been suggested that a weighting of importance should be placed on the determinands so that an overall aesthetic index could be established (NRA, 1996).

The perception of the beach user should be taken into account in award schemes (see Chapter 6) of which many exist (Williams and Morgan, 1995). Cognisance of such perception is sadly lacking in all current award schemes (see Chapter 6). Perception by the general public of the beach aesthetic appearance and water quality has become increasingly important (David, 1971; Williams, 1986; House, 1993; NRA, 1996; Williams and Nelson, 1997). The problems of beach litter are being tackled with respect to the physical (Williams and Simmons, 1997 a, b) and psychological well-being of the consumer (Williams and Nelson, 1997). Emphasis is being applied increasingly to development of aesthetic health indicators which will aid in the implementation of planning measures to deal with beach health hazards (Philipp *et al.*, 1997). The presence of sewage-related debris (SRD) and medical items tend to evoke stronger feelings of unpleasantness with respect to beach aesthetics than items such as cans or plastic bottles but the tolerance level on a world-wide basis has yet to be quantified. The former items attract media attention because of the potential health risks associated with stepping on syringes, ingesting SRD or other contaminated material (Walker, 1991; Rees and Pond, 1995a). Herring and House (1990) concluded that sewage-derived debris had a greater social impact than any other aesthetic pollution environmental parameter. Williams and Nelson (1997) (Box 12.2) showed that the general public are more affected by mixtures of generic debris categories (e.g. cans, bottles and SRD such as condoms and sanitary towel backings), and it appears that females are more sensitive to beach debris (in particular SRD) than males (which could be due to a higher recognition of these particular items). It has been stated by the UK House of Commons Committee that "*while the risk of infection by serious disease is small, the*

Box 12.2 Public perception analysis at Barry Island, South Wales, UK

Public perception to litter was investigated by questionnaire during August 1995 and 1996 at Barry Island beach, South Wales, UK. Results showed that beach users were acutely aware of land-based and marine coastal pollution. A high percentage of beach users (69 per cent) thought the water to be polluted and a large percentage reported a list of litter items as being present on the beach including food packaging (83 per cent) and excrement (27 per cent).

The most prominent items of debris noted on the beach at the end of the day were food packaging, plastic bottles, aluminium cans, excrement and hygiene items. A composition of general litter and sewage-related debris was found to be the more offensive than individual generic items. The most sensitive groups of people to beach litter were females, people in the age range 30–39, and local people when compared with visitors who travelled greater than 10 miles to their destination.

A high degree of concern about the water condition was expressed by the public and a large number, 69 per cent, decided not to enter the water because they believed it to be polluted. Chi-square analysis at the 0.05 level showed females, and also people in the age category 30–40 years old, to be more sensitive to perception of pollution. However, parents still chose to visit the beach for the sand and amenity value without allowing their children into the water. Water quality was the main reason for not swimming (55 per cent), followed by temperature (23 per cent). Floating objects were considered to be the most obtrusive forms of marine debris by 53 per cent of the respondents. Such objects included anything from food packaging and hygiene items to faecal matter. The colour of the water was reported to be unfavourable by 21 per cent of those surveyed, while 14 per cent of respondents commented that the water had a "foul smell" and oil was perceived to be a problem by 12 per cent.

Source: Williams and Nelson, 1997

visible presence of faecal and other offensive materials carried by the sewerage system can mean serious loss of amenity and is therefore an unacceptable form of pollution" (HCEC, 1990 p. xvii).

Dinius (1981) found that water discoloration was a factor that led respondents to make a judgement about the level of pollution of an area. Any visually unpleasant pollutant has the potential to have a negative impact on tourism, whether or not it poses an actual health risk. The aesthetic quality of the Mediterranean has been affected where eutrophication and algal blooms have occurred. There is also evidence of nutrient enrichment in the Baltic Sea, Kattegat, Skagerrak, Dutch Wadden Sea, North Sea and Black Sea (Saliba, 1995). Izmir Bay, Turkey, has been suffering red algal tides and, in 1993, pollution-related illness caused an estimated 10,000 lost working days amongst local swimmers and fishermen using the Bay (Pearce, 1995).

Eutrophication has been reported as a problem along virtually every country bordering the Mediterranean. One of the consistently worst affected areas is the northern Adriatic where algae affecting areas of sea water up to 50 km^2 have been reported (Pearce, 1995).

One model for aesthetic standards defines the aesthetic value of recreational waters as (MNWH, 1992):

- Absence of visible materials that may settle to form objectionable deposits; absence of floating debris, oil, scum and other matter.
- Absence of substances producing objectionable colour, odour, taste and turbidity.
- Absence of substances and conditions (or combinations) which produce undesirable aquatic life.

It is imperative that future beach management plans consider the beach users perception of the coastline. Although poor visual appearance of the beach does not necessarily infer danger to health, results of other surveys (see Box 12.1) strongly suggested a link between the presence of certain items of debris and higher bacterial counts in water.

There are a number of human health risks posed by marine debris. Injuries caused by marine debris include entanglement of scuba divers (Cottingham, 1989), cuts caused by broken glass and discarded ring tabs from cans, skin punctures from abandoned syringes and exposure to chemicals from leaking containers washed ashore (Dixon and Dixon, 1981). In addition, munitions and pyrotechnics such as smoke and flame markers have been recovered on beaches (Dixon, 1992). Fishermen and those involved in dredging operations are at particular risk from such items although there are numerous reports of injuries to holidaymakers who have inadvertently picked up such items (Dixon, 1992).

Horsnell (1977) has documented the actual safety hazards arising from individual or small numbers of individual chemical packages. Studies by Dixon (1992) have shown a 63 per cent reduction in dangerous or harmful substances in England and Wales between 1982 and 1992. The reduction was most marked for flammable liquids, oxidising substances and corrosives. Koops (1988) analysed chemical cargoes lost off the Dutch coast that included the gases ethylene oxide and chlorine and the corrosive, sulphur dichloride.

Less obvious health risks are posed by items of SRD and medical waste. Clinical waste represents the potential vectors of infectious diseases such as Hepatitis B and Human Immunodeficiency virus (Walker, 1991). In addition, other visible pollutants, such as discarded food, dead animals, oil, containers and tyres, commonly found along the coastline have been associated with microbiological hazards (Philipp, 1991). Where visible litter is present there are also likely to be high counts of *Escherichia. coli* (Philipp, 1991) which are commonly associated with human faecal material. Long-term monitoring of marine debris can therefore become an important part of the process to identify suitable indicators (Pond, 1996).

12.2 Litter survey techniques

There are a number of uses for data gathered by beach survey, including the application of the data to assess the effectiveness of remedial measures; appraisal prior to management programmes; tourism guides (e.g. MCS, 1996); as part of an integrated coastal zone management programme; identification of health hazards and/or particular threats; identification of trends; raising public awareness through public involvement; investigations for identifying the source of the litter, ageing litter and identifying the dynamics of litter in the environment. In all cases the data collected must be of quality suitable for the purpose and, where comparisons are to be made, the data must be standardised. The use of photography as the basis for routine comparisons, training and communication may be important, particularly because the perception of litter is a visual and aesthetic process. Education has a major role to play in respect to the above, both at the formal and informal level.

Litter surveys are conducted to assess types, amounts, distribution and source of litter (Rees and Pond, 1995b) and in turn to assess the effectiveness of legislation. Human health, litter and tourism are intimately connected issues and surveys, such as enumerated below, are needed in order to monitor progress in cleaning up litter (or the lack of it) through time.

Monitoring parameters, sampling stations and sites and sampling frequencies, should all be considered when establishing beach quality monitoring programmes. The problems associated with microbiological sampling of seawater have been well documented (Fleisher, 1985, 1990; Jones *et al.*, 1990; EC, 1995; Rees, 1997) and a number of these factors will apply to aesthetic quality sampling, namely variation between analysts, methods, culture mediums, choice of sampling location, number of transects from which samples are taken, number of sampling points on any stretch of beach, time of sampling (spring to neap tides), frequency of sampling, as well as wind, tide, currents and sunlight. All of these can contribute to inconsistency in results (see Chapters 2 and 9) which raises the question of whether it is possible to take a representative sample. Some of the factors above could certainly apply to beach quality monitoring and it is uncertain whether existing award schemes (see Chapter 6) show realistic representations on which to base any quality assessments.

12.2.1 Survey objectives

Of particular importance is the identification of realistic objectives that must be clearly stated and understood before the survey begins. The objectives of any litter study will define the timing of the sampling period. However, all surveys should encompass varying seasons in order to obtain a representative sample. Baseline studies (to identify the types of material found) are generally carried out over large geographical areas using a low sampling frequency. Assessment studies (to identify density of debris and changes over time) are usually carried out over more intense sampling periods and in smaller geographical areas. Temporal changes, physical characteristics of a beach, tidal patterns and use of the beach can have dramatic influences on the composition of debris found at any one time. It is therefore important that the programme design suits the study aims. Resources may also be a factor determining survey timings. Where sampling is carried out for health reasons it may be desirable to survey throughout the year but the availability of resources may restrict sampling to the bathing season.

12.2.2 Methods

Surveys can be focused on the beaches, seas or rivers where beach debris is used as an indicator of oceanic, riverine, estuarine or lake conditions. One of the earliest litter surveys to be undertaken was by Garber (1960) and this approach has been used by others (NRA, 1992). The main disadvantage of the method used by Garber (1960) was its subjectiveness. For example, only presence or absence of certain visual characteristics relating to water quality was recorded in section A of the official form whereas in section B scales ranged from "absence" to an amount that was sufficient to be objectionable. No definitions of these categories were given.

Surveys can be on a small scale, such as that by Gilligan *et al.* (1992) in Chatham County, Georgia, USA where four types of site were selected to obtain samples representative of tidal influence; or they can be large scale such as those carried out by the Coastwatch Europe network (Dubsky, 1995) and the Tidy Britain Group (TBG) in the UK (Dixon and Hawksley, 1980).

A number of guides and reviews exist to help survey design (Gilbert, 1987; Ribic *et al.*, 1992; Rees and Pond, 1995b; Earll, 1996). To date, it has been inappropriate to apply a standardised methodology to assess riverine and marine debris, due to the different objectives of the surveys and the diverse nature of coastlines world-wide (Faris and Hart, 1995; Verlander and Mocogni, 1996). Faris and Hart (1995) concluded that monitoring studies can be carried out in a variety of ways provided standardised sampling protocols are established at the beginning and basic requirements are followed. Study objectives will determine the ultimate project design. Studies may be simple enumeration studies, assessing types and litter quantities, or they can be more detailed indicating age and origin of items. They can cover large geographical areas or they can relate to detailed information about specific regions or places (Williams and Simmons, 1997a). The time element, personnel needs and the costs are restricting factors. Details, such as site

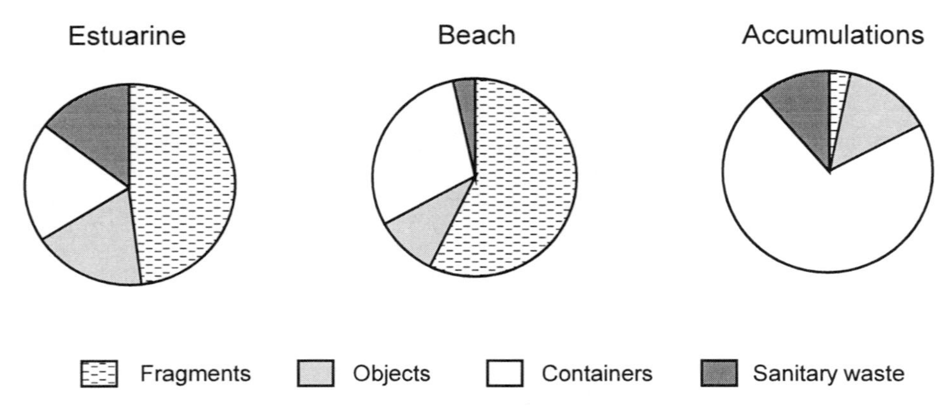

Figure 12.1 Plastic litter found at Merthyr Mawr beach, South Wales, UK (After Williams and Simmons, 1997a)

description, map reference, category definition, wave, wind, current patterns, site topography, physical characteristics of the beach, measurement units, survey frequency and date of survey, all need to be identified and recorded.

It is important to recognise that undertaking a beach survey can be hazardous. In addition to detailed instructions on how to complete the survey, special attention should be paid to safety aspects. Surveyors should wear appropriate footwear as well as gloves if it is necessary to handle the litter.

Litter can be categorised according to size (Ribic, 1990), composition (Dixon and Dixon, 1983) or weight (YRLMP, 1991). Three main methods of assessing type, amount and distribution of marine debris have been documented:

- Record solid waste generated by ships or pleasure crafts (Dixon and Dixon, 1981).
- Collect or observe litter floating in coastal waters (Cuomo *et al.*, 1988).
- Estimate litter during beach surveys (Rees and Pond, 1995b, 1996; Williams and Simmons, 1997a) (Figures 12.1 and 12.2).

All surveys should be repeated to show temporal changes in amount and type of litter and, hopefully, to determine its source, accumulation rate, standing stock, etc.

Individual items of beach marine debris are usually counted and classified or recorded as presence or absence (Pollard, 1996; Rees and Pond, 1996). The sampling size unit is a function of the survey aims. Examples include:

- The whole beach can be surveyed from splash zone to waters edge (Dubsky, 1995).
- Transects may be used of varying width. The optimum transect width is one which provides a reliable sample.
- Transect line quadrats or randomly dispersed quadrats (Dixon and Hawksley, 1980).

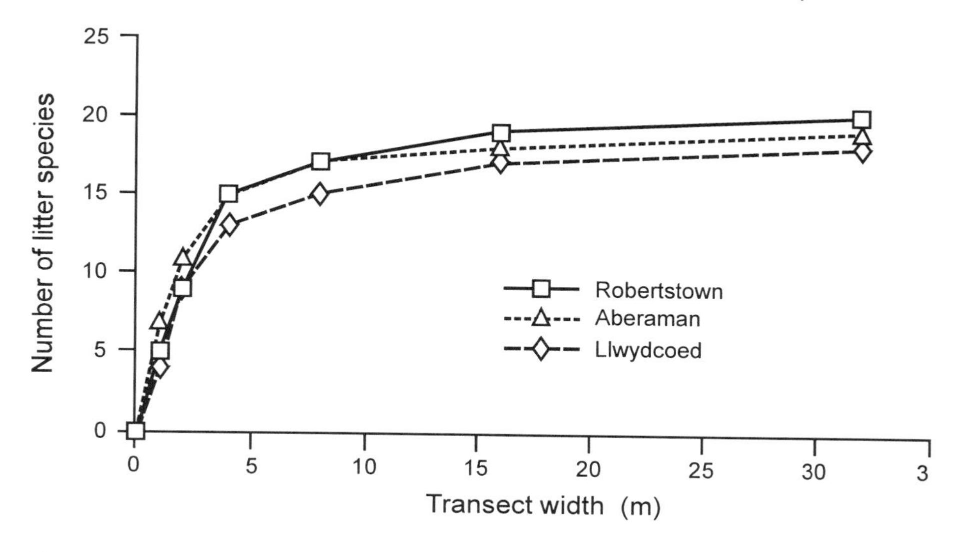

Figure 12.2 A minimal area curve for beach sites on the River Cynon, South Wales, UK (After Earll *et al.*, 1997)

- Strand line counts (Williams and Simmons, 1997a).
- Postal surveys (Dixon, 1992).
- The offshore water column can be sampled (Williams *et al.*, 1993).

The advantages and disadvantages of various methods of litter surveys are shown in Table 12.1. Survey design and methodological development are considered to be of paramount importance and many statisticians and environmental scientists have provided guidelines to aid formulation of sound survey designs (Gilbert, 1987; Ribic *et al.*, 1992). Common to each approach is an emphasis on formulation of realistic objectives that must be stated and clearly understood before work can progress. No precedent exists regarding the optimum type of data, i.e. qualitative or quantitative. Two approaches to aesthetic surveys are described below: transect surveys and questionnaire surveys.

Transect surveys

Gilbert (1987, p. 7) stated,"*the target population is the set of N population units about which inferences will be made. The sampled population is the set of population units directly available for measurement.*". The target population must be limited to litter at sites deemed accessible for sampling purposes, e.g. litter on riverbank sites where both banks can be sampled up to the bankfull position (highest possible water level), beach strandlines, etc. Due to logistical problems of assessing all litter at a site, representative sampling units are needed to provide an accurate portrayal of the whole site. For rivers

Table 12.1 The principle advantages and disadvantages of various methods of litter survey

Method	Advantages	Disadvantages
Five strand lines, excluding the vegetation line	Covers a large area of beach where items may accumulate in algae or as mats of debris left as the tide recedes	Can give biased results as the areas between the strand lines are not surveyed and the method only counts the most recent tidal borne debris (1) Only covers surface litter; some litter may be buried (2) It may be difficult to identify strand lines; these vary daily and seasonally (3) Not suitable for beaches with large boulders (4)
Five strand lines, plus the vegetation line	Area covered includes a good cross section; both accumulated and fresh litter is surveyed	As above (1–3)
Top, bottom and vegetation lines	Easy to use; quicker than the above methods	As for the first method (1–3)
Five metre wide strip transect	Covers a large area of beach where items may accumulate in algae or as mats of debris left as the tide recedes	As for the first method (1–3)
One metre wide strip transect	Covers a large area of beach where items may accumulate in algae or as mats of debris left as the tide recedes	As for the first method (1–3)
Random 2x2 quadrats	Fast method of sampling; sampling is not influenced by location of litter and is therefore statistically valid	Results may be variable and depend on the amount of litter present
Random dispersed quadrats	Fast; economic	Possibility of missing litter clumping
Whole beach	Covers all sections of the beach; avoids bias	Time consuming; care needs to be taken not to miss items Best suited to pocket beaches
Postal surveys	Can cover a large geographical area	May be relatively high percentage of non-response

and beaches, a series of continuous quadrats can be laid starting at the water's edge and finishing at the natural limit of the bank or beach (sites chosen with predominantly natural characteristics). Within each quadrat, litter abundance can be measured in the form of density counts, i.e. the number of individuals of particular litter types within a quadrat. Not every litter type exists on any one particular river or beach.

A useful tool adapted to determine whether transect sampling is an appropriate method for river and beach litter assessments (species) and, if found suitable, to assess the optimum transect width size, is classic minimal area analyses (also known as species area curves) derived from the Braun-Blanquet school of phytosociology (Braun-Blanquet 1921, 1932; Cain, 1932, 1935; Gilbertson *et al.*, 1985).

Narrow belt transects are more easily studied and enable work to be achieved quicker, but wider transects probably yield more reliable data (Burnham *et al.*, 1980). Therefore, the optimum transect width is one which provides a reliable representation of the litter present, for the minimum amount of work. For example, for investigating riverine litter (Figure 12.2), starting from the site's centre point, a tape is placed up the river bank perpendicular to the river flow. A second tape is then placed parallel to the first, at the smallest possible distance apart (in this case approximately 10 cm). The number of litter types is counted and recorded. The exact initial distance decided upon is unimportant, provided it is small enough to contain only one or two items, because recordings are made in relation to a doubling of transect width and not as a function of the exact width measurement. The transect width is doubled and the number of litter types present counted. The doubling and counting procedure is repeated until the number of litter types at each doubling of the transect width has shown no further increase. The resultant data curve starts to level off at the point that resembles the minimal width necessary to obtain representative samples.

Figure 12.2 shows how three different sites produced similar curves, with the curve gradient indicating the number of litter types found, and the curve beginning to level off after 5 m transect widths. On an objective basis it is very difficult to determine the exact optimum transect width. At a 5 m transect width some 13–15 litter types were identified; but at 15 m width 15–17 types were identified. Detailed field work showed that 20 litter types were present at these sites, i.e. 5 m transect widths covered some 65–75 per cent of the litter present whereas 15 m widths covered some 75–85 per cent. The 5 m transect width has been used in many litter surveys (Dixon and Dixon, 1981; Davies, 1989) but there is no clear scientific justification for this. On applying a pre-specified relative error (Gilbert, 1987), results indicated that any between-site comparisons should only be carried out using litter types known to have a more uniform within-site distribution. Commonly occurring litter types, such as plastic sheeting and sewage-derived articles, could be represented realistically using only three transects. Other litter types needed 64 transects, e.g. packing crates. It is meaningless to compare sites of different litter types, because the within-site variation can be greater than the differences between two sites.

Questionnaire surveys
An alternative approach to transect sampling is to survey the entire beach area. This approach has been followed by the Coastwatch Europe network and has been described in detail by Pond (1996) and Rees and Pond (1995b; 1996). Essentially, the study area is divided into manageable units of 0.5 km in length. The method uses standard questionnaires, translated as necessary (see Rees and Pond, 1996). All surveyor groups are provided with detailed instructions on how to complete the survey, including which items of litter should be included in each category, detailed safety notes and the contact telephone number of a national and local co-ordinator. The survey is conducted over a common time period so that the results can be compared between participating countries and between sites within countries. A six-figure map reference of each unit of coastline is recorded and stored on a database in order to ensure that the same units are surveyed in subsequent surveys.

The survey is conducted as soon as possible after high tide. Surveyors are asked to walk along the intertidal area and to return along the splash zone recording the presence or absence of 17 "general" litter categories, such as sewage-related debris, cans, plastic bottles, etc. and nine "major items of debris", such as household refuse. Quantities of some items of litter are also requested. Surveyors are also asked to record potential threats to the area, to investigate the aesthetic quality of inflows (streams and rivers) and to record other information regarding management aspects of the coastal unit. Once the questionnaire has been completed surveyors return it to a national co-ordinating office for data analysis and report writing.

This approach has a number of advantages: the method is simple to use and can be undertaken by relatively inexperienced surveyors under instruction (see section 12.2.4) and the questionnaire can be adapted to focus on particular areas of interest, for example the Coastwatch Europe network has developed a section of the questionnaire to focus on medical and sanitary waste. It can also be adapted to collect qualitative data. Large areas of coast-line can be covered, thus making the sample more representative. The method is also economical and does not require any special equipment or knowledge and can be undertaken in all weather conditions. The main disadvantage is that the method is time-consuming.

12.2.3 Qualitative versus quantitative data analysis

Qualitative data
The problems associated with the techniques used to assess litter and the resulting statistical analyses are comparable with those experienced by ecologists (Ludwig and Goodall, 1978; Ludwig and Reynolds, 1988). Qualitative data can give quick assessments (Hubalek, 1982). Many different pattern

types can exist within communities, including spatial dispersion of litter types (species) "within" a site, and relationships "between" sites.

Ludwig and Reynolds (1988) recommended the variance ratio (VR) test of Schluter (1984) to measure association for more than a single pair of litter types. The Null Hypothesis (H_0) is that no association exists between litter types and the expected VR is 1. If an association occurs, then it is either positive (VR < 1), i.e. the pair of litter types occurred together more often than expected if independent; or negative (VR > 1), i.e. the pair of litter types occurred together less often than expected if independent. Litter types may show no association if independent, or when positive and negative associations between litter types cancel out each other.

The chi-square test can detect pair-wise associations of litter types, with H_0 indicating that the litter types were independent. In riverine litter examples quoted by Simmons and Williams (1997) a $\chi^2_t > 3.84$ at the 95 per cent level rejected H_0, and 31 litter pairs were significantly associated. In the context of ecological monitoring, three qualitative (present/absent) binary techniques are common i.e. Ochai, Jaccard and Dice. Jaccard's technique is particularly robust and when using this technique Simmons and Williams (1997) showed that within-site litter transects were generally no more strongly associated than those between-sites. From these qualitative results it appears that within-site litter variations can be as great as between-site variations. If this is the case the representativeness of transects for each site may be questioned because the results from one transect could be dramatically different from another transect, even at the same site. In the Simmons and Williams (1997) study, no strong associations were apparent between sites; the highest index value reached 0.7 with the majority of indices < 0.5. It appears that either significant differences in litter patterns did not exist between sites, or that the sample size was too small to show differences, or that the statistical test was not able to detect the differences.

Several major limitations negate the benefits of collecting by qualitative data alone. A lack of data versatility is a major problem, with few statistical analyses being appropriate. Even the statistical packages available (Ludwig and Reynolds, 1988) require very time-consuming data manipulation to carry out relatively simple calculations. If data are being compiled for several river catchments or marine sites, it is felt that data manipulation problems alone make qualitative analysis an unfavourable proposition.

Quantitative data

Quantitative data makes it amenable to a broader spectrum of analysis methods giving greater versatility. Three basic patterns may be recognised in litter communities, random, clumped and uniform, and the mean, variance and pattern of individuals within a quadrat are quite different between these

Table 12.2 Significantly correlated litter pairs

Correlated litter pairs	Level of significance[1]
Sanitary towel: panty liner	0.000
Panty liner: tampon	0.019
Sanitary towel: tampon	0.007
Plastic sheeting < 30 cm: plastic sheeting 30–60 cm	0.006
Plastic sheeting < 30 cm: plastic sheeting > 60 cm	0.030
Plastic sheeting 30–60 cm: plastic sheeting > 60 cm	0.000

[1] Equivalent to the probability of the correlation arising purely by chance

Source: Simmons and Williams, 1997

patterns. Once a pattern has been identified, a test must be proposed concerning the community structure. Initially, it is important to determine if sampling units are discrete (natural) or continuous (arbitrary) because this influences the type of spatial pattern analysis. Based on the continuous nature of sampling units, the quadrat variance method of Ludwig and Goodall (1978) can be undertaken enabling spatial patterns to be observed by sampling a series of continuous quadrats. Data may be collected at all litter sites by a series of 1 m^2 quadrats extending up a river bank or along a beach transect. Quadrat-variance methods are based on examining the changes in the mean and variance of the number of individuals per sampling unit, for a range of sampling units.

Two types of quadrat variance methods can be used: paired-quadrat variance (PQV) and two-term local quadrat variance (TTLQV). The former uses (PQV) changes in quadrat spacing to provide spatial pattern information, whilst the latter (TTLQV) uses changes in quadrat size, through the blocking or combining of adjacent quadrats, to determine pattern intensity and range of densities present. Quantitative analyses can be done using the SPSSx® statistical software package (Norusis, 1983). If results from individual quadrats are combined to produce data representing a 1 m wide belt transect, the data could still be used to indicate whether certain statistical tests would be of future use. Data limitations can be due to small sampling areas and the use of only one transect to represent a site.

The normality of litter data sets should be tested using the Kolmogorov-Smirnov one sample non-parametric test (Miller and Miller, 1988), followed by finding the covariation between litter types, e.g. using the Spearman rank correlation. For example, Simmons and Williams (1997) found correlations between sanitary towels, panty liners and tampon applicators (Table 12.2). Plastic sheeting appeared to be correlated with other plastic sheeting of differing sizes, but not with sewage-derived litter (as indicated in the qualitative litter association analyses described above). This result may highlight one of the problems of using qualitative data in this sort of survey. Associations between plastic sheeting and sewage debris may have been calculated because of their common occurrence at sites. Associations shown by qualitative data may have led to the hypothesis that plastics were introduced to the system from the same sources as the sewage-derived litter, hence their association. However, it appears that although both items are present at the same site, their abundance is not correlated significantly. Major limitations of this type of analysis appear to arise from the number of zeros recorded in the data set; consequently large data sets are needed. A second problem is the realistic interpretation of results. When an expanse of coefficients has been calculated, multivariate methods of pattern recognition, such as cluster and principal component analysis, can and should be used (Derde and Massart, 1983).

Cluster analysis may be used to place similar objects or variables into groups or "clusters", in order to produce a hierarchical tree-like structure known as a dendogram. Dendograms demonstrate graphically, in two dimensions, similarities between variables by the varying distances from the x-axis at which the groups are formed. The closer a group is formed to the x-axis, the stronger the similarities between its constituent parts (Simmons and Williams, 1997). The cluster analysis approach appears to be a very useful tool for indicating patterns within a data set and reduction of the numbers of zeros recorded is the main key.

Principal Component Analysis (PCA) is an alternative method of pattern recognition that aims to identify principal components that explain correlations among a set of variables (litter types, see Table 12.3). The method condenses information on litter types from many dimensions (N sites) to two or three dimensions that may be more easily interpreted. In addition, it calculates "loadings" to indicate the significance of each of the variables in determining the data structure. The higher the loading, the greater the importance the variable has in determining that component. In Figure 12.3, the first three factors accounted for 20, 16, and 11 per cent of the data variation respectively (Simmons and Williams, 1997).

12.2.4 Volunteers versus "professionals"

There has been considerable debate about who should conduct litter surveys (Dixon, 1992; Amos *et al.*, 1995; Pollard, 1996; Rees and Pond, 1996). The use of volunteers was discussed at length at the Third International Conference of Marine Debris in 1995 (Faris and Hart, 1995). No conclusions were made, but the Conference recommended that where volunteers are used, clear instructions must be given and good quality assurance procedures must be established. There are both advantages and disadvantages in this approach. The use of volunteers to conduct litter surveys means that a large sample size

Table 12.3 A litter identification key

Source	Category	Type of litter
Sewage derived	Feminine hygiene	Sanitary towels Panty liners Tampon/applicator
	General	Toilet paper Cotton buds Other/unidentified
Housing materials	Combustible	Fencing Hardboard/wood Other/unidentified
	Non-combustible	Brick/rubble Floor coverings Other/unidentified
Household (large)	Brown goods	Furniture Mattress/foam
	White goods	Other/unidentified
Household (small)	Metal	Cans/tins
Commercial/industrial	Metal	Container drums Sheeting Other/unidentified
	Plastic	Polystyrene Sheeting < 30 cm Sheeting 30–60 cm Sheeting > 60 cm Plastic bags Sweet papers Bottles
	Glass	Bottles Other/unidentified
Transport-associated	Motor vehicles	Cars/parts Motorbikes/parts Other/unidentified
	General	Signs/cones
General	Packaging	Cardboard
	Miscellaneous	Cloth/shoes Rope/fishing line Other/unidentified

can be achieved at low cost. However, concerns exist that reporting rates between groups may not be consistent. Trials have shown that volunteers frequently identify litter items incorrectly (Dixon, 1992). This has been investigated recently through the Coastwatch UK programme and it was found that these concerns were largely unfounded (Pond, 1996).

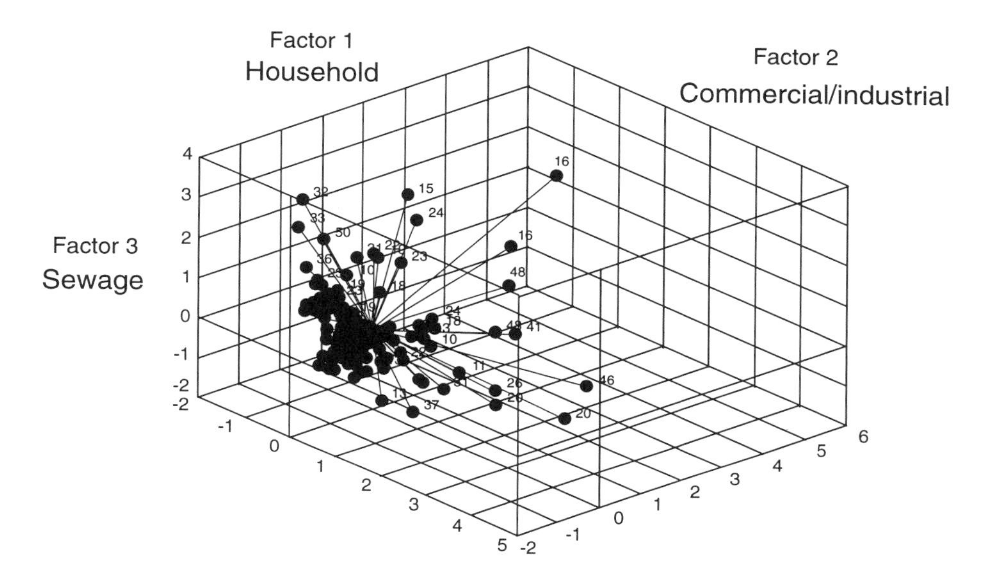

Figure 12.3 Principal Component Analysis of litter sites along the River Cynon in summer and winter. Numbers represent litter types described in Table 12.3

12.2.5 Beach quality questionnaires

Beach quality questionnaires should be objective. Several rating systems are based on a limited number of parameters (Table 12.4) and it should be an axiom that ratings must cover physical, human and biological parameters. Nevertheless, many existing systems have been found wanting in this respect. Virtually all the following do not take into consideration the beach user's perception of his or her environment.

The majority of beach quality schemes look at one or only a few of the parameters associated with beach ratings (Table 12.4). Beach aesthetics cannot be rated effectively on one facet alone, e.g. biological parameters. Table 12.4 shows a summary of the scope of a variety of beach awards and rating systems currently in place. Chapter 6 deals with beach award schemes in greater detail.

12.3 Beach cleaning

Increasingly, environmental management systems are being used to assess the routine performance of management approaches to the environment. Sequences of planning, objective setting, implementation, audit and review are becoming commonplace. The audit process for such systems often requires field measurements to assess whether systems are working. Monitoring in terms of "cleanup" is often the response to litter. The cleaning of

Table 12.4 An overview of the scope of selected beach awards and rating systems

Component	European Blue Flag	Seaside Award (TBG)	Good Beach Guide (MCS)	NRA (sw region)	Chaverri, 1989	Williams *et al.*, 1993	Beach Quality Rating Scale
Water quality	*	*	*		*	*	*
Education and information	*	*					
Access	*	*			*	*	*
Lifeguards/first aid	*	*				*	*
Litter	*	*			*	*	*
Sanitation	*	*					*
Sewage debris	*	*	*	*	*	*	*
Bathing water safety					*	*	*
Climate					*	*	*
Landscape quality					*	*	*
Beach material					*	*	*
Water temperature					*	*	*
Flora and fauna					*	*	*
Refreshments and facilities	Some	Some				Some	*
Beach regulation (dogs, vehicles, etc.)	*	*				*	*
Weighting of factors							*
Scoring based on preferences priorities of beach users							*
Quantification of most or all factors			*	*		*	*
Difference between resort/ undeveloped beaches		*					*

TBG Tidy Britain Group MCS Marine Conservation Society NRA National Rivers Authority (south-west region)

beaches provides a way of collecting data on the types and quantities of marine debris. However, the primary value of these methods is as public participation exercises and as a way of raising public awareness. The cleaning of beaches cannot solve the problem of marine debris permanently because they do not reduce the quantity of debris at source (Simmons and Williams, 1993). Physical “cleanups” are generally carried out by local authorities (Gilbert, 1996), local voluntary groups or volunteers co-ordinated by national voluntary bodies (Pollard and Parr, 1997). However, cleanups are really only useful at the local level, and they are expensive if undertaken by mechanical means or else they are labour intensive. Conversely, if volunteers are employed the costs are minimal. Site selection for beach cleaning programmes is biased towards areas with easy access, tourist locations and depositing beaches. Where volunteers are used, the collection of litter is the primary task and therefore the recording of the items is likely to be less of a priority (Rees and Pond, 1995a). Amos *et al.* (1995) have shown that volunteers participating in beach cleaning programmes undercounted individual items of debris by 50 per cent.

There are two methods of beach cleaning: mechanical and manual. Mechanical cleaning usually involves motorised equipment using a sieve effect that scoops up sand and retains the litter; therefore it is not selective. Litter retention is a function of the sieve. Most sieve machines are coarse grained allowing small items to pass through. The passage of such vehicles over the beach interferes with the beach ecology and the method is costly (Davidson *et al.*, 1991; Kirby, 1992; Acland, 1994; Llewellyn and Shackley, 1996). In addition this technique cannot be used on pebble beaches. Pressure to clean a beach is intense, especially where authorities wish to promote tourism. The advantages of such mechanical cleanups are that the result is achieved quickly, and large areas can be covered and they can provide an apparently pristine beach for visitors. Mechanical cleanups reduce the need for personal contact thus reducing health risks to individuals.

Manual beach cleaning programmes share many advantages and disadvantages with mechanical cleaning. They can help to raise community awareness of the litter problem and enable the sourcing of the litter (Earll, 1996) from a scientific perspective rather than scooping it up “en mass” for deposition in a landfill site. Manual beach cleanups organised as community events on small areas can ensure that the beach is cleaned of small items missed by mechanical methods (Pollard, 1996).

12.3.1 Economic aspects

Cleaning a country’s coastline costs the responsible authorities large amounts of money each year. In the UK, Suffolk District Council estimated that GB£ 50,000 was spent each year on cleaning the grounds around the coast and picking up litter from the foreshore. Authorities in Kent estimated that between GB£ 32,000 and GB£ 48,000 was being spent annually per beach and the direct and indirect costs of dealing with litter on the Kent coast has been estimated at over GB£ 11 million (Gilbert, 1996), which places a strain on the Gross National Product (GNP) of the area. Woodspring District Council reported that GB£ 100,000 was spent on managing litter and sand on just two beaches in the district of Weston Super Mare (Acland, 1994). Nevertheless this expenditure is necessary for tourist beaches.

In 1993, it cost GB£ 937,000 to clean the Bohuslan coast of Sweden (Olin *et al.*, 1994) and more than US$ 1 million was spent in 1988–89 cleaning up the coasts of Santa Monica and Long Beach in California (Kauffman and Brown, 1991). At Studland, Dorset, UK, one million visitors per year along a 6 km stretch of beach results in 12–13 tonnes of litter, collected each week during the summer months at a cost of GB£ 36,000 per annum. Additional costs are incurred when hazardous containers are found and have to be recovered from beaches (Dixon, 1992).

12.3.2 Measurable standards of cleanliness

The public perception of litter is intrinsically linked with standards of cleanliness. A number of issues become pertinent when setting standards or grades of cleanliness and these have been identified by Earll *et al.*, (1997).

- Will the public notice the standards set?
- If the public notice or recognise this material does it matter?
- At what level of littering do these issues become important to the public?
- Are the levels of litter indicative of other pollution, health and environmental hazards?

At present, it appears that very few standards of cleanliness exist regarding beach litter (see Chapter 6). It has been recognised that adequate information is required to support improvements in the cleaning of coastal waters and beaches (Anon, 1972). Coastal authorities, especially in Southern England, responded to the increasing amount of marine litter by extending the cleansing operation beyond the bathing season. The Royal Commission on Environmental Pollution (RCEP, 1984, 1985) noted the substantial costs incurred in beach cleaning operations. In the UK, the Environmental Protection Act (1990) sets standards of cleanliness under the Code of Practice issued under section 89(7) which are considered reasonable to meet. Local Authorities are encouraged to identify as “Category 5 Zone” areas, those beaches in their ownership or control that might reasonably be described as “amenity beaches”. For such designated amenity beaches, the minimum standard is that they should be generally clear of all types of litter and refuse between May and September inclusive. This standard applies, not only to items discarded by beach users, but also to items or materials originating

Table 12.5 Relative contribution of different sources of the marine debris found in St Brides Bay, Wales, 1997

Source	Summer (% of total)	Winter (% of total)	Average over the year (% of total)
Tourism	15.7	1.8	8.8
Shipping	9.1	13.4	11.2
Sewage-related debris	2.9	3.4	3.2
Fishing	8.9	19.5	14.2
Fly tipping	None found	None found	0.0
Medical items	None found	None found	0.0
Non-sourced	63.0	61.7	62.6

from disposal directly to the marine environment. The Code also notes that, in establishing a cleansing standard for beaches, careful consideration should be given to the practical difficulties encountered in collecting and removing litter, and to the damage to sensitive habitats which may result from such operations (DoE, 1991).

12.4 Debris sourcing

An objective of collating and analysing litter is to identify the source because it is only when the source is known that really effective action can be undertaken to remedy the situation. It is essential to have robust quantitative information to enable litter types to be monitored in a systematic manner and to enable assessments and judgements to be made. Sources could be marine (ships), tourists on a beach, fly tipping or a river.

Dixon (1992) concluded that beach litter in the UK consisted mainly of waste generated by ships, sewage discharges and material discarded by the general public, and that discharges of rubbish from ships and other crafts constituted 70 per cent of litter. This is almost the opposite viewpoint to that expressed by Faris and Hart (1995) in the USA, Gabrielides *et al.* (1991) in the Mediterranean, Ross *et al.* (1991) in Canada and Pond and Rees (1994). Litter sourcing seems to be highly site specific and generalisations should be avoided. For example, work carried out at St Brides Bay, Wales in 1997 showed that fly tipping and medical waste were not sources from which litter originated (Table 12.5). Non-sourced litter accounted for 62 per cent of the litter generated. Frost and Cullen (1997) attempted to categorise debris on northern New South Wales beaches by dividing debris into that which has the potential to float and debris that sank. It was then assumed that floatable debris was marine-based and sinking debris was from land-based or *in situ* sources.

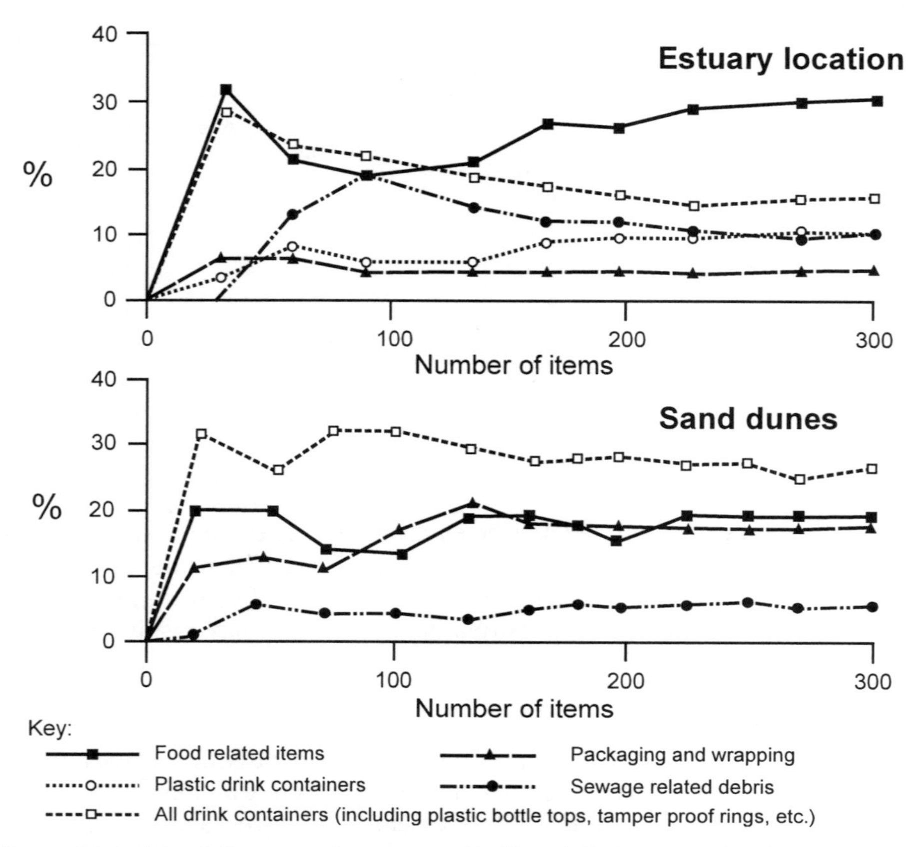

Figure 12.4 Cumulative percentage scores for litter: **A**. Estuary location; **B**. Sand dunes (After Earll *et al.*, 1997)

In Auckland City, attempts have been made to assess the scale of litter discharged from the City into the coastal marine area (Arnold, 1995). This involved monitoring material trapped by a 19 mm wire net placed on three storm-water catchments representative of each land use type. Comparison of the number of items associated with land-use types showed that industrial areas were the major source (9.69 items per hectare per day) followed by commercial (3.33 items per hectare per day) and residential areas (1.22 items per hectare per day).

An interesting approach for litter is to try to identify the people dropping the litter and to make inferences regarding their life style from the types of litter groups. This could help with direct prevention work. However, the number of items that should be collected in order to characterise life style groups still has to be resolved. One way would be to collect, for example, 200 items (in batches of 40–50 items) and to list them by function rather than by material and to carry out a similar analysis to that shown in Figure 12.2, but by plotting the number of categories against percentage occurrence and/or the number of sampled items against the number of litter categories. The common items that should take priority would show up very clearly and a long "tail" would be shown in the plot (Figure 12.4) (Earll *et al.*, 1997). The above approach could be adopted easily and could be carried out on a routine basis and in a cost-effective way. Photography would be an invaluable aid in this approach.

12.5 Elements of good practice

The following are considered to be the main elements of good practice for monitoring aesthetic parameters.

- Monitoring for specific aesthetic pollution parameters should be undertaken where hazards to human health and well being are suspected.
- Selection of aesthetic pollution parameters for monitoring should take into account local conditions and should consider parameters such as surface accumulation of tar, scums, odours, plastic, macroscopic algae or macrophytes (stranded on the beach and/or accumulated in the water) or cyanobacterial and algal scums, dead animals, sewage-related debris and medical waste.
- Sampling of aesthetic pollution indicators should take into account the perception and requirements of the local and any visiting populations in relation to specific polluting items as well as in relation to the feasibility of their monitoring.

12.6 References

Acland, B. 1994 Resort management. In: R.C. Earll [Ed.] *Coastal and Riverine Litter: Problems and Effective Solutions*. Marine Environmental Management and Training, Candle Cottage, Kempley, Glos., GL18 2BU, 18–20.

Amos, F.A., Wickham, A., Rowe, C. and Amos, L. 1995 An evaluation of volunteer beach cleaning and data-recording methods. In: J. Faris and K. Hart [Eds] *Seas of debris: A summary of the Third International Conference on marine debris.* UNC-SG-95-01, North Carolina Sea Grant, Florida, 54 pp.

Anon 1972 National Programme for Clean Water. *Marine Pollution Bulletin,* **3**(2), 22.

Arnold, G. 1995 Litter associated with stormwater discharges in Auckland City. In: J. Faris and K. Hart [Eds] *Seas of debris. A summary of the Third International Conference on marine debris*. UNC-SG-95-01, North Carolina Sea Grant, Florida, 54 pp.

Bonaiuto, M., Breakwell, G.M. and Cano, I. 1996 Identity processes and environmental threat: the effects of nationalism and local identity upon perception of beach pollution. *Journal of Community & Applied Social Psychology,* **6**, 157–175.

Braun-Blanquet, J. 1921 Prinziopen einer Systematik der Phflanzengesellschaften auf floristischer Grunglage, St, Gall. *Naturwiss, Gesell, Jabrb.,* **57**, 305–351.

Braun-Blanquet, J. 1932 *Plant Sociology: the study of plant communities.* McGraw Hill, New York.

Burnham, K.P., Anderson, D.R. and Laake, J.I. 1980 Estimation of density from line transect sampling of biological populations. *Wildlife Monographs,* **72**, 202 pp.

Cain, S.A. 1932 Concerning certain phytosociological concepts. *Ecological Monographs*, **2**, 475–508.

Cain, S.A. 1935 Ecological studies of the vegetation of the Great Smokey Mountains, Il. *Amer. Midl., Nat.,* **16**, 566–584.

Chaverri, R. 1989 Coastal Management; the Costa Rica Experience. In: O.T. Magoon [Ed.] *Coastal Zone '87, Proceedings of the 5th Symposium on Coastal and Ocean Management,* Volume 5, American Society of Civil Engineering, New York, 1112–1124.

Collins Concise Dictionary 1995 3rd edition, Harper Collins, Glasgow, 1583 pp.

Cottingham, D. 1989 Persistent marine debris, Part 1– the threat. *Mariners Weather Log,* **33**, 8–15.

Cuomo, V., Pane, V. and Contesso, P. 1988 Marine litter: where do the problems lie? *Progress in Oceanography,* **21**, 177–180.

David, E.L. 1971 Public perceptions of water quality. *Water Resources Research*, **7**, 453–457.

Davidson, N.C., D'Alaffoley, D., Doody, J.P., Way. L.S., Gordon, J., Key, R., Drake, C.M., Pienkowski, M.W., Mitchell, R and Duff, K.L. 1991 *Nature Conservation and Estuaries in Great Britain. Estuaries Review*. Nature Conservation Council, Northminster House, Peterborough, UK.

Davies, G. L. 1989 *The Investigation of the River Taff Litter Problem*. NRA Welsh region report No. PL\EAE\89\2, National Rivers Authority, Cardiff, UK, 16 pp.

Derde, M.P. and Massatt, D.L. 1983 Use of pattern recognition display techniques to utilise data contained in complex data bases. A Case Study. *Journal of Automatic Chemistry,* **5**, 136–145.

Dinius, S.H. 1981 Public perceptions in water quality evaluation. *Water Resources Bulletin,* **17**(1), 116–121.

Dixon, T. 1992 *Coastal Survey of Packaged Chemicals and Other Hazardous Items.* PECD Reference Number 7/8/188, ACOPS, London, 111 pp.

Dixon, T. and Dixon, T. 1981 Marine litter surveillance. *Marine Pollution Bulletin,* **12**(9), 289–295.

Dixon, T.J. and Dixon, T.R. 1983 Marine litter distribution and composition in the North Sea. *Marine Pollution Bulletin,* **14**, 145–148.

Dixon, T. and Hawksley, C. 1980 *Litter on the Beaches of the British Isles.* Report of the First National Shoreline Litter Survey. Sponsored by The

Sunday Times. Marine Litter Research Programme, Stage 3, The Tidy Britain Group, Wigan, UK, 70 pp.

DoE (Department of the Environment) 1991 *Environmental Protection Act 1990. Code of Practice on Litter and Refuse*. Her Majesty's Stationery Office, London, ISBN 0 11 752363 1.

Dubsky, K. [Ed.] 1995 *Coastwatch Europe. Results of the International 1994 Autumn Survey.* Coastwatch Europe, Dublin, ISBN 0-9523324-6-9, 128 pp.

Earll, R.C. 1996 *Measuring and Managing Litter in Rivers, Estuaries and Coastal Waters. A Guide to Methods*. Working document, 78 pp.

Earll, R., Williams, A.T. and Simmons, S.L. 1997 Aquatic litter, management and prevention–the role of management. In: E. Ozhan [Ed.] *MedCoast '97.* MedCoast, Middle East Technical University, Ankara, Turkey, 383–396,

EC 1995 Seawater Microbiology. Performance of methods for the microbiological examination of bathing water. Part I. EUR 16601 EN, European Commission, ISBN 1018-5593, 113 pp.

Faris, J. and Hart, K. 1995 *Seas of Debris. A Summary of the Third International Conference on Marine Debris.* UNC-SG-95-01, North Carolina Sea Grant, Florida, 54 pp.

Fleisher, J.M. 1985 Implications of coliform variability in the assessment of the sanitary quality of recreational water. *Journal of Hygiene,* **94**, 193–200.

Fleisher, J.M. 1990 Conducting recreational water quality surveys: Some problems and suggested remedies. *Marine Pollution Bulletin,* **21**(12), 562–567.

Frost, A. and Cullen, M. 1997 Marine debris on northern New South Wales beaches (Australia): sources and the role of beach usage. *Marine Pollution Bulletin,* **34**(5), 348–352.

Gabrielides, G.P., Golik, A., Loizides, L., Marino, M.G., Bingel, F. and Torregrossa, M.V. 1991 Man-made garbage pollution on the Mediterranean coastline. *Marine Pollution Bulletin,* **23,** 437–441.

Garber, W.F. 1960 Receiving Waters Analysis. In: E.A. Parson [Ed.] *Proceedings of International Conference on Waste Disposal in the Marine Environment.* Pergammon Press, 372–403.

Gilbert R.O. 1987 *Statistical Methods of Environmental Pollution Monitoring.* Van Nostrand Rheinhold, 320 pp.

Gilbert, C. 1996 The cost to local authorities of coastal and marine pollution–a preliminary appraisal. In: R.C. Earll [Ed.] *Recent Policy Developments and the Management of Coastal Pollution,* 12–14. Marine Environmental Management and Training, Candle Cottage, Kempley, Glos., GL18 2BU, 32 pp.

Gilbertson, D.D., Kent, M. and Pyatt, F.B. 1985 *Practical Ecology for Geography and Biology.* Unwin Hyman, 331 pp.

Gilligan, M.R., Pitts, R.S., Richardson, J.P. and Kozel, T.R. 1992 Rates of accumulation of marine debris in Chatham County, Georgia. *Marine Pollution Bulletin,* **24**(9), 436–441.

HCEC 1990 House of Commons Session 1989–90. Environment Committee. Fourth Report. Pollution of Beaches, Volume 1. Report with appendix together with the Proceedings of the Committee relating to the report, 11 July 1990, Her Majesty's Stationery Office, London, 365 pp.

Herring, B.A. and House, M.A. 1990 Aesthetic pollution public perception survey. Draft report, Water Research Centre, Medmenham, UK, 35 pp.

Horsnell, J.S. 1977 *Hazardous Chemical Washed Ashore: The Case for Government Action*, Isle of Wight County Council, Newport, UK, 7 pp.

House, M. 1993 *Aesthetic Pollution and the Management of Sewage-Derived Waste.* Flood Hazard Research Centre, Middlesex University, London, ISBN 1-85924-054-2, 12 pp.

Hubalek, Z. 1982 Coefficients of association and similarity based on binary (presence-absence) data. An evaluation, *Biological Reviews,* **57**, 669–689.

Jones, F., Kay, D., Stanwall-Smith, R. and Wyer, M. 1990 An appraisal of the potential health impacts of sewage disposal to UK coastal waters. *Journal of the Institution of Water and Environmental Management,* **4**, 295–303.

Kaufmann, J. and Brown, M. 1991 California marine debris action plan. In: O.T. Magoon, H. Converse, V. Tipi, L.T. Tokin, and D. Clark [Eds] *Coastal Zone '91.* Proceedings of the Seventh Symposium on Coastal and Ocean Management. Long Beach, California, July 8–12, 1991, 3390–3406.

Kirby, P. 1992 *Habitat Management for Invertebrates. A Practical Handbook.* Published by Joint Nature Conservation Committee, Peterborough, UK.

Koops, W. 1988 Policy in the Netherlands with respect to response of chemical spills. In: P. Bockholts and I. Heiderbrink [Eds] *Proceedings First International Conference on Chemical Spills and Emergency Management at Sea, Amsterdam, Netherlands.* Kluwer Academic Publishers, London, 103–114.

Llewellyn, P.J and Shackley, S.E, 1996 The effects of mechanical beach cleaning on invertebrates. *British Wildlife,* **7**(3), 147–155.

Ludwig, J.A. and Goodall, D.W. 1978 A comparison of paired and blocked quadrat variance methods for the analysis of spatial pattern. *Vegetation,* **38**, 49–59.

Ludwig, J.A. and Reynolds, J.F. 1988 *Statistical Ecology. A Primer on Methods and Computing.* Wiley, 337 pp.

MCS (Marine Conservation Society) 1996 *Readers Digest Good Beach Guide.* David and Charles, 255 pp.

Mendes, B., Urbano, P., Alves, C., Lapa, N., Norais, J., Nascimento, J. and Oliveira, J.F.S. 1997 Sanitary quality of sands from beaches of Azores islands. *Water Science and Technology,* **35**(11–12), 147–150.

Miller, J C. and Miller, J N. 1988 *Statistics for analytical chemistry.* Second Editon. Wiley, 227 pp.

MNHW 1992 *Canadian Recreational Water Quality Guidelines.* Canadian Government Publishing Centre, Ottawa, Cat., H49-70/1991E, 101 pp.

Morgan, R., Jones, T.C. and Williams, A.T. 1993 Opinions and perceptions of England and Wales Heritage Coast beach users: some management implications for the Glamorgan Heritage Coast, Wales. *Journal of Coastal Research,* **9**(4), 1083–1093.

Norusis, M.J. 1983 *SPSSX Introductory Statistics Guide,* McGraw-Hill, 276 pp.

NAS 1975 Marine litter. In: *Assessing Potential Ocean Pollutants.* A report of the study panel on assessing potential ocean pollutants to the Ocean Affairs Board. Commission on Natural Resources, Natural Research Council, National Academy of Sciences, Washington DC, 405–433.

NRA 1992 *Two year study of beaches in the South West (1990–91).* Report No. JD 00801191, Her Majesty's Stationery Office, London, 5 pp.

NRA 1996 *Development and Testing of General Quality Assessment Schemes.* R & D Project Report 27, Her Majesty's Stationery Office, London, ISBN 0-11-310109-0, 24 pp.

Oldridge, S. 1992 Bathing water quality: a local authority perspective. In: D. Kay [Ed.] *Recreational Water Quality Management, Vol. 1, Coastal Waters.* Ellis Horwood Ltd., Chichester, 33–47.

Olin, R., Carlsson, B. and Stahre, B. 1994 The west coast of Sweden–the rubbish tip of the North Sea. In: R. C. Earll [Ed.] *Coastal and Riverine Litter. Problems and Effective Solutions.* Marine Environmental Management and Training, Kempley, Glos., UK, 12–14.

Pearce, F. 1995 Dead in the water. *New Scientist,* 4th February 1995, 26–31.

Philipp, R. 1991 Risk assessment and microbial hazards associated with recreational water sports. *Reviews in Medical Microbiology,* **2**, 208–214.

Philipp, R. 1993 Community needlestick accident data and trends in environmental quality. *Public Health,* **107**, 363–369.

Philipp, R., Pond, K., and Rees, G. 1997 Research and the problems of litter and medical wastes on the UK coastline. *British Journal of Clinical Practice,* **51**(3), 164–168.

Pollard, S. 1996 *Beachwatch '96. 1995 Nationwide Beach-Clean & Survey Report.* The MCS Ltd, Ross on Wye, England, 71 pp.

Pollard, S. and Parr, J. 1997 *Beachwatch '96. 1997 Nationwide Beach Clean & Survey Report.* The MCS Ltd. Sponsored by The Readers Digest, Ross, UK, 81 pp.

Pond, K. 1996 An appraisal of a practical participative environmental project as a tool in coastal zone management. Unpublished Ph.D. thesis, University of Surrey, 334 pp.

Pond, K. and Rees, G. 1994 *Norwich Union Coastwatch UK 1994 Survey Report.* Published by Farnborough College of Technology, Hampshire, UK, 102 pp.

RCEP 1984 *Royal Commission Environmental Pollution. Tackling Pollution: experience and prospects,* Tenth Report. CMMD 149, Her Majesty's Stationery Office, London, 232 pp.

RCEP 1985 *Managing Waste: The duty of care.* Eleventh Report. CMMD 9675, Her Majesty's Stationery Office, London, 214 pp.

Rees, G. 1997 Lies, damned lies and beach awards. In: *Current Quality.* Edition 1, p 1. The Robens Centre for Public and Environmental Health, University of Surrey, UK, and the World Health Organization, ECEH, Rome, Italy.

Rees, G. and Pond, K. 1995a Impacts: Aesthetics, health and physical clearance. In: R.C. Earll [Ed.] *Proceedings of a two-day workshop–Coastal and Riverine Litter: Problems and Effective Solutions.* Published by Marine Environmental Management and Training, Candle Cottage, Kempley, Glos., UK, 5–7.

Rees, G. and Pond, K. 1995b Marine litter monitoring programmes–a review of methods with special reference to national surveys. *Marine Pollution Bulletin,* **30**(2), 103–1089.

Rees, G. and Pond, K. 1996 *Coastwatch UK 1995 Survey Report,* Sponsored by The Daily Mirror, Published by Farnborough College of Technology, Farnborough, Hampshire, England, ISBN 1-899080-50-3, 64 pp.

Ribic, A. C., Dixon, T.R. and Vining, I. 1992 *Marine Debris Survey Manual.* NOAA Technical Report. NMFS 108, US Department of Commerce, 92 pp.

Ribic, C. 1990 Report of the working group on methods to assess the amount and types of marine debris. In: R.S. Shommura and M.I. Godfrey [Eds] *Proceedings of the Second International Conference on Marine Debris.* 2–7th April 1989, Honolulu, NOAA Technical Report. NOAA-TM-NMFS-SWFSC-154, NMFS, US National Oceanic and Atmospheric Administration, Washington, D.C., 1201–1206.

Ross, J. B., Parker, R. and Strickland, M. 1991 A survey of shoreline litter in the Halifax Harbour 1989. *Marine Pollution Bulletin,* **22**(5), 245–248.

Saliba, L.J. 1995 *Assessment of the State of Microbiological Pollution of the Mediterranean Sea (Draft document of consultation on microbiological monitoring of recreational and shellfish-growing waters).* WHO/UNEP joint project, MED POL Phase II, Athens, Greece, 29 November–2 December 1995, EUR/ICP/EHAZ/MTO2(7), WHO Regional Office for Europe, Copenhagen, 127 pp.

Schluter, D.A. 1984 Variance ratio test for detecting species association with some example applications. *Ecology*, **65**, 998–1005.

Simmons S.L. and Williams A.T. 1993 Persistent marine debris along the Glamorgan Heritage Coast: A Management problem. In: H. Sterr, J. Hofstide and H.P. Plag [Eds] *Interdisciplinary Discussions of Coastal Research and Coastal Management Issues.* Peter Frank, Frankfurt, 240–250.

Simmons S.L. and Williams, A.T. 1997 Quantitative versus qualitative litter data. In: E. Ozhan [Ed.] *Medcoast '97, Proceedings of the third international conference on the Mediterranean coastal environment.* METU, Ankara, Turkey, 397–406.

University of Surrey 1987 *The Public Health Implications of Sewage Pollution of Bathing Water.* The Robens Institute of Industrial and Environmental Health and Safety, England, 25 pp.

Valle-Levinson, A. and Swanson, R.L. 1991 Wind-induced scattering of medically-related and sewage-related floatables. *Marine Technology Society Journal,* **25**(2), 49–56.

Verlander, K. and Mocogni, M. 1996 Sampling strategies - statistical validity. In: R.C. Earll [Ed.] *Measuring and Managing Litter in Rivers, Estuaries and Coastal Waters. A Guide to methods.* Marine Environmental Management and Training, Kempley, Glos., UK, 62–67.

Walker, A. 1991 Swimming – the hazards of taking a dip. *British Medical Journal,* **304**, 242–245.

WHO 1989 *European Charter on Environment and Health.* ICP/RUD 113/Conf.Doc/1 Rev.2, 8303r, World Health Organization Regional Office for Europe, Copenahgen, 7 pp.

Williams, A.T. 1986 Landscape aesthetics of the River Wye. *Landscape Research,* **11**(2), 25–30.

Williams, A.T and Morgan, R. 1995 Beach awards and rating systems. *Shore and Beach,* **63**(4), 29–33.

Williams, A.T. and Nelson, C. 1997 The public perception of beach debris. *Shore and Beach,* **65**(3), 17–20.

Williams, A.T. and Simmons, S.L. 1997a Estuarine litter at the river/beach interface. *Journal of Coastal Research,* **13**(4), 1159–1165.

Williams, A.T and Simmons, S.L. 1997b Movement patterns of riverine litter. *Water, Air and Soil Pollution,* **98**, 119–139.

Williams, A.T., Simmons, S.L. and Fricker, A. 1993 Offshore sinks of marine litter: A new problem. *Marine Pollution Bulletin,* **26** (7), 404–405.

YRLMP 1991 *Yorkshire Rivers Litter Monitoring Project, 1991.* Devised by the TBG and Sponsored by the NRA, Tidy Britain Group, Wigan, UK, 12 pp.

Chapter 13*

EPIDEMIOLOGY

Epidemiological data are frequently used to provide a basis for public health decisions and as an aid to the regulatory process. This is certainly true when developing safeguards for recreational waters where hazardous substances or pathogens discharged to coastal and inland waters may pose a serious risk of illness to individuals who use the waters. Epidemiological studies of human populations not only provide evidence that swimming-associated illness is related to environmental exposure, but also can establish an exposure–response gradient which is essential for developing risk-based regulations. Epidemiology has played a significant role in providing information that characterises risks associated with exposure to faeces contaminated recreational waters. The use of epidemiological studies to define risk associated with swimming in contaminated waters has been criticised because the approach used to collect the data is not experimental in nature. This perception is unlikely to change, given the highly variable environments where recreational exposures take place. Although the variables may be difficult to control, it is possible to carry out credible studies by following certain standard practices that are given below. This Chapter discusses the place of epidemiological investigations in providing information to support recreational water management and the scientific basis of "health-based" regulation.

13.1 Methods employed in recreational water studies

Epidemiology is the scientific study of disease patterns in time and space. Epidemiological investigations can provide strong evidence linking disease incidence and environmental or other exposures. However, this statistical inference does not provide absolute proof of a direct cause and effect, although the combination of strong statistical association with biological plausibility offers strong evidence of causality. Epidemiological methods can quantify the probability that observed relationships occurred by "chance" factors. The methods employed can range from the study of recorded outbreaks of illness (i.e. seeking to infer the causes of morbidity from existing patterns of recorded individuals who are ill (and called "cases") and

* *This chaper was prepared by D. Kay and A. Dufour*

possible "controls" who are not ill), through to carefully designed studies in which volunteers are exposed to a hazard (such as faecally contaminated bathing water) and then followed up for a suitable period to investigate the incidence of illness. The type of study employed is dependent on:

- The objectives of the study, i.e. the required use of the data to be acquired.
- The nature of the exposure and illness under study.
- Available epidemiological and biostatistical expertise, together with economic constraints.

It is vital that these three elements are considered at the outset of any investigation. The first element, "objective(s) of the study", is perhaps the most important aspect because the available types of study discussed below each provide data with distinct potential uses. It is vital that the data produced by the more rudimentary epidemiological designs are not over-interpreted and that their limitations are understood clearly by the scientific and policy communities.

There have been a many epidemiological investigations for the health impacts of exposure to faecally polluted recreational waters reported in the scientific literature since 1953 (Table 13.1). These investigations fall into three main design categories described in the following subsections.

13.1.1 Retrospective case-control studies

Retrospective case-control studies are used to determine whether a particular personal characteristic or environmental factor is related to disease occurrence. Cases refer to persons who have a specific illness or disease. Controls, who do not have the illness or disease, are selected. The selection may seek to "match" for variables such as age and sex, or an unmatched design can be employed in which possible confounders are controlled during the analysis phase. Cases and controls are queried to determine if their exposure to environmental hazards have been similar. For example, cases of typhoid fever and their matched controls may be questioned about their past activity with respect to swimming events. The results of questioning may, for example, show that swimming activity is more likely to have occurred with typhoid cases than with controls, indicating a potential association between swimming and the disease. This type of study is most useful in disease outbreaks where a retrospective case-control study may be conducted to determine if certain activities or exposures were related to the disease or illness under investigation. This approach also may be useful in establishing the relationship between serious illness, such as hepatitis, and exposure to bathing waters. The advantage of conducting retrospective case-control studies is that they are not very costly and are reasonably easy to carry out. Their disadvantage is that, while the linkage between disease and exposure can be determined, it is seldom possible to determine the magnitude of the exposure.

Table 13.1 A summary of epidemiological studies

Country	Water body	Indicator	Symptom(s)	Reference
Australia	Sea	Faecal coliform Faecal streptococci	E/ENT/R	Corbett *et al.*, 1993
	Sea	Faecal coliform Faecal streptococci *C. perfringens*	GI/R/O	Harrington *et al.*, 1993
Canada	Fresh	Total staphylococci	R/GI	Lightfoot, 1989
	Fresh	Total staphylococci Faecal coliform Faecal streptococci	R/GI	Seyfried *et al.*, 1985a,b
Egypt	Sea	Enterococci *E. Coli*	GI	El Sharkawi and Hassan 1982
France	Fresh	Total coliforms Faecal coliforms Faecal streptococci *Aeromonas spp.* *P. aeruginosa*	All + S S GI S S	Ferley *et al.*, 1989
	Sea	Faecal streptococci Total coliforms Faecal coliforms	E/S/GI	Foulon *et al.*, 1983
Hong Kong	Sea	*E. coli* *Klebsiella* Faecal streptococci *Enterococci* *Staphylococci* *P. aeruginosa* *Candida albicans* Total fungi	GI+S GI+S GI+S GI+S ENT GI+S O E+O	Cheung *et al.*, 1990
Israel	Sea	*S. aureus* Enterococci *E. coli* Total staphylococci *P. aeruginosa*	GI GI GI	Fattal *et al.*, 1991
Israel	Sea	Enterococci *E. coli* *S. aureus* *P. aeruginosa*	GI	Fattal *et al.*, 1986
Netherlands	Fresh	*P. aeruginosa*	ENT	Asperen *et al.*, 1995
South Africa	Sea	Faecal coliform *E. coli* Faecal streptococci Total staphylococci	GI/R/S	Schirnding *et al.*, 1993
	Sea	Faecal coliform *E. coli* Faecal streptococci Total staphylococci	GI/R/S	Schirnding *et al.*, 1992
Spain	Sea	Faecal streptococci	S/E/ENT/GI	Mujeriego *et al.*, 1982

Continued

Table 13.1 Continued

Country	Water body	Indicator	Symptom(s)	Reference
UK	Sea	Faecal streptococci Faecal coliform	E/S/ENT/R	Fleisher *et al.*, 1996
	Sea	Faecal streptococci	GI	Kay *et al.*, 1994
	Fresh	Total coliforms Faecal coliforms Faecal streptococci Total staphylococci *P. aeruginosa*	E/S/ENT/R	Fewtrell *et al.*, 1993
	Sea	Faecal streptococci	GI	Fleisher *et al.*, 1993
	Fresh	Total coliforms Faecal coliforms Faecal streptococci Total staphylococci *P. aeruginosa* Enterovirus	E/S/ENT/R	Fewtrell *et al.*, 1992
	Sea	Total coliforms Faecal coliforms Faecal streptococci Salmonella Enterovirus	ENT/GI/S/O	Alexander and Heaven, 1991
	Sea	Total coliform Faecal coliform Faecal streptococci	E/GI/ENT/R	Balarajan *et al.*, 1991
	Sea	Total coliforms Faecal coliforms Faecal streptococci	E/S/ENT/R	Jones *et al.*, 1991
	Sea	NR	GI	Brown *et al.*, 1987
	Sea	Total coliform	O	PHLS, 1958
USA	Fresh	Enterococci *E. coli*	GI	Dufour, 1984
	Sea	Enterococci	GI	Cabelli, 1983
	Both	Total coliform	ENT/GI/R	Stevenson, 1953

E Eye symptoms
S Skin complaints
GI Gastro-intestinal symptoms
ENT Ear nose and throat symptoms
R Respiratory illness
O Other
NR Not reported

13.1.2 Prospective cohort study

The second type of investigation is a prospective cohort study. In this study, individuals are recruited immediately before or, more commonly, after participation in some form of recreational water exposure. A control group is similarly recruited and both cohorts are followed up for a period of time. The exposure status of the bathers and non-bathers is self-selected and not randomised in this type of study. During the follow-up period, data are acquired on the symptoms experienced by the two cohorts using questionnaire interviews, either in person or by means of telephone inquiry. The quality of the recreational water environment is defined through environmental sampling on the day of exposure. The exposure data are often combined to produce a "daily mean" value for the full group of bathers using a particular water on any one day. Many days of exposure are required to define adequately the relationship between "exposure day" water quality and disease. Thus, data on "exposure" are available which can be related to "illness" outcome through an exposure–response curve predicting illness from indicator bacterial concentration. However, this approach will not provide a unique exposure measure (i.e. microbial indicator concentration) for each exposed individual and may lead to systematic misclassification bias. In addition, indicator organism counts are an indirect, and very often very inadequate, estimate of exposure to pathogens.

13.1.3 Randomised trials

A third approach, the randomised trial, is also a "prospective" study design but it differs from the cohort study outlined above in several respects. First, volunteers are recruited at the outset of the experiment. This group is generally interviewed prior to exposure, given medical examinations and then randomised into bather and non-bather groups. Both groups report to a test beach on a predetermined day. Typically, the bathers undertake a brief period of water exposure whilst the non-bathers remain on the beach. In well conducted studies, supervisors monitor each group and note the time and place of exposure undertaken by each bather. Both groups may also be given similar food, in the form of a packed lunch, and may undertake an identical short interview on the study day. Typically, volunteers report for a post-exposure interview and medical examination a week after exposure and complete a further postal questionnaire to examine any illnesses with longer incubation periods. During the exposure period, samples should be acquired from the bathing area in sufficient number to characterise fully water quality at the time and place of exposure for each bather, so that exposure can be assigned to each individual based on the time and place of exposure.

13.2 Major studies

13.2.1 UK Public Health Laboratory Service retrospective studies

The most widely quoted example of a retrospective case-control study was completed by the UK Public Health Laboratory Service between 1953 and 1958 (PHLS, 1959). This study was designed to identify any link between bathing in sewage-polluted waters and cases of poliomyelitis or enteric fever (paratyphoid). All cases of these two notifiable illnesses reported in seaside

District Council areas were used in the study. Controls (matched for age and sex and selected, where possible, from the same street) were also identified. The availability of water quality records for the beaches used by the identified "cases" was also investigated, but microbiological information was rarely available for the appropriate times and locations of the exposure event(s). All cases and controls were interviewed by local medical staff to determine their bathing history. The authors concluded that there was no evidence linking the incidence of poliomyelitis and a history of sea bathing.

Role of retrospective case control studies
Although the retrospective design is not useful for developing exposure–response relationships, it is appropriate for linking illnesses to environmental exposures.

13.2.2 US Environmental Protection Agency prospective studies
The United States Environmental Protection Agency (US EPA) conducted a series of retrospective studies in the mid-1970s (Cabelli, 1983). At the time, these studies were the most extensive and carefully conducted prospective epidemiological investigations ever attempted. The main elements of the design are described briefly below.

Trials were conducted on Saturdays and Sundays, i.e. weekend-only bathers and non-bathing beachgoers were recruited in the hope of avoiding the multiple exposure problem. Demographic data were acquired during the initial beach interview and during a subsequent telephone interview. Bathing activity was recorded during the initial beach interview; bathers were defined as those who had experienced full head immersion, thus risking ingestion of water via the nasal and oral orifices. Wet hair at the time of recruitment was used as an indicator of head immersion and defined the exposure status of the participant.

The recruitment of study participants targeted family groups that included non-bathing controls. A letter was posted 1–2 days after the beach contact to remind all participating families that they should be recording all illness. At 8–10 days after recruitment the respondents for each family group were contacted by telephone to record any symptoms that had developed since the day they were at the beach.

Only gastro-intestinal (GI) symptoms were considered to be related to both swimming and pollution and the water quality indicators chosen. A subgroup of highly credible gastro-intestinal (HCGI) symptoms was defined as vomiting, diarrhoea accompanied by fever or which was disabling, or nausea and/or stomach ache accompanied by fever.

A range of bacterial indicators was used to characterise water quality at different locations, namely 11 indicators at New York City beaches, 5 at Lake Pontchartrain and 2 in Boston. Water quality was measured at times of maximum swimming activity, with 3–4 samples collected from 2–3 sites at chest depth, 12 inches below the surface of the water. This sampling design was used to characterise water quality for each test day. The geometric mean of these samples was used as the exposure estimate for all bathers on that day.

A rate difference between the bather and non-bather groups was used to quantify the swimming-associated morbidity rate for any particular symptom. Each trial day was associated with a specific water quality and an attack rate. However, individual trial days were not used as the raw data of subsequent analyses because the non-bathing control group for each day was of insufficient size. Analyses, therefore, was performed on water quality data gathered by summer and beach, or data from beach trial days that formed natural clusters of similar indicator densities.

Bivariate log-linear least squares regression was used to define the relationship between the seasonal swimming associated rate for GI symptoms (i.e. the rate difference between bathers and non-bathers) and water quality (as $\log_{10}$ geometric mean faecal indicator concentration). Cabelli (1983) reported statistically significant relationships between enterococcus density and swimming associated GI symptoms (GI $r^2 = 74\%$; HCGI $r^2 = 52\%$ i.e. 52 per cent of the variance in HCGI symptom reporting was explained by the predictor variable enterococcus density). These relationships have been used to quantify the risk implied by the previous (NTAC, 1968) standards leading to the formulation of the most recent US Federal standard systems (US EPA, 1986).

Role of non-randomised prospective approach

Precise measurement of water quality at the time of exposure for each swimmer is not possible. The advantages of the non-randomised prospective approach include:

- Selection of participants after voluntary swimming activity allows a broad range of swimmers to be studied.
- Swimming-associated illness is expressed in terms of exposed populations rather than the probability of individual infection. Under some circumstances this approach may be meaningful from a public health perspective.

13.2.3 UK prospective randomised trial studies

The randomised trial involved recruitment of healthy adult volunteers at seaside towns with adjacent beaches that had traditionally passed EC Imperative Standards (Jones *et al.*, 1991; Kay *et al.*, 1994). After initial interviews and medical checks, volunteers reported to the specified bathing location on the trial day where they were randomised into bather and non-bather groups.

Bathers entered the water at specified locations where intensive water quality monitoring was taking place under the supervision of a marshall who recorded their activities. All bathers immersed their heads on three occasions. On exiting the water bathers were asked if they had swallowed water. The locations and times of exposure were known for each bather and, thus, a more precise estimate of "exposure" (i.e. indicator bacterial concentration) could be assigned to each bather (Fleisher *et al.*, 1993; Kay *et al.*, 1994). A control group of non-bathers came to the beach and had a picnic of identical type to that provided for all volunteers. One week after exposure all volunteers returned for further interviews and medical examinations and later they completed a final postal questionnaire, three weeks after exposure.

Detailed water quality measurements were completed at marked locations that defined "swim zones". Samples were collected synchronously at locations 20 m apart every 30 minutes and at three depths (i.e. surf zone, 1 m depth and at chest depth, 1.3–1.4 m). In the case of the latter two sampling depths, samples were collected at approximately 30 cm below the surface of the water. Five bacterial indicators were enumerated. Faecal coliforms and faecal streptococci were analysed using triplicate filtrations for each of three dilutions (i.e. nine plates per bacterial enumeration) to narrow the confidence limits on enumeration of total coliforms. Total staphylococci and *Pseudomonas aeruginosa* were also enumerated.

The initial analysis of the data from the UK randomised trial experiments centred on the links between water quality and gastro-enteritis (see Fleisher *et al.*, 1993 and Kay *et al.,* 1994). The data were analysed for relationships between water quality, as indexed by any of the five bacterial indicators measured at any of the three depths (i.e. 15 potential predictor variables) and gastro-enteritis. Only faecal streptococci, measured at chest depth, provided a statistically significant relationship between water quality and the risk of gastro-enteritis. This result was replicated at three of the four sites examined and at the fourth site concentrations of faecal streptococci were generally below the threshold level at which an effect was observed at the other three locations. No site specific differences were observed in terms of the exposure response curve. The relationship between faecal streptococci concentrations in recreational waters and the excess probability of gastro-enteritis in the exposed population is shown in Figure 13.1. This trend was not apparent with any other bacterial indicator enumerated and the volunteers reporting symptoms (or the research team) could not have known the concentrations of faecal streptococci to which each bather was exposed.

Multiple logistic regression analyses also allowed for the assessment of the effects of concomitant factors (i.e. other predictors of GI symptoms) on the relationship between water quality and illness. The analysis showed that other factors were significant predictors of GI illness. These included

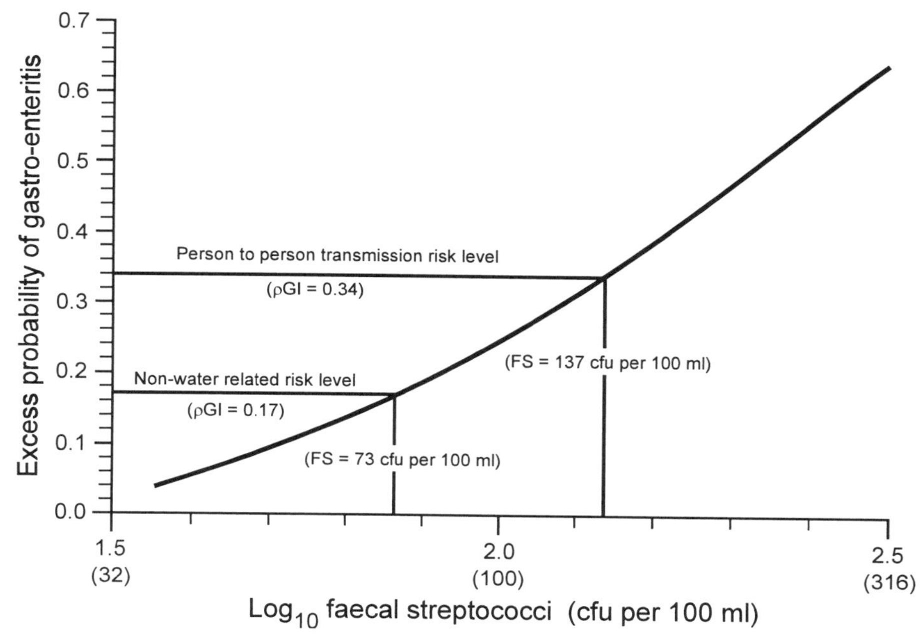

Figure 13.1 The dose response curve linking faecal streptococci with excess probability of gastro-enteritis

non-water-related risk (NWR) factors such as certain food types (see Table 13.2) and person-to-person transmission (PPT) from sick household members. These factors were independent of, and did not confound, the relationship between water quality and gastro-enteritis and can therefore provide markers against which the risk of GI illness from sewage contaminated sea water can be measured (see Figure 13.1). For example, exposure to NWR related risk factors in Table 13.2 represents a risk equivalent to a faecal streptococci concentration of approximately 70 organisms per 100 ml, whereas sharing a household with a person exhibiting GI symptoms represents an equivalent risk to a single exposure to recreational water containing approximately 140 faecal streptococcci per 100 ml.

This UK study, therefore, provides two sources of information. It presents an exposure–response relationship that defines the risk attributable to different levels of faecal streptococci exposure and, it quantifies the risk of other commonly experienced risks in society. As a result it provides scope for a system of risk-based standards.

These trials have also examined the relationships between non-enteric illnesses and exposure to sea water. Significant exposure–response relationships have been reported between acute febrile respiratory illnesses and faecal streptococci concentrations and between ear infections and faecal

Table 13.2 Non-water-related risk factors for gastro-enteritis

Age (grouped by 10-year intervals)
Gender
History of migraine headaches
History of stress or anxiety
Frequency of diarrhoea (often, sometimes, rarely or never)
Current use of prescription drugs
Illnesses within 4 weeks prior to the trial day lasting more than 24 hours
Use of prescription drugs within 4 weeks prior to the trial day
Consumption of any of the following foods in the period from 3 days prior to 7 days after the trial day: mayonnaise, purchased sandwiches, chicken, eggs, hamburgers, hot dogs, raw milk, cold meat pies or seafood
Illness in the household within 3 weeks after the trial day
Alcohol consumption within the 7-day period after the trial
Frequency of usual alcohol consumption
Taking of laxatives within 4 weeks of the trial day
Taking of other stomach remedies within 4 weeks of the trial day
Additional bathing within 3 days prior and 3 weeks after the trial day[1]

[1] This was included in order to control for possible confounding due to multiple exposures among bathers and exposure among non-bathers prior to or after the trial day

coliform concentrations. In addition, eye ailments were elevated in the bather group but unrelated to faecal indicator concentrations in the water (Fleisher *et al*., 1996). Standard systems, to date, have centred on gastro-enteritis as the main outcome. However, as more evidence mounts on these non-enteric illnesses it may be prudent to consider their inclusion into future standard systems incorporating the concept of total disease burden through the use of Disability Adjusted Life Years (DALYs) or other means of cross-comparison between illnesses.

Limitations of the randomised study protocol

The scope of UK randomised trial protocol is limited and should not be over-interpreted. The limitations include the fact that the studies were conducted in north European marine waters with a high tidal range where all waters commonly passed EU Imperative coliform criteria and the US EPA enterococci criteria. It may not be as applicable, however, in the standards design process for Mediterranean bathing waters where solar radiation is more intense, turbidity is lower and tidal activity is limited. Similar comments could be made concerning the application of these results to fresh-water environments.

Furthermore, the results apply only to healthy adult volunteers, and may not be applicable directly to infants or chronically sick people. This is of particular relevance in the consideration of sampling depth. Adult chest depth predicted gastro-enteritis in adult bathers, but the UK operational sampling depth of 1 m (or less) might be more appropriate for child bathers. Another limitation is that the results may not be applicable to special interest groups such as surfers, sail-boarders and other high exposure activities that may involve prolonged contact with the water (often at some distance from the beach).

Role of the randomised prospective design

Randomisation of the exposed (bather) and non-exposed (non-bather) groups removes the problem of self-selection bias that is always potentially present where the exposure status is self-selected. More precise definition of water quality to which each bather was exposed (ideally through measurements taken at the place and time of exposure) provides better definition of the exposure for each bather. The multiple interviews facilitate data acquisition on a broad range of potential risk factors for the illness outcomes and allow accurate case definitions to be made.

13.3 Choice of study design

The primary criteria to be considered in the choice of an appropriate epidemiological study protocol are the objectives of the study and the validity of the findings, both of which determine the use to which the data acquired can be put. A secondary consideration is the scientific capacity and resource availability of the society in which the study is to be conducted. If, for example, the primary objective is to define an exposure–response relationship with maximum precision, then the randomised trial provides the most appropriate protocol. Its suitability derives from its tight control and relatively precise measurement of exposure (i.e. the water quality experienced by each bather) and the extensive data acquired on NWR risk factors. However, there are circumstances, even in affluent developed nations, where the implementation of the randomised trial is not feasible. For example, epidemiological investigations conducted to investigate the health implications of water sports activities, such as white water canoeing, would find a randomised design almost impossible to apply. It would clearly be inappropriate (and irresponsible) to expect a cohort of volunteers recruited from the general public to participate in a potentially dangerous activity for which they would not be skilled. Furthermore, randomisation of existing water sports participants would imply that the non-exposed group would agree willingly not to participate in their sport for a given period. In such circumstances an improved prospective cohort design (Cabelli, 1983) is the most appropriate. The application of this protocol to special interest groups of water sports enthusiasts in marine and fresh waters has been reported by Fewtrell *et al.* (1992, 1993, 1994).

Other circumstances may limit the application of the randomised design; for example, ethical constraints can preclude the inclusion of young people in the volunteer group. This was true for the UK studies but it was not found to be a problem in the randomised trial pilot investigation conducted in the Netherlands during 1996 (van Asperen *et al.*, 1997). The Netherlands randomised trial involved a volunteer group of children with no ethical constraints and studied their exposure to fresh recreational waters. Clearly, the identification and resolution of any potential ethical problems are an important preliminary step in the application of a randomised design.

Where a non-randomised prospective protocol design is applied, i.e. the activity status of the participants is not determined by the researcher, a number of points should be noted. The measure of exposure (i.e. water quality) should, as far as possible, be attributed to small groups of exposed individuals. In effect, daily mean values applied to large groups exposed on one day will mask variability in indicator concentration and underestimate the numbers exposed to high indicator concentration. If the misclassification is random, it will tend to underestimate the slope of the dose–response curve through systematic misclassification bias. Attributing an exposure level to as small a group as possible requires intensive spatial and temporal water quality sampling.

Similarly, such studies should seek to acquire extensive data on NWR risk (and other potential confounding) factors. This might imply recruitment and follow-up interviews of considerable length with well-trained specialist health professionals conducting the data acquisition exercise. As with all epidemiological investigations, tight statistical control and early inputs to protocol design are essential to ensure that the information derived from the data acquired is maximised.

Through appropriate logistic regression analysis, the non-randomised prospective design can produce exposure–response relationships with information on NWR risk factors. However, such relationships should be treated with caution in standards design because of the possibility of misclassification bias and the potential underestimation of effect. Thus, if data on special interest groups or watersports activities are required, then a non-randomised prospective design may be appropriate, provided that any exposure–response relationships are treated with caution in their application to standards design of the type outlined in the *Guidelines for Safe Recreational Water Environments: Coastal and Fresh Waters* (WHO, 1998).

The retrospective case control design is clearly the only possible outbreak investigation tool and it does provide a means for establishing a link between specific pathogens and water exposure. It can provide guidance on maximum

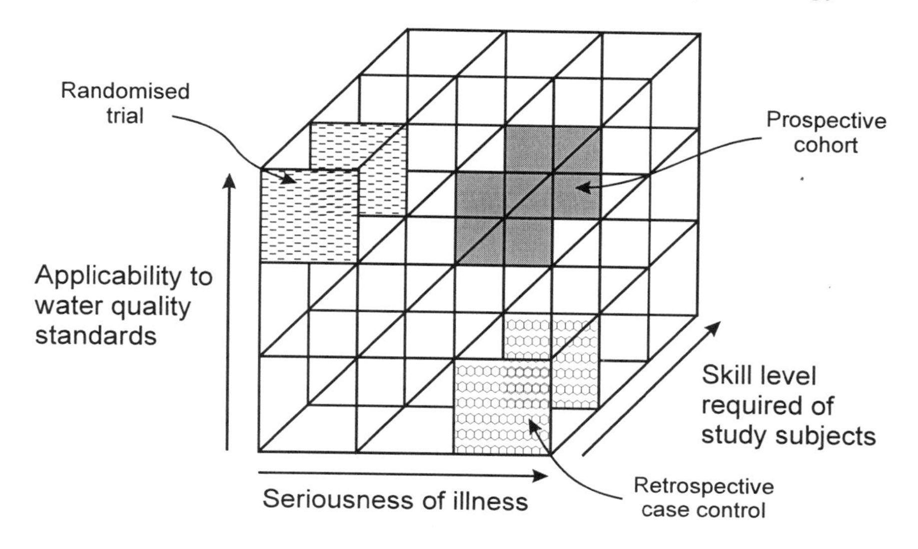

Figure 13.2 Choice of epidemiological protocols in recreational water epidemiology

acceptable concentrations of pollution in recreational waters, provided that exposure data are available, but it is not designed to produce a credible dose–response curve of the type required in health-based standards design.

In summary, the choice of epidemiological protocol design will be driven by the study objectives: i.e. the requirements of risk managers for precise exposure–response relationships, as well as logistical and ethical constraints on project implementation and the resources available. Figure 13.2 illustrates these choices in three dimensions as a guide to appropriate choice of epidemiological protocols in recreational water epidemiology.

13.4 Elements of good practice

- The design of any epidemiological component is critical because it affects every aspect of the recreational water study. It should address why the study is being done, i.e. what is the objective, and how it will be conducted. For example, it should address whether the research will use a case-control, cohort or cross-sectional approach to collect the data. It should also consider how the data will be analysed. These elements should be thoroughly described in the description of the design of the study.
- Health outcomes and exposure should be clearly defined. The endpoint result of exposure to microbiological hazards, as well as the exposure itself are key factors in describing the results of epidemiological studies. The endpoint might be self-reported symptomatology, indicative of exposure to a potentially broad spectrum of pathogens or it may be more specific, as with the isolation of an etiological agent or the reactivity of subject sera to

known antigens. Efforts should be made to make the response to exposure endpoint as specific as possible.

- The population to be studied should be well defined in terms of the participating individuals. This will include demographic information, the means of selecting the population sample and the nature of exclusions, e.g. pregnant women or individuals being treated with steroids or immuno-suppressive agents.
- The numerical size of exposed and non-exposed groups is another critical factor that must be considered in the conduct of epidemiological studies. The sizes of these groups are governed by the frequency of occurrence of the health effect under study. Illnesses or infections that occur at higher frequencies require smaller groups. The size of the required populations is also affected by the magnitude of the differences in the frequency of illness or infections between exposed and non-exposed groups. The smaller the differences to be detected between exposed and non-exposed groups the larger the number of subjects required in each group. Expert advice should be sought with regard to population size before conducting an epidemiological study.
- The approaches for collecting exposure and health effects data should be described in detail. This includes the use of questionnaires and other sources of health data, as well as methods used for collecting exposure data, such as microbiological analytical methods for enumerating micro-organisms in water.
- Data analysis should include the steps taken to control selection, misclassification and confounding bias. The statistical evaluation procedures should be fully described.
- All of the measures taken to ensure the quality of the data should be described including the technical qualifications of all scientists participating in the study.
- The study plan should be submitted to a Human Investigations Committee, or its equivalent, to ensure that any regulatory limitations regarding human studies will be met, especially confidentiality restrictions and informed consent procedures.

13.5 References

Alexander, L.M. and Heaven, A. 1991 *Health Risks Associated with Exposure to Seawater Contaminated with Sewage: the Blackpool Beach Survey 1990*, Environmental Epidemiology Research Unit, Lancaster University, Lancaster, UK, 67 pp.

Asperen, I.A. van, Rover, C.M. de, Schijven, J.F., Bambang Oetomo, S., Schellekens, J.F.P., Leeuwen, N. J. van Havelaar, A.H., Komhout, D. and Sprenger, M.J.W. 1995 Risk otitis externa after swimming in recreational fresh water lakes containing *Pseudomonas aeruginosa. British Medical Journal,* **311**, 1407–1410.

Asperen, I.A. van, Rover, C.M. de, Colle, C., Schijven, J.F., Bambang Oetomo, S., Schellekens, J.F.P., Leeuwen, N.J. van Havelaar, A.H., Komhout, D. and Sprenger, M.J.W. 1995 *Een zwemwatergerelateerde epidemie van otitis externa in de zomer van 1994*. RIVM, Bilthoven, 133 pp.

Balarajan, R., Soni Raleigh, V., Yuen, P., Wheeler, D., Machin, D. and Cartwright, R. 1991 Health risks associated with bathing in sea water. *British Medical Journal,* **303,** 1444–1445.

Brown, J.M., Campbell, E.A. Rickards, A.D. and Wheeler, D. 1987 *The Public Health Implications of Sewage Pollution of Bathing Water*. Robens Institute, University of Surrey, Guildford, Surrey, UK.

Cabelli, V.J. 1983 *Health Effects Criteria for Marine Recreational Waters.* EPA-600/1-80-031, US Environmental Protection Agency, Cincinnati, OH.

Cheung, W.H.S., Chang, K.C.K. and Hung, R.P.S. 1990 Health effects of beach water pollution in Hong Kong. *Epidemiology and Infection,* **105**, 139–162.

Corbett, S.J., Rubin, G.L., Curry, G.K. and Kleinbaum, D.G. 1993 The health effects of swimming at Sydney beaches. *American Journal of Public Health,* **83,** 1701–1706.

Dufour, A.P. 1984 *Health Effects Criteria for Fresh Recreational Waters*. EPA 600/1-84-004. United States Environmental Protection Agency, Cincinnati, Ohio 45268.

El Sharkawi, F. and Hassan, M.N.E.R. 1982 The relation between the state of pollution in Alexandria swimming beaches and the occurrence of typhoid among bathers. *Bull High Inst. Pub. Hlth Alexandria,* **12,** 337–351.

Fattal, B., Peleg-Olevsky, E. and Cabelli, V.J. 1991 Bathers as a possible sources of contamination for swimming associated illness at marine bathing beaches. *International Journal of Environmental Health Research,* **1**, 204–214.

Fattal, B., Peleg-Olevsky, T. Agurshy and Shuval, H.I. 1986 The association between sea-water pollution as measured by bacterial indicators and morbidity of bathers at Mediterranean beaches in Israel. *Chemosphere,* **16**(2–3), 565–570.

Ferley, J.P., Zimrou, D., Balducci, F., Baleux, B., Fera, P., Larbaigt, G. Jacq, E., Mossonnier, B. Blineau, A. and Boudot. J. 1989 Epidemiological significance of microbiological pollution criteria for river recreational waters. *International Journal of Epidemiology,* **18**(1), 198–205.

Fewtrell, L., Godfree, A.F., Jones, F., Kay, D., Salmon, R.L. and Wyer, M.D. 1992 Health effects of white-water canoeing. *The Lancet,* **339**, 1587–1589.

Fewtrell, L., Kay, D., Newman, G., Salmon, R.L. and Wyer, M.D. 1993 Results of epidemiological pilot studies. In: D. Kay and R. Hanbury [Eds]

Recreational Water Quality Management. Vol II Fresh Water. Ellis Horwood, Chichester, 75–108.

Fewtrell, L., Kay, D., Salmon, R.L., Wyer, M.D., Newman, G. and Bowering, G. 1994 The health effects of low contact water activities in fresh and estuarine waters. *Journal of the Institution of Water and Environmental Management*, **8**, 97–101.

Fleisher, J., Jones, F., Kay D., Stanwell-Smith, R. Wyer, M.D. and Morano, R. 1993 Water and non-water related risk factors for gastroenteritis among bathers exposed to sewage contaminated marine waters. *International Journal of Epidemiology,* **22,** 698–708.

Fleisher, J. M., Kay, D., Salmon, R.L., Jones, F., Wyer, M.D. and Godfree, A.F. 1996 Marine Waters Contaminated With Domestic Sewage: Nonenteric Illness Associated with Bather Exposure in the United Kingdom. *American Journal Public Health,* **86**, 1228–1234.

Foulon, G., Maurin, J., Quoi, N.N. and Martin-Bouyer, G. 1983 Etude de la morbidite humaine en relation avec la pollution bacteriologique des eaux de baignade en mer. *Revue française des Sciences de L'eau,* **2**(2), 127–143.

Harrington, J.F., Wilcox, D.N., Giles, P.S., Ashbolt, N.J., Evans, J.C. and Kirton, H.C. 1993 The health of Sydney surfers: an epidemiological study. *Water Science and Technology,* **27**(3–4), 175–182.

Jones, F., Kay, D., Stanwell-Smith, R. and Wyer, M.D. 1991 Results of the first pilot scale controlled cohort epidemiological investigation into the possible health effects of bathing in seawater at Langland Bay, Swansea. *Journal of the Institution of Water and Environmental Management,* **5**(1), 91–98.

Kay, D. 1994 *Summary of Epidemiological Evidence from the UK Sea Bathing Research Programme.* Appendix 5: Select Committee on the European Communities, Bathing Water, House of Lords Session 1994-5, 1st Report with evidence 6th December, Her Majesty's Stationery Office, London, 32–37.

Kay, D. Fleisher, J.M., Salmon, R.L., Jones, F., Wyer, M.D., Godfree, A.F., Zelenauch-Jacquotte, Z. and Shore, R. 1994 Predicting the likelihood of gastroenteritis from sea bathing: results from a randomised exposure. *The Lancet,* **344,** 905–909.

Lightfoot, N.E. 1989 A prospective study of swimming related illness at six freshwater beaches in Southern Ontario. Unpublished PhD Thesis, 275 pp.

Mujeriego, R., Bravo, J.M. and Feliu, M.T. 1982 Recreation in coastal waters public health implications. *Vier Journee Etud. Pollutions, Cannes, Centre Internationale d'Exploration Scientifique de la Mer*, 585–594.

NTAC (National Technical Advisory Committee) 1968 *Water Quality Criteria.* Federal Water Pollution Control Administration, Department of the Interior, Washington, DC.

PHLS (Public Health Laboratory Service) 1959 Sewage contamination of coastal bathing waters in England and Wales: a bacteriological and epidemiological study. *Journal of Hygiene Cambs,* **57**(4), 435–472.

Schirnding, Y.E.R., Fkir, R., Cabelli, V., Franklin, L. and Joubert, G. 1992 Morbidity among bathers exposed to polluted seawater. A prospective epidemiological study. *SAMJ,* **81**, 543–546.

Schirnding, Y.E.R., Straus, N., Robertson, P., Kfir, R., Fattal, B., Mathee, A., Franck, M. and Cabelli, V.J. 1993 Bather morbidity from recreational exposure to sea water. *Water Science and Technology,* **27**(3–4), 183–186.

Seyfried, P.L., Tobin, R., Brown, N.E. and Ness, P.F. 1985a A prospective study of swimming related illness. I. Swimming associated health risk. *American Journal of Public Health,* **75**(9), 1068–1070, 1071–1075.

Seyfried, P.L., Tobin, R., Brown, N.E. and Ness, P.F. 1985b A prospective study of swimming related illness. II. Morbidity and the microbiological quality of water. *American Journal of Public Health,* **75**(9), 1071–1075.

Stevenson, A.H. 1953 Studies of bathing water quality and health. *American Journal of Public Health,* **43,** 529–538.

US EPA 1986 *Ambient Water Quality Criteria for Bacteria - 1986.* EPA440/5-84-002, Office of Water Regulations and Standards Division. United States Environmental Protection Agency, Washington, D.C. 20460, 18 pp.

WHO 1998 *Guidelines for Safe Recreational Water Environments: Coastal and Fresh Waters*. Draft for consultation. World Health Organization, Geneva.

Index

This index is compiled according to recognised indexing standards. The suffix b denotes a reference to material in a text box. The suffix t denotes a reference to a figure or a table.